MECHANICS OF MATERIALS

A Modern Integration of Mechanics and

Materials in Structural Design

MECHANICS OF MATERIALS

A Modern Integration of Mechanics and
Materials in Structural Design

CHRISTOPHER H. M. JENKINS

Professor of Mechanical Engineering
South Dakota School of Mines and Technology
Rapid City, South Dakota

SANJEEV K. KHANNA

Associate Professor of Mechanical and Aerospace Engineering
University of Missouri
Columbia, Missouri

ELSEVIER
ACADEMIC
PRESS

AMSTERDAM • BOSTON • HEIDELBERG • LONDON • NEW YORK
OXFORD • PARIS • SAN DIEGO • SAN FRANCISCO • SINGAPORE
SYDNEY • TOKYO

Elsevier Academic Press
30 Corporate Drive, Suite 400, Burlington, MA 01803, USA
525 B Street, Suite 1900, San Diego, CA 92101-4495, USA
84 Theobald's Road, London WC1X 8RR, UK

This book is printed on acid-free paper. ∞

Library of Congress Cataloging-in-Publication Data
Application submitted.

British Library Cataloging-in-Publication Data
A catalogue record for this book is available from the British Library

ISBN: 0-12-383852-5

For all information on all Elsevier Academic Press Publications
visit our web site at www.books.elsevier.com

Printed in the United States of America
05 06 07 08 09 10   9 8 7 6 5 4 3 2 1

Working together to grow
libraries in developing countries

www.elsevier.com | www.bookaid.org | www.sabre.org

ELSEVIER   BOOK AID International   Sabre Foundation

Table of Contents

Preface

This text is concerned with two interconnected activities:

- Providing a sound foundation in teaching the art and science of structural design
- Bridging the divide between applied mechanics and materials science

First, a few words about structures and structural design are in order. A *structure* is any physical body that must carry loads (usually other than its own weight), and hence develops stresses and strains. The primary engineering disciplines that design structures are aerospace, civil, and mechanical engineering. It is somewhat unfortunate that in the practice of engineering, the name *structural engineer* is often taken to mean someone involved only in the design of civil structures. Aerospace structural engineers design airplanes, rockets, satellites, and the like. Civil structural engineers design buildings, highways, and bridges. Mechanical structural engineers design machinery, vehicles, and consumer products. From a structures perspective, there is much more in common in what aerospace, civil, and mechanical structural engineers do than there is different. A structure is a structure is a structure . . . .

This text attempts to provide a unifying treatment of structural design and should prove useful to any engineer involved in the design of structures. We consider structural engineering in the broadest and most general sense, and it is important that structural engineering students learn from the design of structures in all applications, in or out of their discipline. A wing truss, bridge girder, or overhead crane trolley are all close relatives of the same family of structures. Certainly the design practices of the specific discipline must be learned, but early on it is much more important to design structures generally. To become a great painter, we don't want to always just paint bowls of fruit!

The other divide to be bridged is that between applied mechanics and materials science. In an earlier time, perhaps there was much less gap to be closed. The onset of specialization and the rapid rise of technology, however, have created separate disciplines concerned with the deformation of solid materials. At the risk of oversimplifying, mechanics has roots in physics, whereas materials science derives from chemistry. But in the deformation of a real body, physics and chemistry are inseparable—the macroscopic loads and geometry intimately interact with the material microstructure—and any pedagogical or operational separation is ours alone, not nature's!

The typical undergraduate engineering curriculum follows along this schism. An introductory course in "mechanics of materials" may be taught by mechanics faculty, whereas the introductory course in "property of materials" may be taught by science faculty. Engineering design is taught altogether somewhere else. Mostly, such courses are taught in isolation from one another, both philosophically (in different "languages" and points of view) and physically (in different buildings). (That the faculty is competent in their respective disciplines and teaches their courses well is assumed.) But such a separation is purely artificial, often divided along physical length scales, and the student misses out in never seeing the intimate connection between the macroscopic and microscopic domains of the problem, as well as its human dimensions. Society loses out on having at their service efficient, high-performance material/structural systems.

To introduce mechanics, materials, and design concepts continuously throughout this text, we follow a very methodological process called *total structural design*. The idea is to seek a solution in "total design space," where the *design degrees of freedom* that bound the space include mechanics, materials, and performance variables. We engage the actual design of structures very early by first providing a *generalized design template*, which is then followed and specified for various structural applications. Every chapter first introduces the mechanics perspective through deformation, equilibrium, and energy considerations. Then the constitutive nature of the chapter topic is presented, followed by a link between mechanics and materials concepts. Details of analysis and materials selection are subsequently discussed. A concluding example design problem is provided in most chapters, so that students may get a sense of how mechanics and materials come together in the design of a real structure. We deviate from this consistent approach only in the last chapter, one that provides the interested reader with a brief look at some advanced structural materials.

Exercises are provided that are germane to aerospace, civil, and mechanical applications, and include both deterministic and design-type problems. The reader is encouraged to visit the text Web site at www.books.elsevier.com for information complementary to this text. The student will find the *virtual laboratories* of particular interest. The four labs—tensile test, fracture test, creep test, and fatigue test—allow for simultaneous viewing of structural response across length scales, from the macroscopic to the microscopic. The course instructor will find many additional homework problems and their solutions.

It would be reasonable to ask: Do you have to give some things up to pack all of this into one book? Like engineering design itself, we make trade-offs to reach our objectives optimally. For one, we do not treat statically indeterminate problems as a separate topic. In Chapter 4, we do treat a redundantly supported shaft in the classical way by enforcing continuity of displacements. Very quickly thereafter, however, we show that the direct stiffness method attacks this problem without hesitation! Modern engineers in practice will rarely, if ever, treat a statically indeterminate problem of even the slightest complexity (of which there are plenty) by hand, but rather rely on numerical methods, notably the finite element method. We engage students early on in a study of the direct stiffness method (which is fundamental to the finite element method and modern structural engineering) so as to prepare them for eventual study elsewhere of the finite element method. The power of the direct stiffness method is quickly apparent to the student, especially in that it treats statically determinate and indeterminate problems equally.

We spend little time on shear and moment diagrams, and on beam analysis by the method of sections. For all but the simplest of beam configurations and loadings, the classical method of producing shear and moment diagrams by "graphical calculus" becomes intractable. What the engineer really wants to know is: Where is the internal bending moment maximum? The method used to provide the answer should be the same for any beam configuration. We choose to use singularity functions for this purpose, as they are easy to use and prevalent in reference handbooks (such as Roark's *Formulas for Stress and Strain*). Finding the maximum moment is then straightforward, by setting the shear function to zero.

We hope several other features of the book will be useful to the reader. For one, we use italics in the text to emphasize key concepts and definitions. These italicized words are keyed in the index for ready reference. We use Mathcad® in completing many of the worked examples. This should give the student an opportunity to see how one goes from theory to modern computational practice.

The material presented in this text is suitable for a first course that encompasses both the traditional mechanics of materials and properties of materials courses. At many institutions, however, a wholesale replacement will be difficult, because of a wealth of factors. Thus the text is keyed to allow its use in a more traditional "mechanics of materials" course. We use the icons 𝚼 ⚜ ▣ to indicate chapter sections that are mostly oriented to mechanics (𝚼), materials (⚜), or design (▣), or some combination of these areas. The text is also appropriate for a second course in mechanics of materials or a follow-on course in design of structures, taken after the typical introductory mechanics and properties courses.

This text can be adapted to several different curriculum formats, whether traditional or modern. Instructors using the text for a traditional course may find that the text in fact facilitates transforming the course over time to a more modern, integrated approach. By way of example, we outline two approaches for using the text in different formats.

Use for a Traditional One-Semester Introductory Course in Mechanics of Materials

Traditionally, often at the sophomore level, engineering students are introduced to a first course in mechanics of materials (actually mechanics of "structures"). Usually, students will take a parallel or subsequent introductory course in properties of materials (material science). This text may readily be used for a first course in mechanics of materials by assigning only the appropriate portions of the mechanics-emphasis chapters (𝚼), while assigning the material-emphasis sections (⚜) as background reading on an as-needed basis. An example course outline follows:

Chapter 1. Mechanics and Materials

- All sections except 1.5
- Section 1.5 as background reading
- Review Modules as required

Chapter 2. Total Structural Design

- Sections 2.1–2.4
- Section 2.5 as background reading
- Review Modules as required

Chapter 3. Design of Axial Structures

- Sections 3.1–3.3
- Sections 3.4–3.5 as background reading
- Sections 3.6–3.7
- Tension Test Virtual Lab as background (at www.books.elsevier.com)

Chapter 4. Torsion Structures

- Sections 4.1–4.3
- Section 4.4 as background reading
- Sections 4.5–4.6 (up to Direct Stiffness Method)

Chapter 5. Flexural Structures

- Sections 5.1–5.4
- Section 5.5 as background reading
- Section 5.6 (up to Direct Stiffness Method)

Chapter 6. Combined Static Loading

- Sections 6.1–6.3
- Section 6.4 as background reading
- Sections 6.5–6.6 (up to Failure Theories)

Other chapters and virtual labs as time and interest may allow.

Use for a Second Course in Mechanics of Materials or a Follow-On Course in Structural Design

Chapter 1. Mechanics and Materials

- All sections
- Review Module R1

Chapter 2. Total Structural Design

- All sections
- Review Modules R2 and R3 as required

Chapter 3. Design of Axial Structures

- All sections
- Review Module R4
- Tension Test Virtual Lab (at www.books.elsevier.com)

Chapter 4. Torsion Structures

- All sections

Chapter 5. Flexural Structures

- All sections

Chapter 6. Combined Static Loading

- All sections

Chapter 7. Fracture

- All sections
- Fracture Test Virtual Lab (at www.books.elsevier.com)

Chapter 8. Slender Compressive Axial Structures

- All sections

Chapter 9. Materials for the Twenty-First Century

- All sections for review

Acknowledgments

We wish to express particular appreciation to Dr. David Roylance of the Material Science and Engineering Department at MIT. Dr. Roylance's seminal book *Mechanics of Materials* (Wiley, 1996) was an early inspiration for our own efforts, and we have appreciated working with him over the years and his many helpful suggestions. Much of his material is available at http://web.mit.edu/course/3/3.11/www/modules/, and we encourage the readers of our book to supplement their learning with Professor Roylance's excellent materials.

Author Jenkins wishes to express his deepest love and gratitude to his wife Maureen, and children Amanda and Kelly, for their unwavering support during the arduous years leading up to the publication of this book. He regrets the time taken away from their lives. Warmest appreciation is also extended to the many students who have muddled through the early stages of this book and helped by their patience to make it a reality.

Author Khanna would like to express his heartfelt thanks and recognition to his wife Vinita, who showed a great deal of tolerance and encouragement during the preparation of this book. He also deeply appreciates the understanding of their school-age children Reshma and Yash for giving up their time that belonged to them for family activities. Through this writing experience they have reminded me of the richness in life beyond the material plane. He also wishes to thank his students Vijai Venkata, Marius Ellingsen, and Xin Long for their assistance with several illustrations in the book.

About the Authors

Christopher H. M. Jenkins is Professor of Mechanical Engineering at South Dakota School of Mines and Technology. He teaches and conducts research in the areas of continuum mechanics, computational and experimental mechanics, mechanical design, and structural dynamics. He received his B.S. in Physics from Florida Institute of Technology, and his M.S. and Ph.D. in Mechanical Engineering from Oregon State University. Since 1986, he has focused on investigating the mechanics, materials, and design of ultra-lightweight and highly compliant structures, founding the Compliant Structures Laboratory at SDSM&T in 1994. Dr. Jenkins is a consultant to a number of government laboratories and private corporations, a registered professional engineer in three states, and an active member of several national professional societies. He has edited 4 technical books, authored or co-authored 12 book chapters and over 130 peer-reviewed journal and conference papers, and presented numerous invited lectures. Professor Jenkins is an Associate Fellow of the American Institute of Aeronautics and Astronautics.

Sanjeev K. Khanna is Associate Professor of Mechanical and Aerospace Engineering at the University of Missouri—Columbia. He received his B.S. and M.S. from the Indian Institute of Technology, Kanpur, and his Ph.D. from the University of Rhode Island. He teaches and conducts research in the areas of solid mechanics, experimental mechanics, quasi-static and dynamic fracture of monolithic and composite materials, composite materials development and characterization, nano-mechanical behavior of materials, residual stress measurement, welding engineering, and design of welded structures. He is an active member of the American Society of Mechanical Engineers. Over the past few years the National Science Foundation under the CAREER program and the automobile industry have extensively funded his research on modeling the spot welding process and associated residual stress generation, and studying the effects of fatigue loading on failure in spot welded structures.

1   Mechanics and Materials

If a contractor builds a house for a man, and does not build it strong enough, and the house he built collapses and causes the death of the house owner, the contractor shall be put to death.

—Code of King Hammurabi of Babylon (ca. 1600 BC)

Objective: This chapter will provide the motivation and background for integrating the study of mechanics and materials in structural design.

What the student will learn in this chapter:

- Why we must consider both mechanics and materials in the design of structures
- An overview of structures, including loads, structural types, and analysis methods
- An overview of materials, including atomic structure, bonding, and properties
- An example of a mechanics ⇔ materials link: probabilistic structural analysis

1.1   Introduction

The design of structures crosses many disciplines and has a long history. Some of the earliest structures were civil structures, mostly shelters such as tents and huts. Simple mechanical structures also were in early use, such as levers, and supports for processing and cooking game. Later, marine structures such as rafts and canoes came into existence. Other structures followed, including carts, construction equipment, and weapons. Then aerospace structures, such as balloons and parachutes, appeared.

Today, the fields of aerospace, civil, marine, and mechanical engineering are all involved in structural design. On a cursory level, we can define a *primary structure* as any physical body that must carry loads (other than its own weight), and hence develops stresses and strains. Often these stresses and strains are trivial, and the body can be considered a *secondary structure*. However, in many cases, inadequate design for carrying loads can lead to significant, even catastrophic, failure in primary structures. Unfortunately, examples of such failure are readily found, including the Tacoma Narrows Bridge disaster, the Kansas City Hyatt Regency skywalk collapse, and the Space Shuttle Challenger explosion. From early on, as in King Hammurabi's days, adequate structural design was recognized as critically important.

This text seeks to present structural design in a very broad and integrated context, what we call *total structural design*. This encompasses not just mechanics issues, such as the analysis of stress and strain, but materials issues and manufacture/assembly issues as well. Hence we want to think of a structure as an integrated material/structural system, realized with cradle-to-grave design.

In this text, you will learn about:

- Fundamental mechanics or macroscopic concepts such as stress, strain, and stiffness
- Fundamental materials or microscopic concepts such as atomic structure and bonding, defect theory, and the physics of material properties
- The intimate linkage between macroscopic and microscopic concepts, and their role in the design of structures
- How to apply a *generalized design template* in the design of structures
- How to manipulate *design degrees of freedom* to achieve an optimal structural design
- The different classes of structures and their special design considerations
- Simultaneous engineering of materials and structures

1.2 Motivational Examples ⊤🏛💻

As performance expectations for products rise, the need for removing separate consideration of structure and material increases. At high levels of performance, we no longer have the luxury of first designing a structural "form," and then searching for an available material to fill in the space. We must think in terms of *material/structural systems*, which are designed in an integrated fashion. Numerous examples can be provided.

A skyscraper leaves very little room for inefficiency in structural design. Consequently, glass walls cannot be designed simply for covering and viewing purposes, but must be part of the load-carrying system as well. Coverings that also carry loads are called *monocoque*.

Another example is the development of ultralightweight structures for space applications. In areas such as imaging, solar propulsion, and communications, increased collecting area provides significant gains in performance. Since launch vehicle capacity is limited in terms of size and lift, the options for stowing, launching, and deploying very large precision structures in space is limited. (The International Space Station is an example of a large mechanically erectable structure—but it is not a precision structure like an antenna.)

One of the concepts being widely investigated at the current time is that of membrane/inflatable "gossamer" structures. These are made from thin, flexible films, unfurled or inflated on orbit, with some components possibly rigidized afterward. Extremely low areal densities must be developed, in order to keep total launch mass below specified limits. (*Areal density* is the total system mass divided by the collecting area.) Consider that the areal density will have to be diminished from that of the current Hubble space telescope (at $\sim 100\,\text{kg/m}^2$), to $0.1\,\text{kg/m}^2$—a decrease of three orders of magnitude! Such increases in system performance will require close cooperation between structural engineers and materials scientists, who each have a good understanding of the other's discipline. A good example of a gossamer spacecraft is shown in Figure 1-1.

Another example of a membrane structure is a tension fabric structure. In civil structures, where large areas need to be covered with clear spans, the "tent" roof provides an impressive solution. The roof of the Denver International Airport shown in Figures 1-2 and 1-3 is a wonderful example. Again, the close coupling of materials, mechanics, and design is crucial for membrane structures.

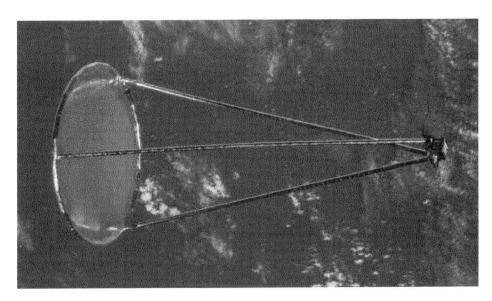

FIGURE 1-1 *Inflatable Antenna Experiment launched in 1996. (Courtesy of NASA.)*

FIGURE 1-2 *Tension fabric roof at the Denver International Airport.*

As still another example, consider the linkage system on a General Electric aircraft turbine engine as seen in Figures 1-4 and 1-5.

This system shows a richness of structural diversity. The system is composed of rods, beams, shafts, and plates, among others. The links (rods) may be under tension or compression, the beams and plates may undergo bending, and the shafts may be placed in torsion. Clamping

FIGURE 1-3 *Tension fabric roof detail at Denver International Airport.*

FIGURE 1-4 *Variable stator linkage system on a GE Aircraft Turbine Engine. (Courtesy of Mr. Andrew Johnson, GE Aircraft Engine Co.)*

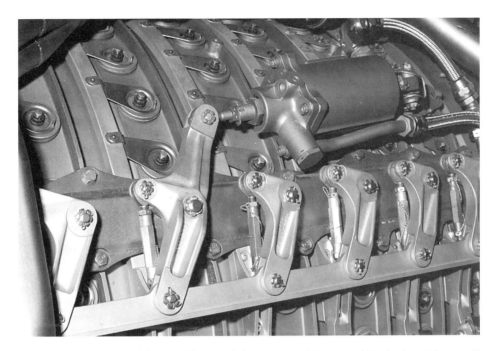

FIGURE 1-5 *Close up of the variable stator linkage system. (Courtesy of Mr. Andrew Johnson, GE Aircraft Engine Co.)*

forces are also apparent in a number of places. Everything has to work under an extreme environment (such as high temperature), and it's got to work right! Again, mechanics–materials–design must all come together in an integrative fashion.

1.3 Linkage between Mechanics and Materials Science in Structural Design ⍦⌖🖳

Applied mechanics and materials science are broad fields of human endeavor with long histories. *Mechanics* deals with the theoretical and experimental analysis of forces on material bodies, and the resultant motions and deformation that follow. *Materials science* is concerned with the atomic structure of materials and the properties resulting there from. For our purposes here, we offer a working, albeit oversimplified, definition: mechanics is physics, materials science is chemistry.

The early "royal philosophers" such as Da Vinci, Galileo, Hooke, and Young were truly renaissance scholars, in that they brought to their investigations a broad background in chemistry, physics, mathematics, and classical studies. The development of analytical mathematics and mechanics in the 18th century increasingly fostered specialists, as contrasted with these earlier generalists. At the same time, chemistry and physics were developing themselves, and with the expanded use of metals in the industrial 19th century, each began to focus into specialized areas (metallurgy, for example). As specialized bodies of knowledge developed, it became increasingly difficult to be inclusive and bridge across various disciplines.

This divergence continues into modern times. Consider, for example, Seely's text *Resistance of Materials* (1925), which contained a 55-page chapter on "Mechanical Properties of

Structural Materials"; by the printing of the 4th edition in 1955, nearly all mention of material properties had disappeared, save for a discussion on fatigue. One can find today very few "mechanics of materials" texts that combine mechanics and materials science (not to mention design) in an integrated and cogent manner (for an exception, see Roylance, 1996).

Today, examples abound that show the need for engineers and scientists who have an integrated, interdisciplinary background bridging mechanics and materials science. Consider, for example, the important and active area of high-performance composite materials. Here, an intimate knowledge of structure–property relations is demanded for technological advancement. Bulk response can be predicted in an averaged sense using a mechanics approach, which is necessary to design a real composite structure; but only knowledge of the fine-scale (micro- to nanoscale) structure–property relations and interactions among the constituents can lead to an optimal "engineering" of these materials for an intended application.

Other topics of current interest include computational modeling of materials with evolving microstructures; advanced manufacturing and processing challenges to mechanics and materials; mechanics and statistical physics of particulate materials; mechanics and materials science of contact; and processing and mechanics of nanoscale, multilayered materials. We will show in what follows that every structural design should be an integration of mechanics and materials technology.

1.4 Overview of Structures 'Y'

1.4.1 Form and Function
Engineers would typically argue that the purpose (function) of an artifact should determine its appearance (form), at least in a general way. On the other hand, artists might consider the form of an object to be the primary thing!

In engineering design, functionality drives the form of a product. However, the number and variety of possible forms that ultimately embody the design may be astounding. As an example, consider the design of a paper clip. The functional requirements might be listed as:

- Provide a clamping force to hold two or more sheets of paper together
- Easily engaged/removed
- Lightweight
- Reusable
- Inexpensive

Consider now the possible forms the clip might take (Figure 1-6). Simply put,
Design is a process that determines the form of an artifact that (optimally) satisfies the functional requirements.

1.4.2 Structural Loads
In structural design, we use the term *loads* to mean forces and moments applied to the structure, either externally on the boundaries (*surface loads*), or within the structure (*body loads*).

Loads are further considered to be either static or dynamic. *Static loads* are loads that do not depend on time, that is, they are of constant magnitude, direction, and location. Although it might seem that certain structures are static, for example, a civil structure such as a building, this is rarely the case. "Live loads" from occupant activity, wind loads, seismic (earthquake)

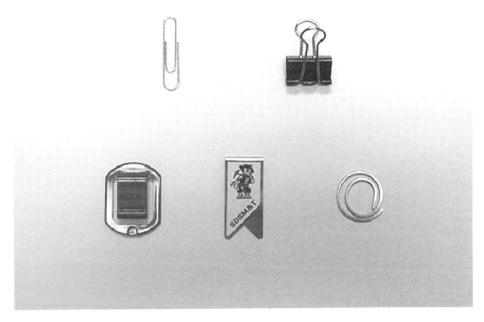

FIGURE 1-6 *Possible forms for a paper clip.*

loads, thermal cycling, and so on, all may give rise to a dynamic load environment. However, if the loads vary 'slowly' with time, they are often considered *quasi-static* and taken as static loads. ("Slowly" usually means that structure inertia forces due to accelerations may be neglected with respect to the difference of the externally applied forces and the internal resistance forces.)

Dynamic loads are divided into two main categories:

1. *Steady-state loads* are loads that maintain the same character (frequency, amplitude, etc.) over the long term.
2. *Transient loads* are loads that change their character (e.g., they may decay) with time.

Common structural loads are summarized in Table 1-1.

The orientation of the load on a structural member is also important. Although this issue will be discussed in more detail in later chapters, a brief summary is given in Figure 1-7.

1.4.3 A Taxonomy of Structures

Humans have always tried to understand complex systems by decomposing them into a number of simpler, more manageable parts. The hope is that when this compartmentalized knowledge is put back together (synthesized), an accurate representation of the whole system results. While this approach has worked well in countless human enterprises, it is based on the linear assumption of superposition, which fails as systems become nonlinear and more complex.

TABLE 1-1 Summary of Common Structural Loads

Surface loads (Common name)	Units SI (US)	Body loads	Units SI (US)
Concentrated force ("Point load")[a]	N (lb)	Gravitational force ("Gravity load" or "weight")	N (lb)
Distributed force ("Line load")[a] ("Pressure")	N/m (lb/in) N/m^2 (psi)	Thermal stress ("Thermal load")	N/m^2 (psi)
Moment or couple ("Bending moment" or "torsion moment")	N-m (lb-in)	Magnetic forces ("Magnetic moment")	N-m/m^3(lb-in/in^3)

[a]In principle, "point" and "line" loads cannot exist, since any load will act over a finite spatial area, no matter how small; however, if that area is small relative to the structure size, then for practical purposes the load may be considered a "point" load or a "line" load.

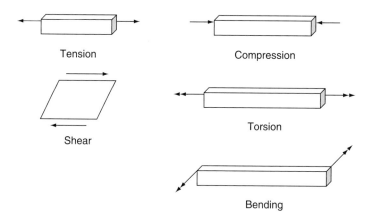

FIGURE 1-7 *Simple load orientations. Single-headed arrows are forces; double-headed arrows are moments (that follow the right-hand rule).*

Forewarned by this knowledge, we will attempt here a *taxonomy* (classification) *of structures* (complex systems) by decomposing them into their structural elements (simpler parts). Most real structures are comprised of a number of different types of structural elements, any one of which may assume a variety of different roles. The primary characteristic used to classify a structural element is how it carries loads.

Carrying loads is the primary function of a structure, and it is this characteristic that largely determines the form of the structural elements. Loads carried include tensile and compressive axial loads; shear loads; torsion moment loads; bending moment loads; distributed loads; gravity loads; and thermal loads.

Structural forms fall into two major categories: *line-forming* and *surface-forming*. Surface-forming elements may be further subdivided into *area-forming* and *volume-forming* elements (see, e.g., Schodek, 2001). A taxonomy of structural elements is given in Table 1-2.

A review of Table 1-2 reveals that beams, plates, and shells are structural elements fully capable of carrying all types of loads. Specialty elements such as cables, rods, and membranes

TABLE 1-2   A Taxonomy of Structural Elements

Function ⇓	Line-forming			Surface-forming			
				Area-forming		Volume-forming	
Loads carried	Cable	Rod	Beam	Plane membrane	Plate	Curved membrane	Shell
Tensile axial	✓	✓	✓	✓	✓	✓	✓
Compressive axial	–	✓	✓	–	✓	–	✓
Direct shear	–	–	✓	–	✓	–	✓
Torsion moment	–	✓	✓	–	✓	–	✓
Bending moment	–	–	✓	–	✓	–	✓
Distributed force	✓	–	✓	✓	✓	✓	✓
Thermal	✓	✓	✓	✓	✓	✓	✓

carry limited types of loads. (Note: *finite element analysis* codes used in structural analysis have libraries of structural elements based upon a similar taxonomy.) Schematics of the elements are shown in Figure 1-8.

Line-forming elements (LFEs) are slender structures having one spatial dimension (length) significantly greater than any other dimension (width, height, thickness, etc.). Schematically, for analysis purposes, LFEs may be represented as lines, either straight or curved as required. LFEs may carry loads in tension, compression, torsion, bending, or some combination of these, depending on the nature of the applied loads, structural geometry, material properties, and boundary conditions. LFEs may form axial, torsional, or bending structures such as rods, cables, and beams.

1. *Rods* carry concentrated axial forces in either tension or compression (no moments or transverse forces). Thus they are *axial structures*. Only *pinned* (simply supported) boundary conditions are required. Cross-sectional geometries may take any shape, but simple shapes are most common, such as circular and rectangular. The cross-section may be solid or hollow.
2. *Cables* carry concentrated axial tensile forces (but not compressive axial forces—you cannot push a rope!), as well as concentrated or distributed transverse forces (but not moments!). In the latter case, cables deform under the action of these transverse loads in such a way that they remain *axial* structures (specifically, no-compression axial structures). (Of course, in the former case of axial loads, cables are by definition axial structures.) As with rods, only pinned boundary conditions are required. Cross-sectional geometries may be of any shape, but solid circular cross-sections are common.
3. *Beams* are the most complete LFEs since they can carry axial compressive or tensile forces as in rods or cables, as well as transverse concentrated or distributed forces as in cables; moreover, they can carry torsion and bending moments. In cases where moments are applied, at least one boundary condition must be able to support moment reactions (e.g., a "fixed" or "built-in" condition). Cross-sectional shapes may be of any geometry, solid or hollow.

Surface-forming elements (SFEs) are thin structures having two spatial dimensions (length and width) significantly greater than the third dimension (thickness). Schematically, for

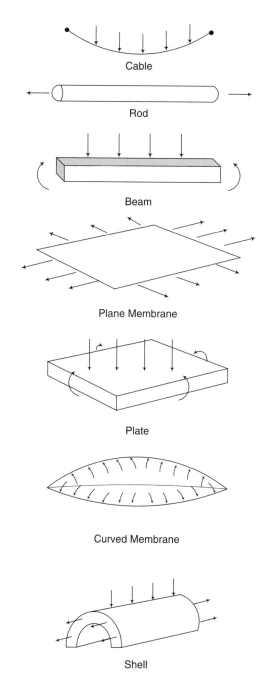

FIGURE 1-8 *Schematic of basic structural elements (not all loads shown).*

analysis purposes, SFEs may be represented as surfaces, either straight or curved as required. SFEs may carry loads in tension, compression, torsion, bending, or some combination of these, depending on the nature of the applied loads, structural geometry, material properties, and boundary conditions. SFEs may form axial, torsional, or bending structures such as *plates* and

TABLE 1-3 Common Boundary Conditions[a]

Name (alternate names)	Translation constraint	Rotation constraint
Free—2D or 3D	None	None
Fixed (clamped)—2D or 3D	No translation $(u = v = w = 0)$	No rotation $(u' = v' = w' = 0)$
Ball and socket (swivel)—3D	No translation $(u = v = w = 0)$	None
Simple (pinned)—2D	No translation $(u = v = 0)$	None
Simple (roller)—2D	No translation $\perp$ to roller surface (e.g., $v = 0$)	None

[a]The prime or $'$ symbol indicates differentiation with respect to a spatial coordinate, i.e., a gradient or slope. u, v, and w are translations in the x-, y-, and z-directions, respectively.

shells. Being more complicated structures, SFEs are outside the scope of this text and will not be described further.

1.4.4 Boundary Conditions

It should be obvious that if a structure is not tied down somewhere, it will not be able to carry loads in many situations. Imagine trying to hang a weight from a hook not connected to the ceiling! Thus we see that the conditions at the structure boundaries, that is, the *boundary conditions*, are critically important in structural design and analysis.

A number of idealized boundary conditions can be defined. (We say "idealized" since boundary conditions on real structures rarely meet these definitions exactly.) In general, boundary conditions either resist translation or rotation, or both, in one or more directions. In order to do this, they must generate resisting forces (translations) or moments (rotations). In other words, translation and/or rotation are constrained at the boundary by the generation of either a force and/or moment, respectively. (Mathematically, boundary conditions provide additional *equations of constraint*.)

If there are $\left\{ \begin{matrix} \text{fewer} \\ \text{more} \end{matrix} \right\}$ constraints than the minimum required, the structure is $\left\{ \begin{matrix} \text{under constrained} \\ \text{constrained} \end{matrix} \right\}$, i.e., the structure is *statically indeterminate* in that there are $\left\{ \begin{matrix} \text{more} \\ \text{fewer} \end{matrix} \right\}$ independent equations available than unknowns to solve for. A number of common boundary conditions are provided in Table 1-3 and Figure 1-9.

An example of a real boundary condition can be seen in the pin joint in the truss shown in Figure 1-10.

1.4.5 Strength and Stiffness

When we think of a structure carrying loads, two primary questions must be asked:

- Is it strong enough?
- Is it stiff enough?

In other words, we wish to know how well the primary functional requirement of a given structure—carrying loads—is performed; this is the primary role of *structural analysis*.

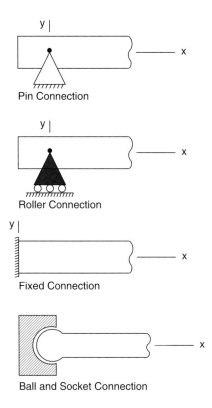

FIGURE 1-9 *Some examples of boundary conditions listed in Table 1-3.*

(As we'll see in Chapter 2, a number of other questions must be asked as well, such as: Is it light enough? Cheap enough? Repairable enough?)

Strength and stiffness are independent quantities (they are, we will find, *design degrees of freedom*, a concept to be discussed in Chapter 2) that depend on the constituents of the material in different ways. We'll come to understand these concepts more in later sections as well. At this point, we just mention that strength considerations are usually of greater importance in structures than stiffness. If you carry a load with a large rubber band, you are usually not too worried if it stretches quite a bit, as long as it doesn't break (i.e., as long as it is strong enough). However, if the band stretches too much, your arms may not be long enough to keep the load from interfering with the ground, and now stiffness (or lack of it!) does become a concern.

An important issue we cannot avoid is: "How much is enough?" Unfortunately, the answer is a somewhat unsatisfying "It depends." It depends on the application, the failure mode, how catastrophic would be the failure (e.g., loss or injury of human life), the past history of similar designs, how well known are the loads and material properties, and so on. In any case, we must always compare what we have (stress, strain, deflection—the load effects) to what we can allow (strength, stiffness—the resistance). This comparison is done generally in one of two ways (a third being a combination of these two): *resistance factor design* and *load factor design* (the third being *load and resistance factor design*).

By way of illustration, one common, elementary approach to determining "enough" is to define a *factor of safety* (FS) (some might say a factor of ignorance!) that tries to account for the effect of many of the issues described above into one quantity. The factor of safety

A

B

FIGURE 1-10 *(A) Pin joints in a truss. (B) Pin joint detail.*

(FS $\geq$ 1) is applied to the resistance side of the comparison, and as such creates a *knockdown factor* resulting in an *allowable resistance*. For example, if the resistance considered is strength, then we have:

$$\frac{S_{\text{failure}}}{\text{FS}} = \sigma_{\text{max}} \qquad (1.1)$$

where
S_{failure} = designated failure strength (tensile yield, tensile ultimate, compressive yield, etc.)
σ_{max} = maximum stress (at the critical section) associated with the appropriate failure mode (shear, tension, etc.).

The most general form of the comparison is through *load and resistance factor design* (LRFD):

$$\alpha R = \sum_{i=1}^{I} \beta_i Q_i \qquad (1.2)$$

where
α = resistance factor (≤ 1)
R = nominal resistance
β_i = ith load factor (usually >1)
Q_i = ith nominal load effect, associated with the ith nominal applied load, I total loads being applied.

We will not pursue this later approach in this text. Organizations such as the American Institute for Steel Construction (AISC) provide tables of load and resistance factors for various applications (AISC, 2000).

1.4.6 Load Path

Since structures carry loads, a useful tool in structural analysis is the concept of *load path*, that is, the path by which the load is carried. It helps if one can imagine a "flow" of stress or load along the path. In many cases, this is trivial and obvious, particularly when there is only a single, unique path. For example, the load from a sign is carried (literally to ground in this case) by the cantilevered arm and the beam-column (post) as in Figure 1-11.

A sign supported by a truss presents a more complicated load path in Figure 1-12.

To continue the discussion, imagine that in Figure 1-11, the vertical beam-column is attached to ground through a bolted flange connection (Figure 1-13 is a detailed section of the lower portion of Figure 1-11). What is the load path to ground now?

We will find as we proceed that the load sharing among multiple paths apportions itself in large part according to the path stiffness. As a simple example, consider a load carried by two parallel springs, one considerably stiffer than the other. It is not hard to imagine that the stiffer spring carries a greater portion of the load. Developing an intuition for load paths will prove to be a very useful asset for the structural engineer.

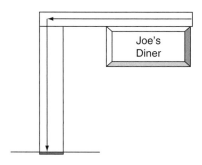

FIGURE 1-11   *Load path for a simple sign, arm, and post.*

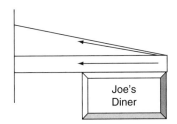

FIGURE 1-12   *Load paths in truss-supported sign.*

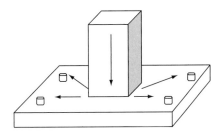

FIGURE 1-13   *Load paths at the base of the beam-column in Figure 1-11.*

EXAMPLE 1-1: Reactions—Exact and Approximate

Consider the following problem of an overconstrained (Why?!) beam of length L subjected to a uniformly distributed load p_0 (Figure E1-1).

1. The exact solution for the reaction loads at A, B, and C (by methods found in Chapter 5) is:

$$R_B = \frac{5}{8}\frac{p_0 L}{2}, \quad R_A = R_C = \frac{3}{16}\frac{p_0 L}{2}$$

In fact:

$$\frac{R_B}{R_A} = \frac{5/8}{3/16} = 3.33$$

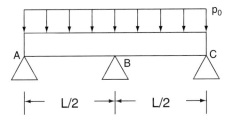

FIGURE E1-1   *Configuration sketch.*

Continued

EXAMPLE 1-1: Reactions—Exact and Approximate–*Cont'd*

2. As a first guess without knowing better, an approximate solution might be simply to take $R_A^* = R_B^* = R_C^* = p_0L/3$. In that case, the error in (say) R_B is

$$\frac{R_B - R_B^*}{R_B} \times 100 = \frac{0.625 - 0.333}{0.625} \times 100 = 47\% \text{ and } \frac{R_B^*}{R_A^*} = 1.00$$

   Not very good! Our estimate is only about one-half the true answer and the reaction at B does not equal the reaction at A.

3. Let's use the concept of load path to come up with another approximate solution. We imagine each of the boundary supports as "collection points" of load arriving there by different load paths. For example, a reasonable choice might be as shown in Figure E1-2.

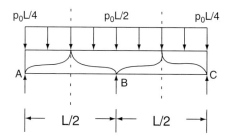

FIGURE E1-2 *Estimate of load paths.*

   Now it can be seen by inspection that

$$\tilde{R}_B = \frac{p_0L}{2}, \ \tilde{R}_A = \tilde{R}_C = \frac{p_0L}{4}$$

and

$$\frac{R_B - \tilde{R}_B}{R_B} \times 100 = \frac{0.625 - 0.50}{0.625} \times 100 = 20\% \text{ and } \frac{\tilde{R}_B}{\tilde{R}_A} = 2.00$$

   (A much better approximation than in 2!)

1.4.7 Characteristic Tasks in Structural Analysis

Structural analysis supports structural design. Structural analysis can include a variety of tasks aimed at understanding the complex response of structures, specifically the resultant stress, strain, and displacement to applied loads, which answer the questions about strength and stiffness. A sample of such tasks is given below:

- Loads analysis
- Strength analysis
- Stiffness analysis

- Natural frequency analysis
- Dynamic response analysis
- Damping analysis
- Thermal (heat transfer) analysis
- Thermoelastic (thermal deformation) analysis
- Structural life (fracture, fatigue) analysis
- Mass property (weight, center of gravity) analysis
- Precision (shape and position accuracy) analysis
- Sensitivity (of the design to perturbations or small changes) analysis

1.4.8 Methods of Structural Analysis

Structural analysis consists of three fundamental parts:

1. *Equilibrium*. Here we consider forces, moments, and the application of Newtonian mechanics (or the potential and kinetic energies and the application of Lagrangian mechanics), and stress. If the body is in global equilibrium, then every local particle of the body is also in equilibrium. Equilibrium implies negligible accelerations (inertia forces), and hence implies static analysis. (Dynamic analysis would consider the motion of the body, requiring inertia forces in the full equations of motion.)
2. *Deformation*. Here we consider the geometry of material particle displacement and the concept of strain. We assume that any material is continuous and fully populated with material particles (the *continuum hypothesis*). We further assume in this text that deformations are small enough that only a linear analysis is required.
3. *Constitution*. Stress and strain are dual quantities that are intimately related within a given material/structural system. These are *stress–strain relations*.

The methods of structural analysis are mathematical analysis and experimental analysis. Mathematical analysis may result in closed-form solutions, series solutions, asymptotic solutions, numerical solutions, and so on. Except for relatively simple structural problems, numerical solutions are usually required. The direct stiffness method, and its descendant the finite element method, are the most ubiquitous and powerful numerical methods for structural analysis today.

Experimental analysis involves the testing of real or prototypical structures and uses various techniques to assess strain and/or displacement. Some of these experimental techniques are:

- Strain gages
- Stress coatings
- Optical interferometers
- Extensometers
- Videography
- Thermography

Quite often, combinations of mathematical and experimental methods are used.

In every case of analyzing real structures, certain assumptions about their behavior or character must be made. This leads to an idealized structure model that will be divorced from the real structure to some greater or lesser degree. **The engineer must always be aware of this discrepancy**.

1.5 Overview of Materials ✎

In the past two decades, industries such as computers, electronics, aerospace, machinery, automotive, metals, energy, chemical, telecommunications, and biomaterials have witnessed major improvements in the performance and reliability of their products. The principal reasons are the development of new materials, novel use of existing materials, better understanding of the structure–property relationships, and incorporation of materials science in the total design of such structures. The listed industries have benefited by billions of dollars every year from the development and use of new materials, and have significantly contributed to the growth, prosperity, security, and quality of human life. Table 1-4 illustrates the types of materials used in these representative industries.

It may be noted that a wide variety of materials are used among the different industries, though many of the materials are common to most of the industries. However, the usage and required properties vary widely. Table 1-5 lists the typical properties required of materials used in different industries.

Thus it is apparent that development and novel use of structural, electronic, and biological materials are the backbone of many industries. It is essential for scientists and engineers to not only understand the science of materials, but also know how to apply that knowledge to the design and creation of new and reliable structural, electronic, and biological products. The properties of structural materials such as toughness, strength, stiffness, and weight, which are determined by the interaction of the atoms and their arrangement in molecules, crystalline and noncrystalline arrays, and defects at the atomic or macroscopic level (see, e.g., Shackelford, 1996). Consequently, the ability to predict and control the material's structure, at both the atomic and macroscopic level, is essential to developing materials that can achieve the

TABLE 1-4   Typical Materials Used in Various Industries

Electronics	Aerospace	Automotive	Machinery
Semiconductors	Metals	Metals	Metals
Polymer composites	Polymer composites	Polymer composites	Polymer composites
Metals	Metal alloys	Metal alloys	Metal alloys
Plastics	Plastics	Plastics	Plastics
Ceramics	Ceramics	Ceramics	Ceramics
Single crystals	Metal-matrix composites	Elastomers	Fluids
	Superalloys		
	Elastomers		
	Single crystals		

Energy	Chemical	Biomaterials
Metals	Metals	Metals
Polymer composites	Polymer composites	Water-soluble polymers
Metal alloys	Metal alloys	Biopolymers
Plastics	Plastics	Polymers
Ceramics	Ceramics	Ceramics
Metal-matrix composites	Metal-matrix composites	Nanomaterials
Superalloys	Fluids	Biotissue
Fluids		Artificial skin

TABLE 1-5 Desired Material Properties for Representative Industries

Desired properties	Industry						
	Elect.	Aero.	Auto.	Mach.	Energy	Chem.	Biomat.
Weight		●	●	●			●
Strength		●	●	●	●	●	●
Stiffness	●	●	●	●	●	●	●
Fracture toughness	●	●	●	●	●	●	●
High temperature resistance	●	●	●		●	●	
Corrosion resistance		●	●	●	●	●	●
Durability	●	●	●	●	●	●	●
Shape formability		●	●	●			●
Reactivity		●			●	●	●
Recyclability			●	●	●	●	

level of performance required in, for example, airplanes, automobiles, or bridges. For example, the launch of space vehicles requires low structural weight. This requires new materials with high strength-to-weight ratios and the ability to withstand a harsh environment such as high temperature, ultraviolet radiation, and abrasive dust. In this regard, some of the materials used are composites, thin films and layered structures, ablative materials, and ceramics. Thus, knowing the fundamental mechanisms of failure such as fracture, fatigue, creep, wear, and corrosion, the goal is to achieve the appropriate material with the desired structure.

Macroscopically, materials may be classified in a number of different ways. One useful choice is to divide them into two classes—isotropic and anisotropic. *Isotropic* materials have identical properties in all directions, whereas *anisotropic* materials have properties that depend on the direction along which the property is measured. The most common materials and their characteristics are listed in Figure 1-14.

For example, aluminum (metal) is an isotropic material because its mechanical properties, such as Young's modulus, do not depend on the direction in which a tensile test specimen is cut out from a large plate. The plate consists of crystals oriented at random. Each direction will thus give a value of about 70 GPa (10×10^6 psi). However, if a single crystal of aluminum was tested in the [100] crystal direction (this will be discussed in detail in Chapter 4), the Young's modulus obtained will be about 64 GPa (9.2×10^6 psi). This is due to the differences in the atomic arrangement in the planes and directions within a crystal. Hence, a single crystal behaves as an anisotropic material, whereas a regular aluminum plate consisting of randomly oriented crystals behaves as an isotropic material.

The common crystal structure of materials and the types of bonds formed in these materials is shown in Figure 1-15. The interactions of the electrons with each other and with the nucleus are responsible for the binding of the atoms into molecules and crystals. The mutual attraction between atoms results in four types of bonds.

The ionic, covalent, and metallic bonds are called *primary bonds*, while the van der Waals bonds are called *secondary bonds*. The primary bonds are strong and stiff and are mainly responsible for the physical and mechanical properties of metals and ceramics, such as high melting points and high Young's modulus. The secondary bonds are relatively weak and are primarily found in polymers. (A more detailed discussion of atomic bonding is given in Review Module 1.)

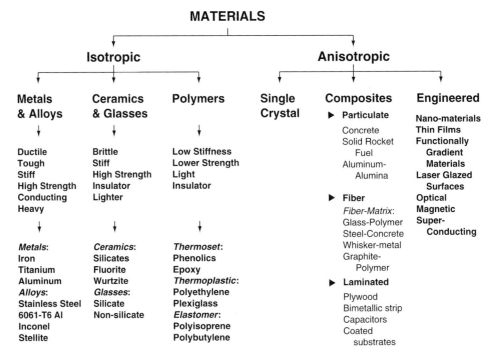

FIGURE 1-14 *Common materials and their characteristics.*

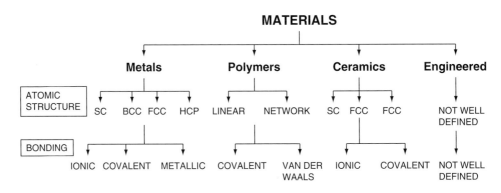

FIGURE 1-15 *Common atomic and molecular structures and bonds.*

The four main crystal structures are *simple cubic* (SC), *body-centered cubic* (BCC), *face-centered cubic* (FCC), and *hexagonal close-packed* (HCP). The deformation response of these crystal structures to applied load is different and accounts for the difference in the properties such as ductility and strength. Materials with the SC and HCP structure are brittle; examples are beryllium, magnesium, zirconium, and cobalt. The FCC materials, for example, aluminum, gold, silver, lead, copper, iron, and platinum, have low strength but high ductility. The BCC materials have high strength but relatively lower ductility. Examples of BCC materials are titanium, tungsten, vanadium, iron, chromium, and zirconium. It may be noted that some materials exist

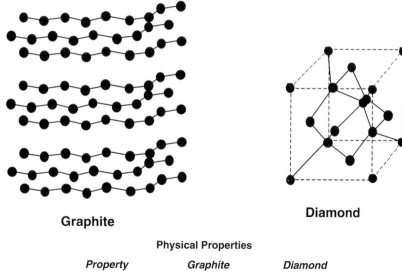

Graphite

Diamond

Physical Properties

Property	Graphite	Diamond
Hardness	Very Low	Very High
Color	Black	Colorless
Density	High	Low
Thermal Conductivity	Good	Excellent

FIGURE 1-16 *Comparison of graphite and diamond property and structure.*

in more than one crystal structure depending on the conditions. For example, at low temperature iron has the BCC structure, but at a high temperature it transforms to an FCC structure.

It is now apparent that atomic and molecular arrangements play an important role in determining the microstructure and mechanical behavior of materials. The rules of bonding then control the resulting properties of materials. For example, let us consider the structure–property relationship in graphite and diamond, both of which are crystalline modifications of carbon. The structure of graphite is shown in Figure 1-16. It is made up of planes of carbon atoms, which are covalently bonded in planar hexagonal arrays, with weak secondary bonds between the planes. Within these planes the atoms are closer together than in the diamond crystal, whereas the binding force is less between the planes, which are more widely separated than in diamond. The diamond crystal is also illustrated in Figure 1-16. In diamond, each carbon atom is surrounded by four other carbon atoms symmetrically arranged at the corners of a tetrahedron, with the surrounded atom at the center. These carbon atoms are bonded by covalent bonds. The difference in bonding and arrangement of carbon atoms results in different structures, which in turn result in very significant differences in the properties, as illustrated in Figure 1-16.

1.6 Mechanics ⇔ Materials Link: Probabilistic Structural Analysis ⅄ 🕸

Although we do our best to determine exactly the magnitudes of loads acting on a structure, or accurately determine material property values, there will always be some uncertainty associated with our knowledge of these quantities. For example, we will never know for certain that the wind load on a structure will have a fixed magnitude, frequency, and duration. If we measure the tensile yield strength of a dozen samples of a given material, we will see scatter in our data.

Thus we can consider that the structural property data we seek has some distribution of values. For our purposes here, we can consider this distribution to be Gaussian or normally distributed about a mean value. (Other distributions, such as Weibull, are possible, but the actual form of the distribution is not that important for our present discussion.) For example, consider the yield strength S_Y data for a steel bar (Figure 1-17).

The combined effects of uncertainty in loads and geometry will manifest themselves in uncertainties in the calculated stress. For example, the distribution of stress values σ might look like Figure 1-18.

A deterministic analysis (using only the mean values) of the factor of safety, without reference to the uncertainty in the strength and stiffness values as shown by their distributions, would show by Eq. (1.1) that:

$$FS = \frac{<S_Y>}{<\sigma>} = \frac{6}{3} = 2$$

That sounds like a good margin! This would be terribly misleading, however, since a more complete, probabilistic analysis would show that there is a region of overlap of values such that $\sigma > S_Y$! Figure 1-19 shows Figures 1-17 and 1-18 plotted side-by-side.

What this means is that there is some probability that a stress value will exist that exceeds a possible strength value—with ensuing failure! (More details of this subject will not be provided here, but see, e.g., Shigley and Mischke, 2001.) The important point to keep in mind is that although for demonstration, calculation, or design purposes we may use a single value to represent the magnitude of a structural property, in reality a range of values exists for that property.

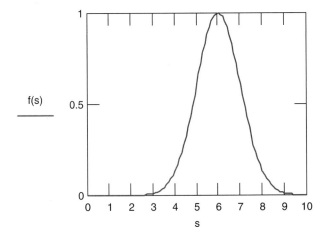

FIGURE 1-17 *In this figure, s = S_Y, and f(s) is the distribution of values of the yield strength. The mean value of S_Y is $<S_Y> = 6$ ($\times 10^5$ psi, perhaps).*

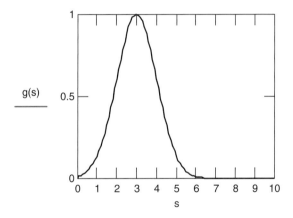

FIGURE 1-18 *Here* s = σ, g(s) *is the distribution of stress values, and the mean value of stress is* <σ> = 3 (× 10⁵ *psi, say*).

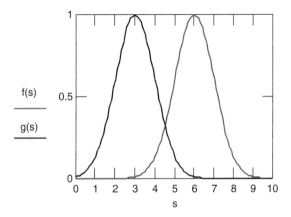

FIGURE 1-19 *The region of overlap between stress and strength is clearly seen in this figure.*

Key Points to Remember

- Structures are physical artifacts that carry loads other than their own weight.
- Every structure is a material as well. Hence the mechanics of a structure is intimately linked with the properties of the structural material.
- Critical elements of structural mechanics include geometry, loads, boundary conditions, and load paths.
- Critical elements of structural materials include atomic bond type and strength, crystal structure, and material properties.
- It should always be kept in mind that mechanics and materials quantities are distributed random variables, even though in many cases we operate with their mean values.

References

AISC (2000). *Load and Resistance Factor Design Specification for Structural Steel Buildings.*

Roylance, D. (1996). *Mechanics of Materials.* Wiley, New York.

Schodek, D. L. (2001). *Structures*, 4th ed. Prentice-Hall, Upper Saddle River, NJ.

Seely, F. B. (1925). *Resistance of Materials.* Wiley, New York.

Shackelford, J. F. (1996). *Introduction to Materials Science and Engineering*, 4th ed. Prentice-Hall, Upper Saddle River, NJ.

Shigley, J. E., and Mischke, C. R. (2001). *Mechanical Engineering Design*, 6th ed. McGraw-Hill, New York.

Problems

1-1. What is a structure? Give examples. Give examples of physical artifacts that are *not* structures.

1-2. What is the difference between a primary and a secondary structure?

1-3. Give two examples for each of the following:

	Aerospace	*Civil*	*Marine*	*Mechanical*
Early structure	1.	1.	1.	1.
	2.	2.	2.	2.
Modern structure	1.	1.	1.	1.
	2.	2.	2.	2.

1-4. Is the skin of an airplane an example of a monocoque structure? Of a car? Of a human?

1-5. Give a simple example of how physics and chemistry provide different points of view of the same phenomenon.

1-6. In the design of a paperclip, how does the requirement for being reusable relate to mechanics issues? Materials issues?

1-7. Give an example of a dynamic load whose magnitude remains constant but whose direction of application changes with time.

1-8. Through what mechanisms do concentrated forces act? Gravity forces? Electromagnetic forces?

1-9. What are the different types of loads acting on the pole due to a steady wind blowing, at points A, B, and C in Figure P1-9?

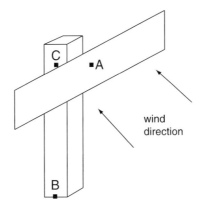

FIGURE P1-9

1-10. What is the type of loading in each of the members A to D in Figure P1-10?

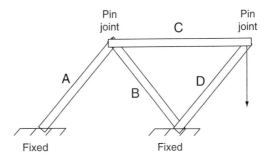

FIGURE P1-10

1-11. What is the loading at points A and B in Figure P1-11?

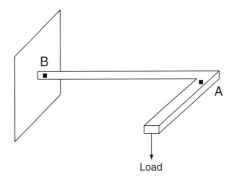

FIGURE P1-11

1-12. Give two examples of structural designs/products that were inspired by new materials, or the novel use of existing materials. (You may wish to conduct a search on the Internet or in the library.)

1-13. Draw the load path in the specimen, shown in Figure P1-13, subjected to a tensile load. Will the specimen bend under the load or remain straight?

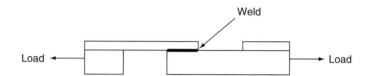

FIGURE P1-13

1-14. Sketch the load paths for the following structures:

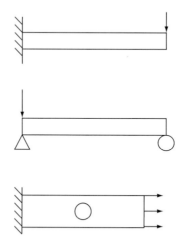

FIGURE P1-14

1-15. Give two examples each from everyday life of real structures undergoing:

  - Tension
  - Compression
  - Shear
  - Torsion
  - Bending

1-16. Give an example from your local environment for each of the structural elements listed in Table 1-2/Figure 1-8.

1-17. When does a beam become a plate and vice versa?

1-18. Give an example from everyday life for each of the boundary conditions shown in Figure 1-9.

1-19. Sketch beams with the following combinations of boundary conditions:

  - Free-clamped
  - Pinned-roller
  - Pinned-clamped

  Draw a correct and complete free body diagram for each end of each beam.

1-20. Which is more conservative, an FS of 2 or 3? Why?

1-21. (a) Define isotropic and anisotropic.
   (b) Give two or three examples of anisotropic materials.
   (c) In common engineering materials, what is the type of relationship between crystal structure and mechanical properties?

1-22. From Table 1-4, select one material for each application and give a brief description of how it is used in each case.

1-23. Give names of at least five elements that have the BCC, FCC, and HCP structures.

1-24. Which of the elements named in Problem 1-23 are ductile and which are brittle? What are their elastic moduli?

1-25. The energy of interaction of two atoms separated by a distance r can be written as

$$U(r) = -\frac{\alpha}{r} + \frac{\beta}{r^8}$$

where α and β are constants.

   (a) State which of the terms is 'attractive' and which is 'repulsive'.
   (b) Show that for particles to be in equilibrium $r = r_0 = \left(\frac{8\beta}{\alpha}\right)^{\frac{1}{7}}$.
   (c) In stable equilibrium show that the energy of attraction is 8 times that of the repulsion.

   (For this problem, the student should refer to Review Module 1.)

1-26. For the data supplied in Table P1-26A:

TABLE P1-26A Data on the Ultimate Tensile Strength of Copper (listed in order of decreasing magnitude)

Sample number	Strength (MPa)	Sample number	Strength (MPa)
1	1175	21	1215
2	1185	22	1215
3	1190	23	1217
4	1193	24	1218
5	1196	25	1219
6	1198	26	1220
7	1201	27	1220
8	1203	28	1222
9	1203	29	1223
10	1206	30	1223
11	1206	31	1226
12	1207	32	1226
13	1210	33	1228
14	1210	34	1228
15	1211	35	1230
16	1211	36	1230
17	1212	37	1236
18	1213	38	1239
19	1214	39	1242
20	1215	40	1252

TABLE P1-26B Frequency Distribution of Ultimate Tensile Strength

Group intervals (MPa)	Observations in the group	Relative frequency	Cumulative frequency
1166-1178	1	0.025	0.025
1179-1192	2	0.050	0.075
1193-1206			
1207-1220			
1221-1234	9	0.225	0.900
1235-1248			
1249-1261	1		
Total	40		

- Calculate the mean strength, and the one, two, and three standard deviation values
- Complete Table P1-26B
- Plot relative and cumulative frequencies versus ultimate tensile strength

2   Total Structural Design

Scientists discover what is, engineers create what never was.

—Theodore von Kármán

Objective: This chapter will introduce the concepts of engineering design, total structural design, the generalized structural design template, and methods of structural analysis and materials selection.

What the student will learn:

- The basic concepts of engineering design
- The extended concept of total structural design
- How these concepts can be implemented using standardized procedures
- Mechanics ⇔ materials ⇔ design link: structural hierarchy

2.1 The Design Space 🖳

Design is a creative human endeavor that is the essence of engineering. *Design* is an activity that turns abstract information about the need for a product or process into concrete knowledge about realizing that product or process (Figure 2-1).

Design sits at the intersection of science, art, business, politics, and psychology. Design solutions are nonunique: different engineers will arrive at different design solutions; some will be better solutions than others, but there is no single, correct solution to any design problem.

This last point is important, because it leads us to define the *design solution space*. Before we describe this abstract space, however, let's step back and think about physical space, for example the space of the environment you're in right now. We usually define three-dimensional physical space by a set of three coordinates, say $x-y-z$. If you're indoors, there's probably an intersection of two walls and the floor that could quite nicely form the coordinate axes of the space (Figure 2-2).

We can now think of the physical space as being defined by the coordinates, and consequently define a variety of solutions in this space, e.g., the trajectory of a particle to given excitation. If that trajectory depends on each of the coordinates independently, then we would

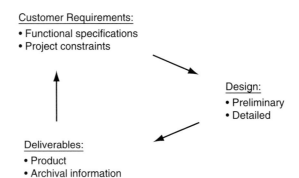

FIGURE 2-1 *The design process at its most fundamental level is satisfying customer requirements.*

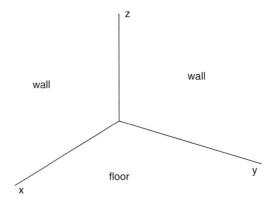

FIGURE 2-2 *Physical coordinate space.*

say the particle had three *coordinate degrees of freedom* (CDOF). A simple example would be the magnitude of the particle's distance d from the origin:

$$d(x, y, z) = \sqrt{x^2 + y^2 + z^2} \qquad (2.1)$$

If the particle were constrained to move only along the $x-z$ wall, then there would be just two coordinate degrees of freedom, since one of the coordinates is constrained by the equation $y = 0$, that is:

$$d(x, z) = \sqrt{x^2 + z^2}, \; y = 0 \qquad (2.2)$$

If the particle is now constrained to move in a circle of fixed radius R along that same wall, located with its center at (x_0, z_0), how many degrees of freedom are there? Well, there is only one coordinate degree of freedom, since x and z are no longer independent, but related by the equation of the circle.

$$x^2 + z^2 = R^2 \qquad (2.3)$$

(The position in this case will be left as an exercise for the student.)

A set of coordinates can thus define a *space*, and this can be a physical space as just discussed or an abstract space, such as the space of real and imaginary numbers. In a like manner, we will use a variety of quantities as generalized coordinates to define the *design space*. For example, we might have a design problem where we consider only elastic modulus E, tensile yield strength S_{yt}, and package volume V as the design variables or *design degrees of freedom* (DDOF). The design space would then look like Figure 2-3.

We would look for a solution inside this design space. Most likely, we would also constrain the solution (often because of practical issues), just as in our particle problem, by placing limits on, for example, volume—say $V < V_{max}$, some maximum volume (Figure 2-4).

Our search for a solution is now constrained to be below the plane defined by $V = V_{max}$. If we similarly constrain the other DDOF, you can imagine we would narrow our search to within a box. The important point is that **each of the design degrees of freedom can be manipulated within limits to achieve the best or optimal design.**

In what follows, we will talk of design degrees of freedom (DDOF) in a manner similar to speaking of coordinate degrees of freedom. Of course, the number of DDOF could be quite

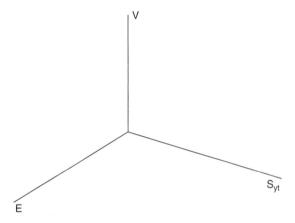

FIGURE 2-3 *Abstract design space.*

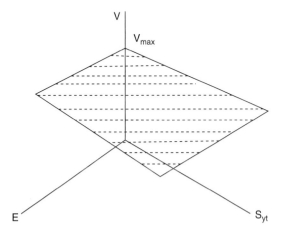

FIGURE 2-4 *Design space constrained by limits on one of the DDOF.*

TABLE 2-1 Examples of Design Degrees of Freedom

Mechanics DOF	Materials DOF	Performance DOF
• Coordinate: translation, rotation	• Weight: mass density ρ	• Cost
• Geometry: length, area, moment of inertia	• Modulus: tensile, shear	• Precision
• Load: type, magnitude, direction	• Strength: yield, ultimate	• Robustness

large, which would no longer let us easily draw the space, but we can nonetheless still think in these terms. Table 2-1 gives examples of a few design degrees of freedom, degrees of freedom that can be manipulated in seeking the design solution (more DDOF will be discussed throughout this text).

2.2   Total Structural Design 'Y'🏯🖥

Modern design encompasses *life-cycle design*, in that the entire product life cycle is considered, "from cradle to grave." For example, in the design of a product, considerations may be given to resources required to:

- Extract and process materials
- Analyze structural response
- Manufacture, assemble, and maintain the product
- Remove the product from service

Modern design is carried out by interdisciplinary teams, with many of the design activities carried out concurrently, or in parallel. This is in contrast to past design practice of sequential design, often called "throw it over the wall" design. Concurrent design relies upon strong communication, such as detailed and shared databases, and groups of diverse people working together for a common purpose.

A *team* is a collection of people who are united by a mutually shared commitment to a common goal, who have complementary skills, and who hold one another mutually accountable for the attainment of that goal. At this level, there are no individual failures, only team failures.

If a structure is required as part of a system, then that structure must be considered within the context of the entire system design process. This is what we refer to as *total structural design*. It is not enough to consider only that the structure carries loads safely. The engineer must also consider how the:

- Structure interfaces with other elements of the system
- Manufacture and assembly of the structure can be made most cost-effective
- Maintenance and repair of the structure can be made most cost-effective
- Structure contributes to cost and weight budgets for the system
- Structure functions in the social context within which it will exist

It is outside the scope of this text to provide a comprehensive discussion of engineering design. The interested student will find many good references available for further study (e.g., see Dieter, 2000; Dym and Little, 2004; Ullman, 1997).

2.3 The Generalized Structural Design Template 'Y' 🏛️🖥️

We now provide a template or pattern for the total design of structures (Figure 2-5). This template is sufficiently general to be used for the design of any structure. It covers three basic activities of the design process: problem definition, preliminary or conceptual design, and detailed design. In subsequent chapters, we will adapt the template to the design of specific structures.

Problem Definition

⇒ Goal: Define the design space
- Define the *service environment*: loads, configurations, physical environment, system interactions, level of uncertainty
- Define *performance requirements*: weight, longevity, safety factor, robustness, …
- Define *project constraints*: cost targets, time targets, physical constraints (manufacturing limitations, assembly limitations, …), system interaction constraints (interfacing constraints, loadsharing, …)

Preliminary Design

⇒ Goal: Manipulate design degrees of freedom (DDOF) to determine a reasonable preliminary solution in the design space
- Define the basic DDOF:

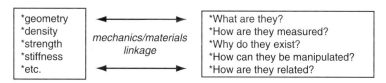

- Perform trade study: investigate various trades (compromises) among the DDOF
- Perform a failure modes effects analysis (FMEA)

Detailed Design

⇒ Goal: Provide the information (specifications, tolerances, materials, drawings, …) necessary to achieve the design in an optimal manner
- Consider:
  - boundary conditions: contact, crushing, stress concentrations, …
  - attachments: joining, stress concentrations, weight, …
  - materials: corrosion, fatigue, "engineer" the material, …
  - life cycle: cost, recyclability, maintainability, repairability, …
  - optimization: with respect to specs., constraints, other trades, …
- Define advanced DDOFs:
  - CTE
  - Fracture toughness
  - Resilience
  - Hardness
  - Fatigue strength
  - Creep compliance/relaxation modulus
  - Ductility
  - Thermal conductivity
  - Corrosion resistance
  - Abrasion resistance

FIGURE 2-5   *The generalized structural design template.*

2.4   Methods of Structural Analysis

There are two parallel approaches to structural analysis: the force method and the displacement method. The *force method* relates unknown forces (internal forces and reactions) to known forces (loads) through the equilibrium equations. This is the familiar method from basic Newtonian statics, where forces (and moments) are summed in order to find the net resultant. This approach requires a *free-body diagram* (see Review Module 3).

If the number of equations equals the number of unknowns, the problem is *statically determinate*, and the analysis proceeds in a straightforward manner. If the number of unknowns exceeds the number of equations, then the structure is *statically indeterminate*, consisting of a number of redundant members, which requires special treatment with the force method.

In the *displacement method* (also known as the *direct stiffness method* and *matrix structural analysis*), displacement variables are the primary unknowns. Modified equilibrium equations are developed that relate displacements to forces through an inherent structural stiffness. No special treatment of statically indeterminate structures is necessary in the method; thus the displacement method provides a universal approach to analyzing structures.

In fact, the displacement method is the basis for most finite element (FE) computer codes available today. Finite element methods (FEM) are powerful tools that have revolutionized the design and analysis of structures since the latter half of the 20th century. In this text, we emphasize the displacement method in order to provide a firm foundation for later study of FEM.

2.5   Failure Mechanisms in Materials

2.5.1   Stress Concentration

Consider a plate with a circular hole subjected to a tensile stress at the ends, as shown in Figure 2-6a. If a portion of the plate is cut out at plane A-A, the stress will be uniformly distributed across the cross-section of the plate as shown in Figure 2-6b, and the system will be in equilibrium.

If a section of the plate is cut out at the plane B-B, the stresses cannot be transmitted across the region where the hole is present. Thus the question arises: what is the distribution of stresses in plane B-B? It has been found that the stresses in the region adjacent to the discontinuity, which is the hole in this example, are higher than those at some distance from the hole. Qualitatively the stress distribution in the plane B-B is as shown in Figure 2-7.

The stress is said to be "concentrated" in the region of the discontinuity; hence the name *stress concentration*. Quantitatively, a *stress concentration factor*, C_σ, can be defined as

$$C_\sigma = \text{Maximum stress at discontinuity/Maximum nominal stress}$$

The stress concentration factor depends on the nature of the applied load, the geometry of the discontinuity, and the dimensions of the specimen in relation to those of the discontinuity. The *nominal* stress is usually the stress based on the reduced area at the discontinuity.

A way of visualizing the effect of an abrupt change in cross-section, such as the hole in the plate, on the stress distribution is the *force flow analogy*. In Section A-A in Figure 2-6a, the force flow lines are uniform as shown in Figure 2-6b. The presence of the hole makes the force

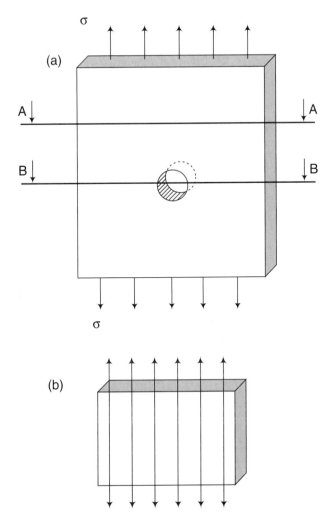

FIGURE 2-6 *Axially loaded plate with a central hole. (a) Basic configuration with nominal stress applied. (b) Section of the plate cut at section A-A.*

flow lines crowd together while passing around the discontinuity, causing an increase in stress around the discontinuity, as shown in Figure 2-7. The maximum stress concentration factor for a large plate with a finite hole is $C_\sigma = 3.0$ under remote tensile loading. The data for this case and other stress concentration factors may be found in references such as Shigley and Mischke, 2001.

It may be noted that the stress concentration effect dies off at a distance of the order of a few hole diameters, and thus the effects of geometric discontinuities are localized. This implies that the local stress around discontinuities in a structure can be found by determining the nominal stress distribution, neglecting local discontinuities, and then multiplying the nominal stress by the appropriate stress concentration factor.

Stress concentration factors are critical in structural design, and neglecting them can result in catastrophic failures. An example is the structural failure of the Comet jet airliner in the

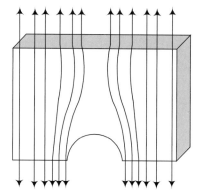

FIGURE 2-7 *Section of the plate in Figure 2-6 cut at Section B-B.*

1950s. The cause of the failure was determined to be the initiation and propagation of cracks from rivet holes 3.2 mm (0.125 in) in diameter (Gordon, 1984). Other examples of stress concentration are fillets, oil holes, sharp corners, keyway grooves, or any location where there is an abrupt change in the shape of the component. The susceptibility of a material to failure in the presence of the stress concentration under a given loading condition is an important consideration. A brittle material may fail more easily than a more ductile material in the presence of the higher stress at the discontinuity because it cannot relieve the stress concentration by material flow (see Chapter 6). Hence a judicious material selection is important.

2.5.2 Fracture

If we take a piece of paper and try to tear it by pulling at the two sides, the paper may tear along a zigzag path in the middle or just tear where we grip it. But if we fold the paper, press it along the fold, and repeat the tearing process, it tears along the fold. The reason is that a stress concentration was created along the fold, which provided a preferred path for tearing. (Also note that paper is a fiber/matrix composite, which may exhibit a different tear response in different directions, that is, paper may exhibit anisotropic behavior. Try tearing a piece of newspaper in the length and width directions and compare the results.)

Consider another example: breaking a glass plate simply supported along two edges and struck in the middle with a hammer—the glass plate will break into many pieces. Now take a diamond-tipped cutter and scratch one surface along a line, in effect creating a sharp-tipped notch or a crack-like defect along the line. Again hit the plate on the face opposite the scratched face. The plate will break along the line created by the diamond cutter. The reason is that the line crack created a region more susceptible to cracking than the rest of the plate. However, if an aluminum plate is scratched with the diamond-tipped cutter and hit by a hammer, no apparent failure may be noticed. This is because different materials have different sensitivities to the presence of a crack-like defect. The brittle glass plate is very sensitive to the presence of a crack and fails when tapped with the hammer, whereas the considerably more ductile aluminum is much more resistant to failure in the presence of the crack when loaded with the hammer.

In this section we introduce some basic aspects of fracture mechanics (a more detailed description is given in Chapter 7). *Fracture mechanics* in a broad sense is that part of solid mechanics that relates to the study of the load-carrying capacity and the remaining life of a structure in the presence of cracks.

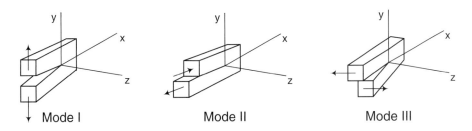

FIGURE 2-8 *Crack-tip loading modes.*

When a crack is introduced in, say, a plate, two free surfaces and a sharp edge are created. The sharp edge is called the *crack tip*. This crack can be primarily loaded in three ways, called *crack-tip loading modes*. Each loading mode is associated with a distinct mode of crack deformation, as shown in Figure 2-8. The crack opening mode, Mode I, is associated with a local displacement in which the crack surfaces move away from each other perpendicular to the leading edge of the crack. The shearing (forward sliding) mode, Mode II, is developed when crack surfaces slide over each other in a direction perpendicular to the leading edge of the crack. The tearing mode, Mode III, is characterized by crack surfaces sliding with respect to each other in a direction parallel to the leading edge of the crack. Superposition of these three modes can fully describe the most general three-dimensional case of local crack-tip deformation and stress field.

In Chapter 7, the stress-field equations will be developed. However, qualitatively it may be mentioned here that the stress in the vicinity of the crack tip is a function of a factor K called the *stress intensity factor*, and this single parameter characterizes the stress field in the vicinity of the crack tip (under a special condition that the material around the crack tip undergoes "small-scale yielding," that is, the size of the plastic zone is small compared to the dimensions of the crack). The factor K also depends on the mode of crack-tip loading and is represented by K_I, K_{II}, K_{III}, associated with Modes I, II, or III, respectively.

In this case, the stresses around the crack tip can be developed on the assumption that the material behaves in a linear elastic fashion. Such an approach is often termed as *linear elastic fracture mechanics* (LEFM). The stress intensity factor has the general form (regardless of the loading mode)

$$K = F\sigma\sqrt{\pi a} \qquad (2.4)$$

where F is a factor dependent on the type of loading and the geometry away from the crack, the gross nominal stress is σ, and a is the *crack length*. When K exceeds a certain intensity value called the *critical stress intensity factor*, K_c, crack extension occurs at a rapid pace. Thus K_c is a measure of the resistance of the material to crack propagation and is also known as the *fracture toughness* of the material. Fracture takes place when any one of the stress intensity factors, K_I, K_{II}, or K_{III}, reaches its corresponding critical value, K_{Ic}, K_{IIc}, or K_{IIIc}. Among the critical stress intensity factors, K_{Ic} is typically most critical. Thus for most materials, Mode I loading is the governing mode of brittle fracture.

Returning to the example of the glass plate with a line crack, it may be mentioned that Mode I loading with a hammer increases the stress intensity factor of the crack to a value greater than K_{Ic} of glass, causing crack extension that ultimately breaks the glass into two pieces.

It should be noted that, in general, it is not necessary for the fracture stress (σ) to exceed the yield strength of the material for fracture to occur. Depending upon the size of the crack and critical stress intensity of the material, the fracture stress may or may not have to exceed the yield strength. For short cracks in a ductile material such as aluminum, the fracture stress may exceed the yield strength, but will be less than or equal to the ultimate strength.

2.5.3 Fatigue

Failure by fracture under fluctuating stress or strain is called *fatigue*. Fatigue failures may be the most common type of failure in engineering structures and result in a loss of several percent of the gross national product every year.

Though generally in practice the repeated loads are random (Figure 2-9a), much fatigue testing and modeling involves harmonic (cyclic) loading between maximum and minimum stress levels that are constant, as shown in Figure 2-9b.

A fatigue failure is essentially the formation and growth of cracks under the action of cyclic loading. Three major factors that affect the fatigue life of a structure are (1) material: hardness, metallurgical structure impurities, stress concentrations and surface condition; (2) environment: temperature, humidity, corrosive and multiaxial stresses; and (3) nature of loading. The effect of these factors on the fatigue life of a component and many other issues are not presented in detail in this text (see, for example, Shigley and Mischke, 2001). However, the following are some basic observations on fatigue failure:

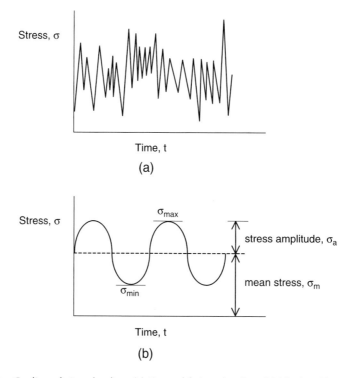

FIGURE 2-9 *Cyclic or fatigue loading. (a) General fatigue loading, (b) idealized harmonic loading.*

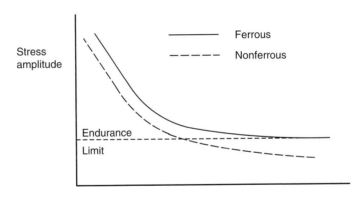

FIGURE 2-10 *Fatigue life as a function of stress amplitude.*

1. As the stress amplitude decreases, the number of loading cycles to failure increases.
2. Below a stress amplitude called the *endurance limit*, ferrous materials exhibit a very long ("infinite") fatigue life as shown in Figure 2-10 (often taken to be greater than 10^6 cycles). However, nonferrous metals do not exhibit such a phenomenon.
3. Hostile chemical environments decrease the fatigue life.
4. Microstructure changes can significantly influence the fatigue behavior. For example, by reducing the size of inclusions and grain size, the fatigue life can be improved.
5. Residual stresses (which are internal stresses in the material) can significantly affect the fatigue life. Compressive stresses are beneficial, while tensile stresses are deleterious.

2.6 Material Selection in Structural Design 🕸️💻

We saw in Sections 1.4 and 1.5 that the classes of general structural forms are relatively small when compared to the classes of materials. Consequently, the universe of materials from which the designer can choose is considerably large. The population includes existing materials that are either naturally occurring (such as wood) or common synthetic materials such as steel or plastic. In some cases, the design may call for a specially engineered material, such as a fiber-reinforced composite. Choosing one material, or even one class of material, from among this bewildering array can be daunting, not to mention inefficient.

Moreover, there are many questions we must ask and answer about the material as we proceed along the design process. Materials, and their processing and manufacturing, are all complex issues. Table 2-2 gives examples of materials information that might be required at various stages in structural design.

2.6.1 Ashby's Method of Material Selection

So, how do we proceed to select materials in a design problem? Ashby has made great improvement to this process by introducing *material property charts*, along with a structured procedure for their use (Ashby, 1999). In our parlance, the charts plot one design degree of freedom against another. This might be elastic modulus E versus density ρ (see Figure 2-11 for a schematic representation), or specific strength (strength divided by density) versus cost. If the ranges of the axes are appropriately chosen (often logarithmic), a large population of materials may be represented on a single chart. Classes of materials, metals for example, will cluster

TABLE 2-2 Examples of Materials Information Required in Structural Design (following *Computer-Aided Material Selection in Structural Design*)

Material identification	Environmental stability	Cost factors
Class, subclass	Toxicity	Raw materials
Industry designation	Compatibility	Special handling
Product form	Corrosion resistance	Special finishing
Condition designation	Recyclability/disposal	Special tooling
Material production history	Flammability	Recycling/disposal
Material properties and tests	**Processability**	**Availability**
Strength	Formability	Sources/vendors
Modulus	Weldability	Sizes
Toughness	Machinability	Shapes
Fatigue behavior	Castability	Amounts
Thermal expansion	Repairability	Delivery time
Density	Previous history	
Conductivity		

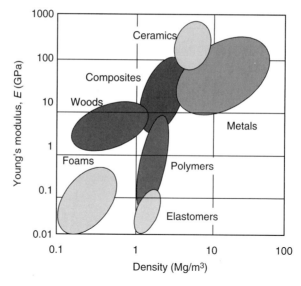

FIGURE 2-11 *Young's modulus versus mass density chart. (Reproduced with permission of Granta Designs.)*

together on the chart and are so noted by a "balloon"; subclasses within a class become "nested" balloons (Figure 2-11).

The utility of a given chart is increased by "lines of constant material indices." A *material index* (MI) groups together into one measure the salient DDOFs. When maximized, the material index optimizes some aspect of the structural/material system. For example, the material index for a minimum weight stiffness design of a tensile rod (for a given load, length, and stiffness specified) is E/ρ. This makes sense, since in this case we want to maximize the inherent stiffness of the material (E) for the minimum weight penalty (ρ). Thus when the material index is maximized, it provides the best solution to the design requirement. Figure 2-12 shows another view of a modulus versus density chart with material index shown.

Ashby and his co-workers have developed many other material indices, depending on the mode of loading: axial, torsion, bending, etc. Cross-sectional shapes can also be included in the index. A basic set used in this text is given in Table 2-3.

On the E versus ρ chart just mentioned, all materials falling on the line of constant $\sqrt{E}/\rho$ (called a "guide line") will perform equally well in a minimum-weight stiffness design, for a rod of those materials loaded in compression or a beam in bending. Materials that lie above the

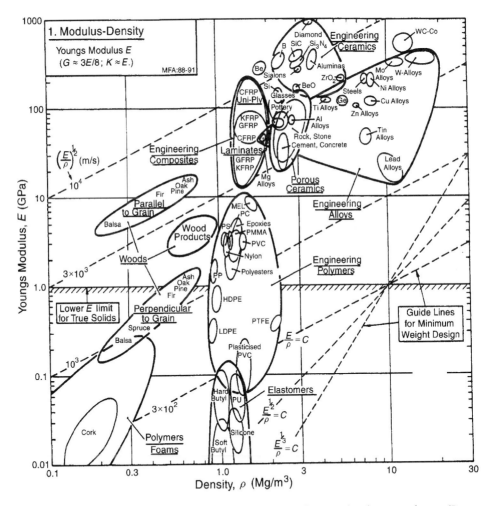

FIGURE 2-12 *Ashby modulus versus density chart. Several material indices are shown. (Reproduced with permission of Granta Designs.)*

TABLE 2-3 Some Material Indices[a]

Structural Element	Parameters Specified	Stiffness	Strength
Rod (tension)	Loading, stiffness, length	E/ρ	S_f/ρ
Rod (compression)	Collapse load, length	$\sqrt{E}/\rho$	S_f/ρ
Torsion bar	Loading, stiffness, length	$\sqrt{G}/\rho$	$S_f^{2/3}/\rho$
Beam	Loading, stiffness, length	$\sqrt{E}/\rho$	$S_f^{2/3}/\rho$

[a]E, Young's modulus; G, shear modulus; S_f, failure strength; ρ, mass density.

line are better than those that lie on the line. One must be careful, however, in making various trades among the indices. Cost is an obvious example. While a certain material may maximize stiffness of a rod, it may be nonoptimum in terms of cost. Strength, ductility, manufacturability, and a variety of other factors must be considered for a truly optimal design.

2.6.2 Material Selection in Total Structural Design

Whenever we design and create a physical object, it has to be made of an appropriate material, such that it enables the object to be manufactured and then perform the desired function over its expected life. Hence the design, manufacture, performance, and durability are all related to the optimum choice of the material.

The design process consists broadly of five steps as shown in Figure 2-13. In modern concurrent design, all of these steps are, at some level, influencing one another continuously and iteratively during the product development cycle.

In the *conceptual design* stage, all the various ideas and schemes are evaluated for the functions that will be performed by the object. Then one or more schemes are selected for further analysis. In the *embodiment design* stage, the approximate dimensions are determined and materials category (metal, plastic, composite, etc.) is selected, for each of the acceptable schemes. In this design stage a basic layout for each design alternative is prepared. In the

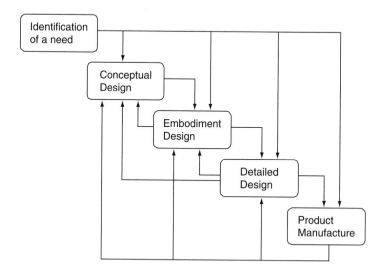

FIGURE 2-13 *A typical methodology for product design.*

detailed design stage, each component of the object is carefully analyzed using engineering principles and the exact material specification is determined. Also, the production process is analyzed and a cost analysis is done. If any changes in the material selection, manufacturing process, or functionality of the object are desired, they must be caught before moving into the production phase.

Thus it is clear from the brief description of the design process that materials selection and evaluation is an integral part of total structural design. In the embodiment stage, the broad category of material is first selected. For example, we may choose a metal and then specify whether it is a ferrous or nonferrous metal—for example, whether it is steel or an aluminum alloy. If the choice was a composite, then the decisions made may be in this order: a metal matrix or polymer matrix composite; the choice of a fiber or particle reinforcement; the type of fiber or particle, such as glass or graphite fibers, or glass or ceramic particles; and a broad specification of the final type of material, such as graphite fiber–reinforced aluminum matrix composite. However, in this design stage no decision is made on the exact nature of the steel or aluminum alloy; for example, we do not yet specify the exact grade of steel with its carbon content, alloy content, and heat treatment. In the case of a composite material, we do not yet specify the exact fiber volume fraction, any fiber coatings or fiber lay-up directions, etc. The exact and final specifications are provided in the detailed design stage.

Even an approximate selection of materials in the conceptual and embodiment stage can be daunting, considering the fact that there are more than 100,000 different engineering materials from which to choose. Moreover, these materials present a very wide range of physical and mechanical properties, and mechanical responses. For example, the elastic modulus could vary from 0.1 to 1000 GPa, the density from 0.1 to 20 g/cm^3, strength from 0.1 to 10,000 MPa, and the fracture toughness over a range of 0.1–300 MPa$\sqrt{\text{m}}$. Also, very different materials can have the same strength, same elastic modulus, or similar fracture toughness.

Hence without any materials selection methodologies and organization of material properties depicting their interrelationships, the choice of an optimum material will at best be difficult. Fortunately we do have help. Several methodologies for materials selection are (Dieter, 2000):

1. Material indices
2. Decision matrices
3. Selection with computer-aided databases
4. Value analysis
5. Failure analysis
6. Benefit–cost analysis

As previously discussed, the organization of the material properties into charts depicting their interrelationships and the use of material indices has been provided by Ashby (1999). These charts are very useful in selecting basic material types (e.g., ceramics, composites, metals, wood) and comparing material choices, if more than one exists. These charts are suitable for evaluating materials in the conceptual and embodiment design stages only. In the detailed design stages, more precise material property data should be used. A variety of Ashby's materials selection charts are available (Ashby, 1999), some of which are also shown in Appendix A2. The entire Ashby method has been encapsulated into the Cambridge Engineering Selector software from Granta Designs (www.grantadesign.com).

Another very useful method of materials selection is the use of *decision matrices*. In the use of this method for materials selection, three important factors must be considered. These factors are: (a) the alternative materials or material–process pairs (the "alternatives"), (b) the material properties or material performance characteristics that will enable the functional requirements of the design to be fulfilled (the "decision criteria"), and (c) the "weighting factors", which are the numerical representations of the relative importance of each criteria. The weight factors should add up to 1; the most important criteria are assigned the maximum weight factor.

The use of decision matrices and Ashby's charts for selecting materials for a desired function in the design of a structural element is illustrated in the chapters that follow through design examples.

EXAMPLE 2-1:

An application of a structural link requires it to function under certain loading and environmental conditions. For example, such a link could be used to operate a gate valve on an HVAC system, rotate the blades on a gas turbine, or adjust the opening of the guide vanes in the water inlet system to a hydroelectric turbine. The choice of a material depends on the application and the mechanical properties of the link called for by the application. Some desirable mechanical properties for a link may be:

1. Elastic modulus, $E \Leftrightarrow$ stiffness
2. Yield strength, $S_Y \Leftrightarrow$ yielding by plastic flow of material
3. Stress concentration effect $\Leftrightarrow$ increased local stresses
4. Durability $\Leftrightarrow$ fatigue life
5. Fracture toughness $\Leftrightarrow$ resistance to failure by crack-like defects

Some additional desirable properties of the link may include:

- Minimum thermal distortion (for gas turbine)
- Durability under impact loading (for hydroelectric turbine)
- Resistance to thermal shock (for gas turbine)

Given next are some design goals for the link under consideration, which are often encountered in engineering design, and the corresponding material property or properties, which should be optimized.

(a) Design goal: To obtain high stiffness and low weight
   Material index: Maximize the ratio of elastic modulus to density, E/ρ
   Possible materials: Carbon fiber reinforced plastics (CFRP), graphite fiber reinforced plastics (GFRP), aluminum alloys, ceramics
(b) Design goal: To obtain high stiffness, low weight, and low cost
   Material index: Maximize the ratio of elastic modulus to density, $E/\rho C_R$, where the relative cost, C_R, is the ratio of the cost of the material to the cost of mild steel
   Possible materials: GFRP, aluminum alloys, steel
(c) Design goal: To obtain high strength and low weight
   Material index: Maximize the ratio of yield strength to density, σ_y/ρ
   Possible materials: GFRP, CFRP, ceramics, aluminum alloys, steel
(d) Design goal: To obtain high strength, low weight, and low cost
   Material index: Maximize the ratio of yield strength to density, $S_Y/\rho C_R$
   Possible materials: GFRP, aluminum alloys, steel

EXAMPLE 2-1: *Cont'd*

(e) Design goal: To obtain high stiffness, high strength, and low weight
Material index: Maximize the ratio of specific strength to specific modulus, $(\sigma_y/\rho)/(E/\rho) = \sigma_y/E$.
Possible materials: Aluminum alloys, CFRP, GFRP, steel

(f) Design goal: To obtain high stiffness, high strength, low weight, and low cost
Material index: Maximize the ratio of specific strength to specific modulus, $(\sigma_y/\rho)/(E/\rho) = \sigma_y/E$
Possible materials: Aluminum alloys, GFRP, steel

(g) Design goal: To obtain high resistance to failure from crack-like defects and low weight
Optimum material
Property function: Maximize the ratio of fracture toughness to density, K_{Ic}/ρ
Possible materials: CFRP, GFRP, aluminum alloys, steel

(h) Design goal: To obtain high resistance to failure from crack-like defects, low weight, and low cost
Optimum material
Property function: Maximize the ratio of fracture toughness to density, K_{Ic}/ρ
Possible materials: GFRP, aluminum alloys, steel

(i) Design goal: To obtain high resistance to failure from crack-like defects and high strength
Optimum material
Property function: Maximize K_{Ic} and σ_y
Possible materials: Aluminum alloys, steel, CFRP, GFRP

(j) Design goal: To obtain high resistance to failure from crack-like defects, high strength, low weight, and low cost
Optimum material
Property function: Maximize K_{Ic} and σ_y
Possible materials: Aluminum alloys, GFRP, steel

Let us assume that the link is to be used to operate a gate valve of the HVAC valve and we set our design goals as high stiffness, high strength, high resistance to failure, low weight, and low cost. From (b), (d), and (j), we find that the materials that satisfy all the requirements well are graphite fiber–reinforced composites, steel, and aluminum alloys. In the next chapter we will solve the link example in more detail.

2.7 Mechanics ⇔ Materials ⇔ Design Link: Structural Hierarchy ⓨ⌂💻

We see that structures may carry a variety of different types of loads, and that some materials may be better suited for carrying certain types of loads than others. For example, wood is better suited to carrying tensile than compressive loads, concrete is better suited for compressive than tensile loads, and steel is relatively good at carrying both.

Concrete is often used in civil structures, because of its good surface properties and relatively low cost. But because of concrete's weakness in tension, it is usually reinforced with steel (Figure 2-14). This is an excellent example of structural hierarchy, in particular the

FIGURE 2-14 *Reinforcing steel bar (rebar) extending from a concrete block wall.*

partitioning of structural functions: we can make the structure more efficient by assigning the tension-carrying function to the reinforcing steel, and the compression-carrying function to the concrete. The point is this: increasing structural hierarchy increases the number of DDOFs, which in turn provides more options for achieving an optimal structural solution.

As you look around, you'll notice other similar examples. A sailboat assigns the tension-carrying function to the sails and guys (cables and ropes), and the compression-carrying function to masts and spars. Early biplanes used a similar partitioning (Figure 2-15).

A beam in bending is still another example. The outermost fibers of the beam see most of the tension and compression but little shear, and the neutral axis sees no bending but maximum shear. Then cross-sectional shapes such as an I or H turn out to be structurally very efficient, since the tension-and compression-carrying functions are relegated to the flanges, while shear is carried by the web. Making these components out of different materials exploits this partitioning even further.

Key Points to Remember

- It is important to define the design space as early in the project as possible. Design degrees of freedom are any of the independent variables that can be manipulated in the design, and they define the design space. The optimum solution is found within (usually) a constrained design space.
- Total structural design takes the modern life-cycle point of view for structural design: cradle to grave.
- The generalized structural design template can be used to break the structural design problem into three phases: problem definition, preliminary design, and detailed design.

FIGURE 2-15 *Tension cable/compression strut partitioning in a biplane wing.*

- Detailed design will include structural analysis concepts such as stress concentrations and cyclic loading. Detailed materials selection is facilitated using material selection charts, material indices, and decision matrices.
- Complex material/structural systems can be designed to achieve efficiency by designing in structural hierarchy, an example being assigning specific load-carrying functions to the material components best suited to carry those loads.

References

Anonymous (1995). *Computer-Aided Materials Selection During Structural Design (NMAB-467)*. National Academy Press, Washington.

Ashby, M. F. (1999). *Materials Selection in Engineering Design*, 2nd ed. Butterworth Heinemann, Boston.

Dieter, George E. (2000). *Engineering Design*, 3rd ed., Chapter 8. McGraw-Hill, New York.

Dym, C. L., and Little, P. (2004). *Engineering Design*. Wiley, New York.

Gordon, J. E. (1984). *The New Science of Strong Materials*. Princeton University Press, Princeton, NJ.

Shigley, J. E., and Mischke, C. R. (2001). *Mechanical Engineering Design*, 6th ed. McGraw-Hill, New York.

Ullman, D. G. (1997). *The Mechanical Design Process*. McGraw-Hill, New York.

Problems

2-1. A mass particle is constrained to a circular trajectory of radius R in x–z space. The center of the circle resides at the location $(x_0,\ z_0)$. Show that the distance of the particle from the origin of coordinates is given by:

$$|\mathbf{r}| = \sqrt{x_0^2 + z_0^2 + 2R(x_0\ cos\ \theta + z_0\ sin\ \theta) + R^2}.$$

2-2. A compound pendulum consists of two rigid massless links of length L (hinged together) and one massive bob of mass M (Figure P2-2). How many coordinate degrees of freedom are involved in this configuration?

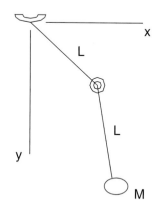

FIGURE P2-2

2-3. A dime rolls on a tabletop without slipping. How many coordinate degrees of freedom are required to completely locate a position on the face of the coin, say the tip of Roosevelt's nose?

2-4. In Table 2-1, add three more DDOFs each to the Materials DOF and Performance DOF lists.

2-5. Are stress and strain independent quantities that can be used simultaneously as DDOFs? Why or why not?

2-6. Discuss the "cradle-to-grave" considerations for design of an automobile.

2-7. Imagine you are to design a new mountain bike. Use the design template to:

- Define the problem
- Define the basic DDOFs
- Outline a trade study among the DDOFs
- Outline possible failure modes for the product

2-8. Determine whether or not the following structures are statically determinate:

- Beam with both ends fixed and central transverse load P (Figure P2-8a)
- Tripod with transverse load P (Figure P2-8b)

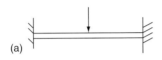

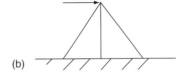

FIGURE P2-8

2-9. Why was the glass plate in Section 2.5.2 hit with a hammer on the side opposite the one containing the scratch?

2-10. How many possible fracture loading modes can be used to tear a sheet of paper by hand?

2-11. A crack is accidentally created on the outer surface of a pipeline used to transport oil. What is the crack loading mode? (Hint: Consider different directions of the crack length.)

2-12. If a crack is cut on the outer surface at 45° to the axis of a closed pressure vessel, what is the loading mode on the crack?

2-13. Consider a ductile and brittle material for a rotating shaft. Which one is expected to have a longer fatigue life?

2-14. Which of the two shafts shown in Figure P2-14 will have a longer fatigue life and why?

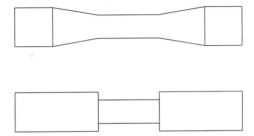

FIGURE P2-14

2-15. Which of the two plates shown in Figure P2-15 will exhibit a greater stress concentration on the x-axis? The diameter of the circular hole is equal to the major diameter of the elliptic hole.

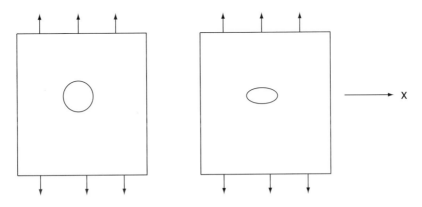

FIGURE P2-15

3   Design of Axial Structures

Ut tensio sic uis. (As the extension, so the force.)

—Robert Hooke, 1679

Objective: This chapter will introduce seven design degrees of freedom—ρ, E_t, E_c, S_{Yt}, S_{Yc}, A, u—in the context of the design of axial structures.

What the student will learn:

- Definition and configuration of axial structures
- Equilibrium and deformation of axial structures
- 1-D elastic constitutive relation
- Linkage of mechanics and materials concepts through the tension and compression tests
- The direct stiffness method as applied to axial structures
- Materials selection for axial structures
- How to design axial structures

3.1   Introduction 𝗬⟨⟩▱

Axial structures carry loads primarily in tension or compression. They are typically long, straight, slender structures, which could be categorized as one-dimensional (1-D) structures. Axial structures are often called by the names *bar*, *link*, *rod*, or *strut*. Whether or not a given structure responds, or can be represented to respond, as an axial structure depends on the nature of the loading, and on the structure's boundary conditions. For example, a truss member will be an axial structure if its ends (joints) are pinned, and if the loading occurs only through these pin joints, as shown in Figure 3-1.

In thinking of these structures as one-dimensional, we are ignoring the fact that all real structures deform simultaneously in three dimensions under the action of any load. We can get away with this neglect, for now, because we will only consider the structure's response to a single load, and that its multiaxial response may not be important for the given design. Later, when we consider combined loading (Chapter 6), we will be required to look at multidimensional response.

FIGURE 3-1 *Although the various members that make up the truss assembly may or may not be axial elements, the rods supporting the hanging lights clearly are axial.*

It is convenient to think of a locus of material particles, parallel with the axial structure axis (*x*), as a material "fiber." The axial structure may then be thought of as comprising a large number of identical longitudinal fibers, with one of the fibers colinear with the long axis of the element.

In addition to carrying axial loads in tension, axial structures can carry loads in compression, that is, axial loads applied so as to compress the element. If the element is rather compact, that is, short and stout (Figure 3-2), the response and analysis is not unlike that of tensile axial structures. Failure is governed largely by strength, since a compact element should be overly stiff. (However, some materials may behave differently in compression as compared to tension, and this will be discussed more fully later on.)

On the other hand, if the element is relatively slender (e.g., a column), compressive failure may be due to lack of stiffness, and a completely different response and analysis is required (this becomes a stability problem which is dealt with in Chapter 8). The axial element, at a given critical load, will buckle or warp (Figure 3-3). Excessive buckling can lead to complete collapse upon continued loading.

An example of slender compression members can be seen in Figure 3-1. Note the columns that carry roof loads down to the truss. We will defer to a later time the study of

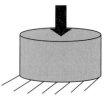

FIGURE 3-2 *Compact compressive axial structure.*

FIGURE 3-3 *Slender compressive axial structure undergoing buckling.*

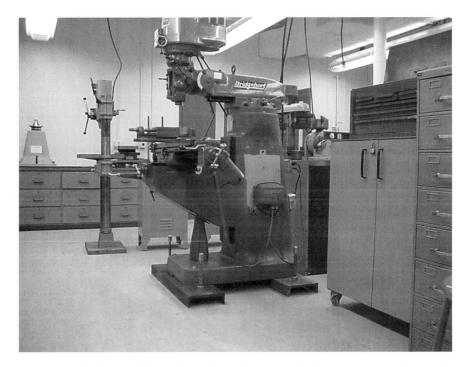

FIGURE 3-4 *The flanges of the "C-channels" used to support this heavy machine tool are undergoing compressive axial loads (as is the floor directly underneath).*

buckling behavior and how to design slender compressive axial structures (we need to first explore flexural structures). This chapter will provide information about design of compact compressive axial structures. An example of a compact compressive member can be seen in Figure 3-4.

It is often found that a *complex axial structure* is made up of axial elements some of which carry tensile loads while others carry compressive loads. The truss in Figure 3-1 is an example. Another example is that of a bolted joint (Figure 3-5). In this chapter, we will lay the foundation for consideration of the design of complex axial structures under combined tension and compression.

FIGURE 3-5 *A variety of bolted joints are displayed on this hydraulic cylinder. The material that has been joined is under compression, while the bolt itself is under tension.*

There are two key concepts to keep in mind as we discuss equilibrium and deformation of axial structures or any structure. The first is that the quantities' stress and strain are measures of the body's response at a point. The second is that any continuous body that is in global equilibrium must have all of its parts in equilibrium, that is, it must be in local equilibrium as well. We take advantage of these concepts to go from the externally applied forces to the internal stresses and strains.

3.2 Equilibrium and Deformation 'Y'

3.2.1 Configuration

In Figure 3-6, an axial structure called a bar, link, rod, or strut of round cross-sectional shape is shown. A force P is applied to one end of the rod, resulting in a displacement u. The other end of the rod has a boundary condition described as either fixed or built in. The rod has geometric properties of length L and constant cross-sectional area A; material properties are *mass density* ρ, *tensile elastic modulus* E_t, and *tensile strength* S_t. (These material properties will be discussed in subsequent sections of this chapter. At this point we make no distinction between tensile "yield" strength and tensile "ultimate" strength.)

(Keep in mind that although we have shown a rod with constant, circular cross-section, primarily for convenience and simplicity, a rod may have a nonconstant and/or noncircular cross-section. However, to remain an axial structure, the axis of every cross-section, which is

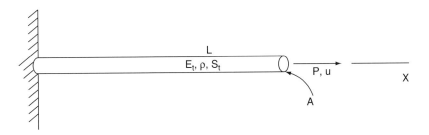

FIGURE 3-6 *Definition sketch of an axial structure.*

the line passing perpendicularly through the centroid of that cross-section, must remain colinear. Structures with constant cross-sectional area, or more generally constant area moment of inertia, are called *prismatic* structures.)

In this chapter, we assume that the compressive structural element is relatively compact, so that buckling considerations need not be introduced. (As we will see later in Chapter 8, compactness is a function of element length, cross-sectional geometry, material modulus, boundary conditions, and imperfections; hence the use of the term "relative" compactness.) For now, a *slenderness ratio* can be defined, which is the effective length of the element divided by $\sqrt{(I/A)}$, where I is the smallest *area moment of inertia* of a cross-sectional area A. A compression member is usually considered compact if the slenderness ratio is less than ~ 10 (see Chapter 8). For our purposes in the current chapter, we then likewise expect any applied compressive load to be less than the critical buckling load. We still, however, deal here with geometries that can be modeled as one-dimensional structural elements (Figure 3-7).

3.2.2 Loads

The loads on an axial structure must be *coaxial* loads (proved later), that is, loads acting coaxially with the significant structural axis (x-axis here), such as the concentrated force applied to the ends of the rods discussed earlier. The force shown in Figure 3-6 is tensile, since it acts to create tension in the bar, that is, acts to displace material particles further away from one another. In Figure 3-7 the force is compressive, so as to create a compressive action on material particles, that is, acting to displace material particles closer together than in their unloaded configuration. No moments are applied to axial structures.

3.2.3 Boundary Conditions

Here we need to be cautious. In the case of tensile axial structures, we readily allow the case of pinned–free boundary conditions. Such a configuration (of coaxial tensile loads and boundary

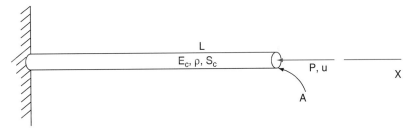

FIGURE 3-7 *Definition sketch of an axial structure in compression.*

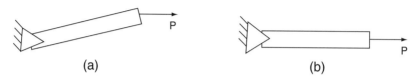

FIGURE 3-8 *Tensile axial structure slightly perturbed from equilibrium returns to the equilibrium position.*

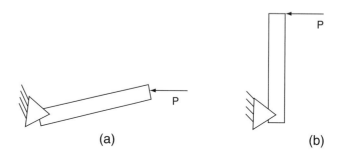

FIGURE 3-9 *Compressive axial structure slightly perturbed from equilibrium does not return to the equilibrium position, but seeks a new equilibrium position.*

conditions combined) is *unconditionally stable*. That is, any slight perturbation (disturbance) of the rod from its equilibrium position (Figure 3-8a) would simply return the rod to the equilibrium position (Figure 3-8b).

Now in the analogous compression case, if we perturb the rod slightly from its equilibrium position (Figure 3-9a), a radically different response occurs. In this case, the rod undergoes large excursions away from the original equilibrium position (until a new equilibrium position is found, as seen in Figure 3-9b). The original equilibrium configuration then must have only been conditionally stable (also called *unstable equilibrium*). We will discuss such issues in more detail in Chapter 8.

For our present discussions, we assume that the configurations (tensile or compressive coaxial load plus boundary conditions) are unconditionally stable. In the compressive case, pinned–pinned boundary conditions are a common example of such configurations.

3.2.4 Equilibrium

We now consider a global *free-body diagram* (FBD) of a rod (which should include the externally applied forces and moments, the internal reaction forces and moments, and the necessary geometry and coordinate descriptions necessary to invoke Newton's laws—see Review Module 3). The right end of the rod is "freed" from the body at some position x from the origin, and the corresponding FBD looks as shown in Figure 3-10.

Shown on the FBD is the internal reaction force F, which from global static equilibrium analysis (bold font indicates vector or tensor quantities)

$$\Sigma \mathbf{F} = \mathbf{0} \tag{3.1}$$

can be shown to be (P is the applied load and F is the internal reaction)

$$P - F = 0 \Rightarrow F = P \tag{3.2}$$

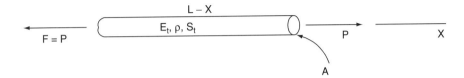

FIGURE 3-10 *Global free-body diagram of an axial structure.*

Furthermore, global moment equilibrium

$$\Sigma M = 0 \tag{3.3}$$

shows that F and P are colinear (coaxial), since they must present no net *couple* (moment) on the rod.

We now look more closely at the stress in an axial structure and the corresponding local equilibrium. Consider the local FBD or *stress element* carved out of the cross-section A, shown in Figure 3-11.

Now an increment of internal force, dF, can be seen to act over an increment of area, dA, which balances the increment of applied load dP (i.e., $dF = dP$). We take as a fundamental postulate (known formally as *Cauchy's stress principle*) that the externally applied load is resisted internally (to enforce local equilibrium) by a stress σ_x such that

$$dF = \sigma_x dA \tag{3.4}$$

Based on the assumptions and discussion for an axial structure given earlier, the stress σ_x is assumed to be constant as long as no intermediate axial loads are applied; hence a simple integration gives:

$$\sigma_x = \frac{F}{A} \tag{3.5}$$

(The stress state and strain state in an axial structure are considered to be relatively simple, in that it is assumed that only normal stresses and strains exist on cross-sections normal to the rod

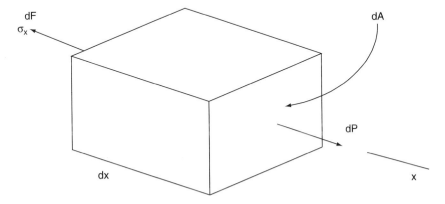

FIGURE 3-11 *Local free-body diagram of an axial structural element.*

axis. The stress distribution is also considered to be relatively simple, except in a very local region around where the load is applied, which we will ignore. Keep in mind that what we really calculate by Eq. (3.5) is an "average" stress acting everywhere over the cross-section of the bar, even though stress is fundamentally a measure of the load response by a structure at a point.)

We might imagine that on the internal face of the stress element, the incremental force is not dF but has changed by some *gradient* (spatial rate of change) of the force, dF/dx,

$$dF + (dF/dx)dx \tag{3.6}$$

Then a local application of equilibrium, that is, summation of (incremental) forces, in the x-direction gives (using dF and dP)

$$dF - [dF + (dF/dx)dx] = 0 \Rightarrow dF/dx = 0 \tag{3.7}$$

since $dx \neq 0$. Now, substituting $\sigma_x dA$ for dF from Equation (3.4) gives

$$d(\sigma_x dA)/dx = 0 \tag{3.8}$$

or, since $dA \neq 0$,

$$\boxed{d\sigma_x/dx = 0}. \tag{3.9}$$

This is a very fundamental and general requirement for local equilibrium, that is, the gradient of the stress equals zero.

(By and large, most applications of axial structures will not see a variation in external load along their length, and hence the incremental forces on opposite faces of any section will be equal. We will, however, find Eq. (3.9) to be an important relation in other structures we consider.)

Local and global equilibrium can be related through the following equations:

$$
\begin{aligned}
F &= \iint_A dF = \iint_A \sigma_x dA = P \\
M_y &= \iint_A z \cdot dF = \iint_A \sigma_x z dA = 0 \\
M_z &= -\iint_A y \cdot dF = -\iint_A \sigma_x y dA = 0.
\end{aligned}
\tag{3.10}
$$

3.2.5 Deformation

Under the action of the tensile force F, a material particle located at position x in the rod will displace an amount $u(x)$ in the $+x$ direction. The corresponding strain is a strictly geometrical quantity with arbitrary definition. The simplest definition we can take is that strain (at a point!) is the comparison of a local change in length between two material particles relative to some "gage length," which is taken as the length between the same particles in the undeformed state, namely dx. If we take the deformed length to be dx^*, then the strain in the x-direction is defined as

$$\varepsilon_x \equiv \frac{dx^* - dx}{dx} \tag{3.11a}$$

$$\boxed{\varepsilon_x = \frac{du}{dx}}$$ (3.11b)

(There is no single, unique definition for strain. We use here the common definition of one-dimensional *engineering strain*.)

This is a very fundamental and general requirement of deformation that relates strain to displacement and is called the *strain–displacement relation* (although it is more correctly the strain–displacement gradient relation).

We see that the strain ε_x is a normal (here tensile or extensional) strain in the axial direction. The axial displacement anywhere along the rod (at any cross-section) can then be found from

$$u(x) = \int_0^x \varepsilon_x dx$$ (3.12)

with the total increase in the rod length given by

$$u(L) = \int_0^L \varepsilon_x dx$$ (3.13)

We emphasize that for a rod modeled as a 1-D structure, all fibers (which are axial fibers) deform identically.

3.3 Constitution

3.3.1 Density and Weight

Students often find the concepts of *density* and *weight* confusing. This is exacerbated by the confusion in systems of units (see Review Module R2), and by the fact that two types of density are often reported: mass density and weight density.

By definition we take density to mean *mass density* ρ, specifically the mass per unit volume of a given material. Mass relates to the amount of material; unit volume could be a cube, say, 1 mm on a side (but it could also be a spherical unit volume, etc.). The mass of a given material is the same anywhere in the universe, and hence is a very fundamental property. Units of mass are, for example, kilogram or slug; units of mass density are, for example, kg/m^3 or $slug/ft^3$.

Weight, w, on the other hand, is a force, specifically the gravitational force. It depends on both the material (its mass) and the intensity (i.e., the acceleration g) of the local gravitational field:

$$w = mg$$ (3.14)

Units of weight are force units, such as newtons or pounds.

We recognize that the weight of an object on the moon is about one-sixth of its weight on earth. Thus we take mass or mass density as a fundamental design degree of freedom. Weight density is also commonly used, that is, the weight per unit volume; unfortunately, the same symbol is often used for both weight density and mass density, and sometimes simply "density" is reported without definition, and one has to take special note of the units used.

It is common to talk about wanting to "minimize the weight" of a structure. That's a bit sloppy, because what we really want to do is "minimize the mass" of the structure. This can be done either by reducing the density (again, we take density in this text to always mean mass density), or by reducing the volume of material—either way, we reduce the mass, and for a given gravitational field, reduce the weight.

3.3.2 One-Dimensional Elastic Constitutive Relation

In 1678, the English mechanician Robert Hooke published a paper titled "De Potentia Restitutiva" ("The Spring"). In it, he detailed the results of experiments he had conducted on materials, such as metallic wire. He showed that the deflection u in the wire was proportional to the applied tensile load P.

$$P = ku \tag{3.15}$$

where k is a proportionality constant (called the *stiffness*) to be defined later. It is critical at this point to realize that every elastic body can be considered as a spring of one sort or another, just as Hooke did. We will come back to this concept time and time again.

Dividing both sides of this relation by AL (the product of the initial area and length of the wire) and rearranging terms, the relation can be rewritten in terms of stress and strain as

$$\sigma = E_t \varepsilon, \tag{3.16}$$

where σ is the *engineering stress* P/A, ε is the *engineering strain* u/L, and $E_t = kL/A$ is the *tensile elastic modulus* of the material. This relation, which shows the proportionality now between stress and strain, is formally called the *1-D Hooke's law*.

(In most materials, the tensile elastic modulus is not very different from the compressive elastic modulus. However, in some very special materials, such as soils, a value of compressive elastic modulus E_c that is different from their tensile elastic modulus E_t may occur. Such materials are called *bimodulus* or *bilinear* materials. However, in the rest of this book, no distinction will be made between the tensile and compressive moduli, and the subscripts t and c will be dropped. The same convention will also apply to material strength unless specifically mentioned otherwise.)

We have just discovered something very fundamental: the mechanics quantities stress and strain are related physically in the material through material quantities (such as the elastic modulus in this case). Such properties can be determined through a tension test or a compression test of the material, to be discussed subsequently.

EXAMPLE 3-1:

The rod shown in Figure E3-1a consists of two coaxial circular cylindrical sections. The material is high-carbon steel ($E = 30 \times 10^6$ psi, $S_Y = 70 \times 10^3$ psi). Determine the overall extension of the rod, and check for strength design.

From the strain–displacement (gradient) relation we can find the total elongation:

$$u = \int_0^L \varepsilon_x dx.$$

But in the present example, due to discontinuities of loading and geometry, we need to break the integration up. First we complete three FBDs associated with each of the discontinuities (Figure E3-1b).

EXAMPLE 3-1: *Cont'd*

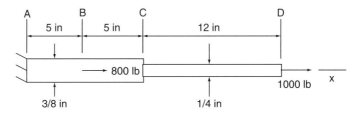

FIGURE E3-1A

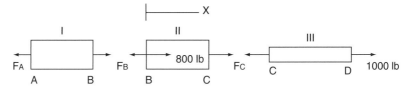

FIGURE E3-1B

Next we determine the loading over each section by applying the equilibrium relations:

- Section III: $\Sigma F_x = 0$: $-F_C + 1000\,\text{lb} = 0 \Rightarrow \underline{F_C = 1000\,\text{lb}}$
- Section II: $\Sigma F_x = 0$: $-F_B + 800\,\text{lb} + F_C = 0 \Rightarrow \underline{F_B = 1800\,\text{lb}}$
- Section I: $\Sigma F_x = 0$: $-F_A + F_B = 0 \Rightarrow \underline{F_A = 1800\,\text{lb}}$

Now we can integrate separately over each section of the rod:

$$u = \int_0^{L_1} \frac{P_1(x)}{EA_1(x)}\,dx + \int_{L_1}^{L_2} \frac{P_2(x)}{EA_2(x)}\,dx + \int_{L_2}^{L} \frac{P_3(x)}{EA_3(x)}\,dx$$

$$= \frac{F_A}{EA_1}\int_0^{5''} dx + \frac{F_C}{EA_1}\int_{5''}^{10''} dx + \frac{F_C}{EA_3}\int_{10''}^{22''} dx$$

$$= \frac{1}{30 \times 10^6\,\text{psi}} \left[\frac{1800\,\text{lb}}{\frac{\pi}{4}(0.375\,\text{in.})^2}(5\,\text{in.}) + \frac{1000\,\text{lb}}{\frac{\pi}{4}(0.375\,\text{in.})^2}(5\,\text{in.}) + \frac{1000\,\text{lb}}{\frac{\pi}{4}(0.25\,\text{in.})^2}(12\,\text{in.}) \right]$$

$$= \underline{0.124\,\text{in.}}$$

As a check on strength design, we can compute the stress in each segment of the rod:

$$\sigma_1 = \frac{P_1}{A_1} = \frac{1800\,\text{lb}}{\frac{\pi}{4}(0.375\,\text{in.})^2} = 16{,}300\,\text{psi}$$

$$\sigma_2 = \frac{P_2}{A_2} = \frac{1000\,\text{lb}}{\frac{\pi}{4}(0.375\,\text{in.})^2} = 9{,}050\,\text{psi}$$

$$\sigma_3 = \frac{P_3}{A_3} = \frac{1000\,\text{lb}}{\frac{\pi}{4}(0.25\,\text{in.})^2} = 20{,}400\,\text{psi}$$

Continued

EXAMPLE 3-1: *Cont'd*

The stress is seen to be less than the strength ($S_Y = 70 \times 10^3$ psi) in every case. However, we have not taken into account the effects of stress concentrations near boundaries and discontinuities—the stresses here could be considerably higher!

3.4   Mechanics $\Leftrightarrow$ Materials Link: The Tension Test

A typical response of a steel tensile specimen to monotonically increasing tensile load is shown in Figure 3-12. (The full details of the tensile test are described at the text Web site.)

To fully utilize the potential of a material in a structure, the designer must understand both the elastic and plastic deformation, as well as the stress–strain response over both regions.

3.4.1   The Elastic Response

We will first discuss the elastic response of a metal. As seen from Figure 3-12, the elastic stress–strain response is a straight line, or in other words the stress–strain response is linear. If the material is unloaded in the elastic region, it returns to its original unloaded state along the straight line. Linear elastic behavior is intrinsically related to the nature of atomic bonding, which is discussed later in this section.

When a uniaxial tensile stress is applied to the specimen, it extends along the tensile axis and undergoes transverse contraction along the two perpendicular axes (see Figure 3-13).

For linear elastic deformation, say in the x direction, the longitudinal strain ε_{xx} and transverse strains ε_{yy} and ε_{zz} are related through

$$\varepsilon_{yy} = \varepsilon_{zz} = -\nu\varepsilon_{xx}, \tag{3.17}$$

where ν is called the *Poisson's ratio*, a property of the material. The negative sign indicates that the dimensions of the specimen along the transverse axes decrease as the length of the

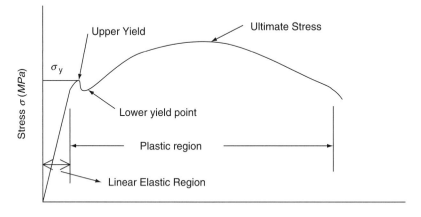

FIGURE 3-12   *Schematic stress–strain diagram for a mild steel.*

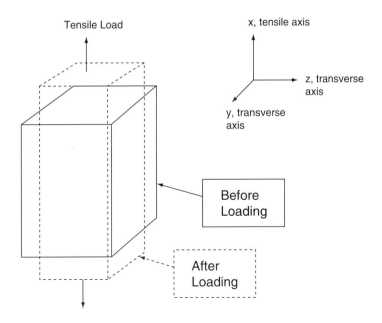

Tensile Load

x, tensile axis

z, transverse axis

y, transverse axis

Before Loading

After Loading

FIGURE 3-13 *Elastic tensile deformation of a specimen.*

specimen along the tensile axis increases. The slope of the straight line in the linear elastic region is called the *Young's modulus* or the *elastic modulus*, *E*, of the material; *E* is a property of the material.

Following elastic deformation, plastic (or permanent) flow commences at a stress equal to the *yield strength*, S_Y, again a property of the material. If the material is loaded to or beyond the yield strength, and then unloaded, it does not return to its original unloaded state, but retains some deformation, explained further in the discussion on the plastic region in Chapter 6. If shear stresses are applied, the shear strain is related to the shear stress by the shear modulus *G*, which is a material property (see Chapter 4).

(It may be noted that the volume of a material increases slightly during elastic tensile deformation and decreases during elastic compression. During plastic deformation the volume of the material is conserved.)

3.4.2 Atomistic Basis for Elastic Behavior

Elastic behavior has its origin in the forces between the atoms of the solid and therefore depends on both the nature of chemical bonding and the atomistic structure of the material (e.g., the crystal structure). The general nature of the atomic bonding and interatomic forces has been described earlier in Section 1.6. We will rely on that here, as well as the material science review in Review Module R1, to qualitatively explain the atomic basis for elastic behavior of crystalline solids.

The material properties obtained from the elastic response discussed earlier, namely, *E*, *v*, and *G*, are also known as "microstructure-insensitive properties", as these quantities are independent of crystalline imperfections, such as interstitial inclusions, vacancies, and dislocations. When an atomic force is applied along a certain crystal axis in a metal, for example along a [100] axis, the interatomic spacing changes (the details of crystal axes will be discussed more fully in Chapter 4). The spacing increases along the [100] direction and decreases along the two perpendicular directions [010] and [001]. For small extensions, Δr, in the atomic spacing, the force increases linearly with $\Delta r / r_o$ (see Review Module R1). The slope of the interatomic force versus interatomic distance curve around the equilibrium spacing, r_o, is related to the Young's modulus in the [100] direction. (If the extension, Δr, is large, the linear relation between force and displacement does not hold.)

The Young's modulus can be calculated for materials with different types of atomic bonding. However, the calculation is easiest for materials consisting of ionic bonds, such as sodium chloride. The final expression is given in Eq. (3.18) below without proof (Eisenstadt, 1971):

$$E = \frac{\sigma_x}{\varepsilon_x} = [2.61 \times 10^9 Z_C Z_A e^2 \left(\frac{n-1}{r_o^4} \right)][1 - 2v] \qquad (3.18)$$

Equation (3.18) is valid only for ionic crystals similar to NaCl in the [100] direction. Different crystal structures will result in different values of the numerical constant (which is 2.61×10^9 here). It should be noted that Young's modulus is strongly dependent on the quantity r_o [expressed in meter, and 1 angstrom ($\mathring{A}$) = 10^{-10} m], and r_o depends on the ionic radii of elements forming the solid. As an approximation, the value of r_o can be the sum of the ionic radii of the two elements forming the ionic crystal. The value of E also depends on the product of the magnitude of the cation and anion charges, Z_C and Z_A, respectively. The other quantities in the expression are as follows: e is the charge in an electron ($e = 1.6 \times 10^{-19}$ coulombs, C), n is the exponent in the repulsive potential term, and v is the Poisson's ratio. Relations similar to Eq. (3.18) can be derived for metallic and covalent bonded materials, but the derivation is more complicated and beyond the scope of this text.

The concept of the elastic modulus can also be qualitatively perceived as being the "spring stiffness" at an atomic level. In this spring model, the interatomic forces are applied to the atoms through extensions or contractions of springs imagined to be connecting the atoms (recall our discussion about springs earlier). If all the atoms are in the equilibrium position, r_o, the spring force (and extension) is zero. If an atom is displaced through a small distance, $\Delta r = r - r_o$, the force exerted on a spring is $F = k \Delta r$ [recall Eq. (3.15)], where k is the spring stiffness. The value of k is in the range 20–200 N/m for covalent bonds, and 15–100 N/m for metallic and ionic bonds. Hence, covalent bonded solids are stiffer. In polymers, which typically have only partial crystallinity at most, the chains consist of atoms held together by covalent bonds, but the interchain bonding is of the weak van der Waals type, resulting in a low spring constant of 0.5–2 N/m. Hence, long-chain polymers have low elastic modulus.

EXAMPLE 3-2:

Calculate the elastic modulus in the [100] direction of MgO. The ionic radii of Mg^{2+} and O^{2-} are 0.78 $\mathring{A}$ and 1.32 $\mathring{A}$, respectively, and $n = 7$ for the two ions.

Using a value of $v = 0.3$ for the Poisson's ration, the value of E is given by

$$E = \left[\frac{(2.61 \times 10^9)(2 \times 2)(1.6 \times 10^{-19})^2 (7-1)}{[(0.78 + 1.32) \times 10^{-10}]^4} \right] (1 - 2 \times 0.3)$$

$$= 0.32 \times 10^{12} \text{N/m}^2 = 320 \text{ GPa } (4.64 \times 10^6 \text{ psi})$$

3.4.3 Engineering Quantities Obtained from a Tension Test

The tension test can provide a wealth of important material properties for use in structural design. A number of these properties are described below.

1. *Engineering strain, ε,* is defined as the change in length per unit original length,

$$\varepsilon = \frac{l - l_o}{l_o}, \tag{3.19}$$

where
   l = final length
   l_o = original length
   Strain is a nondimensional quantity (length/length) with units such as mm/mm or in/in.

2. *Engineering stress, σ,* is

$$\sigma = \frac{P}{A_o} \tag{3.20}$$

where
   P = applied load
   A_o = original area of cross-section of the tensile specimen over the gage length
   Stress has dimensions of force/length^2 with units such as N/m^2 (pascal or Pa) or lb/in^2.

3. *Modulus of elasticity* or *Young's modulus, E,* is defined as the ratio of the uniaxial engineering stress to the corresponding engineering strain:

$$E = \frac{\sigma}{\varepsilon}. \tag{3.21}$$

Over a small range of strain (<0.2% strain for metals), the material generally deforms elastically. If the load is removed the material of the specimen returns to its original dimensions. This is called the *linear elastic behavior*. The modulus of elasticity is defined only in the linear elastic range. E is the slope of the $\sigma-\varepsilon$ curve in the elastic range.
   Modulus of elasticity has the same dimensions as stress and the same units.

4. *Poisson's ratio, ν:* The elongation of the specimen is generally accompanied by a contraction of its lateral dimensions. The absolute value of the ratio of the lateral strain to longitudinal strain is known as the Poisson's ratio:

$$\nu = -\frac{\text{lateral strain}}{\text{longitudinal strain}} \tag{3.22}$$

(Note: A separate strain measurement has to be made on the specimen to measure the lateral strain.)

5. Modulus of resilience: This is the area under the stress–strain curve up to the yield point S_Y of the material. It is a measure of the energy the material can store elastically:

$$\text{Modulus of resilience} = \frac{S_Y \varepsilon_Y}{2} = \frac{S_Y^2}{2E} \tag{3.23}$$

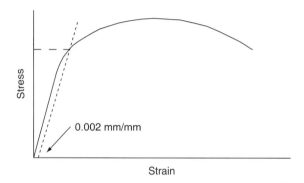

FIGURE 3-14 *Use of 0.2% offset strain for determining yield strength of gradually yielding materials that do not depict a distinct yield point.*

where

S_Y = yield strength of material

ε_Y = strain at the yield point

Modulus of resilience has dimensions of energy per unit volume, and units such as $N\text{-}m/m^3$ or $in\text{-}lb/in^3$

6. *Yield stress*: As the load is increased, the material reaches a point where it can no longer deform elastically and it begins to undergo plastic (or permanent) deformation. The stress at which the material develops the plastic deformation is known as the *yield stress*, σ_Y and the strain at this stress level is called the yield strain, ε_Y.

   (Note: Ferrous materials exhibit a distinct yield stress; nonferrous materials do not display a distinct yield stress, hence a 0.2% or 0.002 offset strain is used to define the yield stress, as shown in Figure 3-14.)

7. *Yield strength*, S_Y: When a standard ASTM tensile test specimen is used to determine the yield stress, σ_Y, then this yield stress is a material property and is known as the yield strength S_Y.

8. *Tensile strength* or *ultimate tensile strength* S_U: As the specimen elongates beyond the yield stress, its cross-sectional area decreases permanently and uniformly through the gage length. However, the change in the cross-sectional area is no longer uniform when the specimen is loaded beyond the ultimate tensile strength (UTS) S_U. Beyond UTS the change in cross-section is localized and concentrated in a small region of the specimen. This region is called the *neck* and the phenomenon is called *necking*. The UTS is a measure of the overall strength of the material. Beyond the UTS the engineering stress in the material drops (why?) until the specimen breaks or fractures, as shown in the engineering stress–engineering strain plot in Figure 3-15.

9. *Fracture stress*: The engineering stress at which the specimen breaks into two pieces is called the fracture stress:

$$\text{Fracture stress} = \frac{\text{Fracture load}}{\text{Original cross-sectional area}} \tag{3.24}$$

10. *Percent reduction of area:*

$$\text{Reduction of area} = \frac{\Delta A}{A_o} \times 100\% \tag{3.25}$$

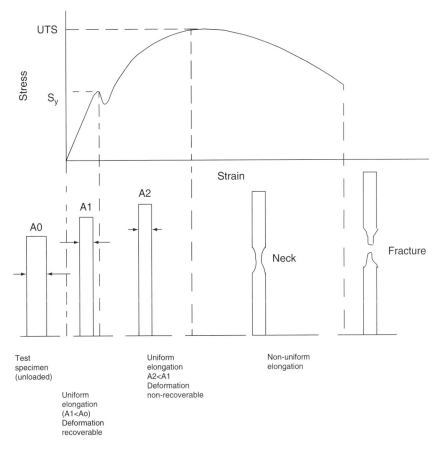

FIGURE 3-15 *Engineering stress–engineering strain plot of a tension test to failure.*

where

A_o = original area of cross-section

A_f = area of cross-section at fracture

$\Delta A = A_o - A_f$ is the change in area of cross-section

This is a measure of the ductility of the material, that is, the magnitude of strain the material can withstand before fracturing.

11. *True stress*: This is the ratio of the load to the instantaneous area supporting the load. During elastic deformation the change in the area of the cross-section may be negligible. However, during plastic deformation the change in cross-sectional area is not negligible and is more pronounced after start of necking.

12. *True Strain* and *Engineering Strain*: True strain, ε_t, is defined as

$$\varepsilon_t = \int_{l_o}^{l} \frac{dl}{l} = \ln\left(\frac{l}{l_o}\right) \tag{3.26}$$

Engineering strain can be written as

$$\varepsilon = \frac{l}{l_0} - 1$$

$$\Rightarrow \frac{l}{l_0} = 1 + \varepsilon$$

$$\Rightarrow \ln\left(\frac{l}{l_0}\right) = \ln(1 + \varepsilon)$$

$$\Rightarrow \varepsilon_t = \ln(1 + \varepsilon)$$

For small strains, the true strain ε_t and engineering strain ε are related by the preceding equation. For large strains this equation does not hold, and only the true strain values must be used. For example, in metalworking calculations, true strain must be used. For large strains the true strain can be calculated using the change in cross-sectional area at a point:

$$\varepsilon_t = \ln\left(\frac{A_0}{A}\right) \tag{3.27}$$

where A is the current cross-sectional area.

13. *True stress–true strain curve*: A typical true stress–true strain curve from a tension test is shown in Figure 3-16.

Unlike the engineering stress–strain curve, the slope is always positive and the slope decreases with increasing strain. Although in the elastic range stress and strain are linearly related, the total curve is approximated by a power law as shown in Figure 3-16.

When the true stress–strain curve is plotted on log-log scales, a straight line is obtained as shown in Figure 3-17, and the slope of the line is equal to the exponent of ε_t in the power law expression representing the curve.

14. *Toughness*: This is the area under the entire true stress–strain curve and is given by

$$\text{Toughness} = \int_0^{\varepsilon_f} \sigma d\varepsilon \tag{3.28}$$

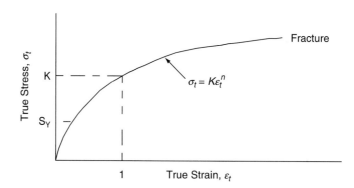

FIGURE 3-16 *A typical true stress–true strain curve obtained during tensile tests of a structural material. K is the true stress at a true strain of unity.*

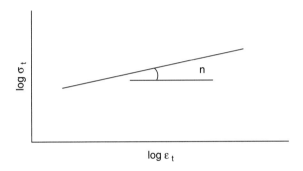

FIGURE 3-17 *A true stress–true strain curve on a log-log plot.*

where ε_f is the true fracture strain.

Toughness is energy per unit volume stored and/or dissipated up to the point of fracture. Note that this energy pertains only to the volume of material at the region of the neck. The volume of the material away from the neck undergoes less strain and hence stores/dissipates less energy.

15. *Types of stress–strain curves*: Figure 3-18 shows the most common mechanical responses of different materials to an axial load.

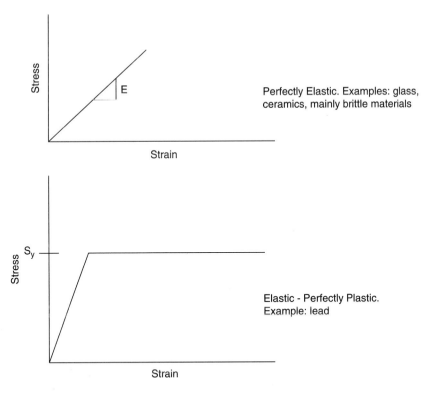

FIGURE 3-18 *(Continued)*

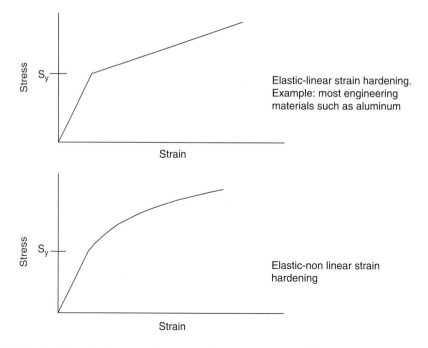

Stress

S_y ─

Strain

Elastic-linear strain hardening.
Example: most engineering
materials such as aluminum

Stress

S_y ─

Strain

Elastic-non linear strain
hardening

FIGURE 3-18 *(Cont'd) Schematic illustration of various types of idealized stress–strain curves.*

EXAMPLE 3-3:

Consider the data shown in the following table, obtained from a tension test on a 0.505-in. (12.85 mm) diameter aluminum bar. Convert this data to true stress and true strain data.

Measurement no.	Load (lb)	Gage length (in)
1	0	2.000
2	1000	2.001
3	3000	2.003
4	5000	2.005
5	7000	2.007
6	7500	2.030
7	7900	2.080
8	8000 (max)	2.120
9	7900	2.170
10	7600 (fracture)	2.205

EXAMPLE 3-3: *Cont'd*

Answer:
Engineering σ and ε:

1. Say at Load = 1000 lb

$$\sigma = \frac{F}{A_0} = \frac{\text{Load}}{\text{Original cross-sectional area}} = \frac{1000}{\frac{\pi}{4}(0.505)^2} = 5000 \, \text{psi}$$

$$\varepsilon = \frac{l - l_0}{l_0} = \frac{\text{Final length} - \text{Original gage length}}{\text{Original gage length}}$$

$$= \frac{2.001 - 2.000}{2.000} = 0.0005 \, \text{in/in}$$

2. Load = 8000 lb (max)

$$\sigma = \frac{F}{A_0} = \frac{\text{Load}}{\text{Original cross-sectional area}} = \frac{8000}{\frac{\pi}{4}(0.505)^2} = 40,000 \, \text{psi}$$

$$\varepsilon = \frac{l - l_0}{l_0} = \frac{\text{Final length} - \text{Original gage length}}{\text{Original gage length}}$$

$$= \frac{2.120 - 2.000}{2.000} = 0.060 \, \text{in/in}$$

The engineering stress and strain values are in shown in the accompanying table.

Measurement no.	Stress (psi)	Strain (in/in)
1	0	0
2	5000	0.0005
3	15,000	0.0015
4	25,000	0.0025
5	35,000	0.0035
6	37,500	0.0150
7	39,500	0.0400
8	40,000	0.0600
9	39,500	0.0850
10	38,000	0.1025

True σ and ε:
The diameter at maximum load is 0.491 in:

$$\text{True stress} = \frac{F}{A_{\text{instant}}} = \frac{8000}{\frac{\pi}{4}(0.491)^2} = 42,251 \, \text{psi}$$

$$\text{True strain} = \ln\left(\frac{l}{l_0}\right) = \ln\left(\frac{2.120}{2.000}\right) = 0.058 \, \text{in/in}$$

Continued

EXAMPLE 3-3: *Cont'd*

Note that true stress is higher while true strain is lower, but only slightly. If the diameter at fracture is 0.398 in, $\sigma_{eng} = 38,000$ psi, $\sigma_{true} = 61,090$ psi, $\varepsilon_{eng} = 0.1025$ in/in, and $\varepsilon_{true} = 0.476$ in/in.

Note that now a lot of error is introduction in the analysis beyond the maximum load when necking starts. Thus after necking starts, true σ and ε must be used for accurate analysis.

A good way to find the instantaneous area is to use the fact that the volume of the specimen is constant during plastic deformation, and approximately constant in the elastic portion.

$$\Rightarrow A_0 l_0 = Al$$

$$\Rightarrow \text{Instantaneous area, } A = \frac{A_0 l_0}{l}.$$

Thus, corresponding to a load of 5000 lb, the instantaneous area is obtained as

$$A = \frac{A_0 l_0}{l} = \frac{\frac{\pi}{4}(0.505)^2 \times (2.000)}{2.005} = 0.1998 \text{ in}^2$$

and the true stress at 5000 lb load is

$$\sigma_t = \frac{5000}{0.1998} = 25,025 \text{ psi},$$

while the true strain at this load is

$$\varepsilon_t = \ln\left(\frac{l}{l_0}\right) = \ln\left(\frac{2.005}{2.000}\right) = 0.0025 \text{ in/in}$$

The student should similarly obtain the true stress and true strain values for the remaining loads in the table, plot the true stress versus strain response, and compare it qualitatively with Figure 3-16.

Other quantities of interest that may be obtained for this problem are given below:

- Modulus of elasticity = slope of elastic portion
  at point 0.0 lb, strain = 0.0
  at 35,000 psi, strain = 0.0035

$$E = \frac{\Delta\sigma}{\Delta\varepsilon} = \frac{35,000 - 0}{0.0035 - 0} = 10 \times 10^6 \text{ psi}$$

- Percent reduction of area at maximum load:

$$= \frac{A_0 - A_f}{A_0} \times 100$$

$$= \frac{\frac{\pi}{4}(0.505)^2 - \frac{\pi}{4}(0.491)^2}{\frac{\pi}{4}(0.505)^2} \times 100$$

$$= 5.5\% \text{ (not a measure of ductility)}$$

EXAMPLE 3-3: *Cont'd*

- Percent reduction of area at fracture:

$$= \frac{(0.505)^2 - (0.398)^2}{(0.505)^2} \times 100$$

$$= 37.8\%$$

$$= \text{a measure of ductility}$$

- Ultimate tensile stress, $S_U = 40,000 \, \text{psi}$

3.5 Mechanics ⇔ Materials Link: The Compression Test 🍸🕸

3.5.1 Compression Testing

The compression test, in which the specimen is subjected to a compressive load as shown in Figure 3-19, gives useful information about the behavior of a material under compression and in the stress–strain curve.

Many operations in metalworking, such as forging, rolling, and extruding, are performed with the workpiece under compressive loads. In all such operations, the phenomena of buckling is not an issue; only the compressive behavior under large compressive deformation is needed, which can be obtained from compact compression tests.

The compression test is typically carried out by compressing a solid cylindrical specimen between two flat platens. However, friction between the specimen and the platens is an important factor to be considered. Significant friction can prevent the top and bottom surfaces from expanding freely, resulting in the phenomena of *barreling*, shown schematically in Figure 3-20. Thus the solid cylindrical specimen deforms into a barrel shape due to friction at the platen–specimen interface.

Effective lubrication is necessary to obtain a nearly uniform cross-sectional area along the height of the specimen, and minimize energy dissipation by friction during the test.

Engineering compressive strain is given by

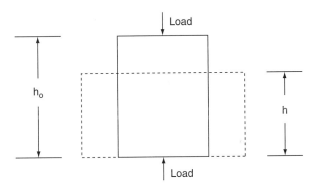

FIGURE 3-19 *Compressive deformation of a short solid cylindrical specimen.*

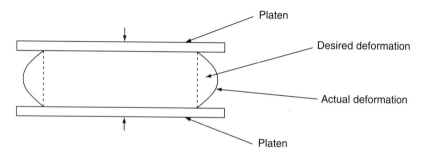

FIGURE 3-20 *Barreling phenomena during compression testing.*

$$\varepsilon = -\frac{h - h_{\mathrm{o}}}{h_{\mathrm{o}}}. \qquad (3.29)$$

An approximate value of true strain is given by

$$\varepsilon_{\mathrm{t}} = -\frac{h - h_{\mathrm{o}}}{h} \qquad (3.30)$$

where h_{o} and h are the original and instantaneous height of the specimen, respectively.

The above definition for true strain [Eq. (3.30)] is valid for small strains and only gives an average value of the strain over the whole deformation range. A more accurate definition of true strain [Eq. (3.31)], which holds true for large strains, such as those encountered when a material is loaded to its ultimate strength and beyond, is given by (see also Section 3.4):

$$\varepsilon_{\mathrm{t}} = \int_{h_{\mathrm{o}}}^{h_{\mathrm{f}}} \frac{dl}{l} \qquad (3.31)$$

where h_{f} is the final height of the specimen.

With effective lubrication and ductile materials, compression tests can be carried out uniformly to large strains, whereas in a tension test, nonuniform deformation starts when the ultimate tensile stress of the material is reached at a strain that is relatively much lower than the compression strains required for nonuniform deformation in the same material. It has been found that for ductile metals, the true stress–true strain curves are identical in monotonic tension and compression. However, this is not true for brittle materials.

Brittle materials such as ceramics and concrete are tested in compression by using a *disk test*. In this test, a disk is subjected to compression between two hard and flat platens as shown in Figure 3-21. When a disk is loaded in compression, tensile stresses develop along the horizontal axis, that is, perpendicular to the vertical loading axis. Thus fracture begins along the vertical axis or perpendicular to the tensile stress direction. Brittle materials are tested for their tensile fracture behavior by this test, since the gripping forces required for direct tensile testing can lead to premature fracture in the grip area.

The tensile stress σ in the disk, which is uniform along the centerline, is

$$\sigma = \frac{2P}{\pi dt} \qquad (3.32)$$

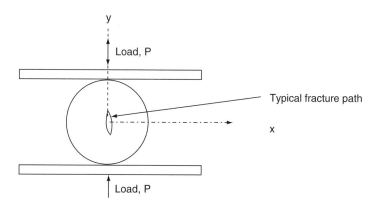

FIGURE 3-21 *The disk test for brittle materials testing.*

where d is the diameter, t is the thickness, and P is the load at fracture. (Interestingly, in the disk test the material failure occurs due to tensile stresses at the center, though the externally applied load is compressive.)

3.5.2 Mechanics–Materials Link: Compressibility of Materials

Consider a hypothetical volume of material made up of equal-sized particles arranged in such a way that spherical symmetry (i.e., everything is the same in all directions) is achieved. The particles interact with each other through interatomic forces. If a *hydrostatic* compression (i.e., an equi-triaxial loading) is applied, the equilibrium distance between the particles will change under the action of the external uniform compressive force field until the repulsive or attractive interatomic forces balance the external forces. (Similar statements could be made for hydrostatic tension, but such a loading is difficult to implement in practice.) Such changes will be transient, and no changes take place in the relative positions of the individual particles by which their energy would be reduced permanently. Thus in such a material a spherically symmetric external force field (hydrostatic) produces a spherically symmetric deformation because the lattice possesses spherical symmetry. In real life, materials with three-dimensional symmetrical cubic crystals, such as manganese, copper, and lead, produce a deformation response as mentioned earlier.

In crystalline bodies with inherent anisotropy in the arrangement of particles in the atomic structure, a different deformational response in different directions is produced, even under the action of a spherical external force field. The imposition of hydrostatic pressure (or tension) on an anisotropic crystalline body will produce not only volumetric changes (produced as well in materials with symmetric arrangement of particles) but also changes in the shape. In such crystalline materials, for example zinc, magnesium, and zirconium, which have hexagonal crystal structures, the difference in compressibility in two directions of the crystal will be quite different.

When homogeneous materials are compressed, there is a reduction in the interatomic distances, a steep increase in the slope of the atomic force versus separation curve with decreasing atomic distance, and a sharp increase in the repulsive energy of the particles (see Figure R1-2 in Review Module R1) resulting in a decrease of compressibility of all isotropic and homogeneous materials. This effect on compressibility is usually represented by the bulk modulus, which increases with increasing compressive force. *Bulk modulus* is the elastic modulus associated with a volume change and is defined as the ratio of the hydrostatic pressure to the fractional change in volume due to the applied pressure.

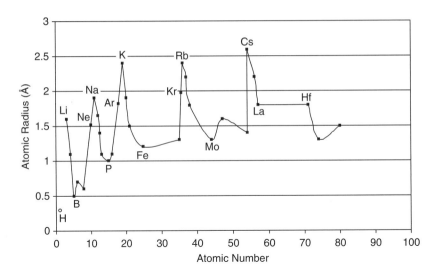

FIGURE 3-22 *Radii of atoms as a function of atomic number (adapted from Freudenthal, 1950).*

The *compressibility* of elements in the periodic table varies with the atomic number and there exists a periodic relationship. On comparing Figures 3-22 to 3-24, it should be noticed that the most compressible solids are those with the largest atomic radii (Figures 3-22 and 3-24) and the lowest density (Figures 3-23 and 3-24), e.g., alkali metals (such as Li, Na, K, Rb, and Cs) as shown schematically in these figures. Compressibility may be visualized in terms of the compression of an aggregate of impenetrable elastic spherical electron shells. The larger the radii of the spheres, the greater the extent of compression by hydrostatic compression. With increasing density and external pressure, the resistance to inelastic deformation increases in both crystalline and amorphous materials. In crystals this increase manifests itself by an increase in the critical shear stress (mainly at high pressures).

When a material is compressed, failure by fracture is much less likely than when it is subjected to tensile loads, since cracks would tend to close rather than open under compressive loading. Moreover, if hydrostatic compression is applied to homogeneous and relatively incompressible metals and amorphous substances, fracture cannot occur. However, in materials possessing microscopic inhomogeneities, such as cast iron, or a macroscopically inhomogeneous structure, such as concrete, a nonhomogeneous response is produced to a hydrostatic compression stress state. Thus failure may be initiated in such materials even under hydrostatic compression. Nevertheless, materials that are very sensitive to the presence of cracks in tension, such as ceramics, concrete, and stone (in other words, very brittle materials), are considerably stronger in compression, by about 8 to 15 times their failure strength in tension.

3.6 Energetics

In this book, we make considerable use of "Newtonian mechanics," which is a vector mechanics of forces, displacements, and displacement derivatives. For one, Newtonian mechanics is a very intuitive approach—it is easy to visualize forces and displacements as arrows! However, there is also considerable reason to take a Lagrangian mechanics approach, which is a scalar mechanics of work and energy and their derivatives. Energy methods are important

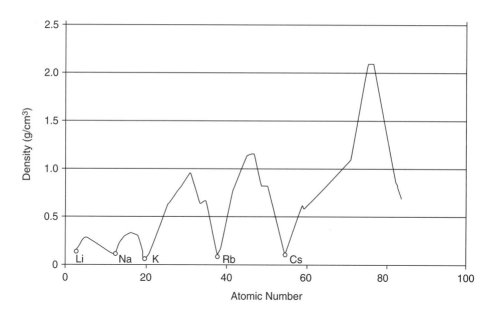

FIGURE 3-23 *Schematic variation of density of elements as a function of atomic number (adapted from Freudenthal, 1950).*

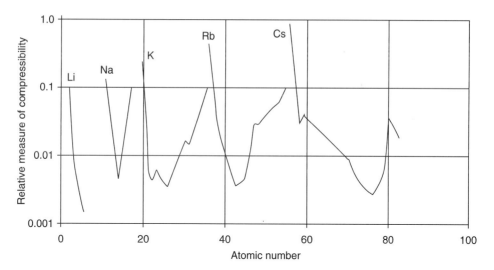

FIGURE 3-24 *Schematic periodic variation of compressibility of elements as a function of atomic number (adapted from Freudenthal, 1950).*

because they provide fundamental principles for formulating the governing structural equations, and as such they are the gateway to more advanced analysis techniques.

We know from earlier studies that mechanical energy can be divided into potential energy and kinetic energy. Since we only deal with static problems here, we can disregard kinetic energy considerations. The *total potential energy Π* of a structure includes both the energy of

FIGURE 3-25 *Linear elastic translational spring.*

deformation stored within the structure (i.e., the *strain energy*) and the potential energy associated with the work of the forces and moments responsible for the deformation.

The strain energy U is equal to the *recoverable work* done on the body. Consider a linear elastic translational spring of stiffness k under the action of a net force P as shown in Figure 3-25 (see Section 3.7 for a more complete discussion of this spring).

From the definition of work (the incremental work dU of a force is the scalar product of the vector force $\mathbf{F}$ with the resulting incremental displacement vector $\mathbf{ds}$ or $\mathbf{F} \cdot \mathbf{ds} = F \cos \theta \, ds$, where θ is the angle between $\mathbf{F}$ and $\mathbf{ds}$), we have for the spring that $dU = P \, du$ (du being the increment of displacement in the x-direction). Integrating over the total displacement u we obtain the strain energy in the spring as

$$U = \int_0^u P d\xi = \int_0^u k\xi d\xi = \frac{1}{2} k u^2 \tag{3.33}$$

(Here ξ is just a dummy variable of integration.)

The potential energy E of the applied loads relevant to the structure is the negative of the work done by the applied loads during the deformation: $E = -Pu$. The total potential energy is

$$\Pi = U + E \quad \Pi = \frac{1}{2} k u^2 - Pu \left(= \frac{1}{2} k u^2 - k u^2 = -\frac{1}{2} k u^2 \right).$$

Note that Π is a quadratic function (quadratic form) of u. A plot of Π versus u looks like Figure 3-26.

Apparently Π has a minimum, which in this case can be analyzed by elementary calculus (more powerful methods of the calculus of variations will be needed for more complex energy functions), that is, $d\Pi/du = ku - P = 0 \Rightarrow ku = P$ and $d^2\Pi/du^2 = k > 0$.

These two simple results are very important:

1. We have derived the equilibrium equation $ku = P$ from the minimization of the total potential energy. That means that the equilibrium configuration achieved under a set of applied loads is the unique one that minimizes the total potential energy. This is the *principle of minimum potential energy*. It is a cornerstone of structural mechanics!
2. The stiffness k is guaranteed to be positive.

One final point can be made. The stored energy of a rod can be calculated in terms of stress and strain through use of the constitutive relations. For the one-dimensional Hooke's Law model, we get a strain energy per unit volume of the rod as $U = 1/2 k u^2 = 1/2 P u = 1/2 \sigma_x A \varepsilon_x L$ or $U = 1/2 \sigma_x \varepsilon_x (AL)$.

Hence the strain energy per unit volume is given by $1/2 \sigma_x \varepsilon_x = 1/2 E \varepsilon_x^2 = 1/2 \sigma_x^2 / E$. In general, for the one-dimensional case the strain energy is given by

$$U = \frac{1}{2} \int_{\text{volume}} \sigma_x \varepsilon_x dV \tag{3.34}$$

Two- and three-dimensional forms of the strain energy will be discussed later.

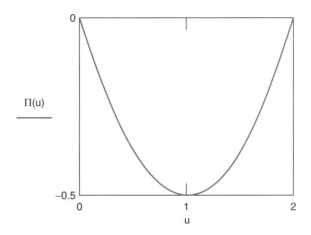

FIGURE 3-26 *Plot of total potential energy for a linear elastic translational spring.*

EXAMPLE 3-4:

Referring to Example 3-1, calculate the total strain energy in the rod.

In general, the strain energy is given by

$$U = \frac{1}{2}\int_{\text{vol}} \sigma_x \varepsilon_x dV = \frac{1}{2}\int_{\text{vol}} \frac{\sigma_x^2}{E} dV, \quad dV = dx\,dy\,dz.$$

Again, due to the discontinuities in stress and volume, the integral must be broken as before:

$$U = \frac{1}{2E}\left(\int_{\text{vol.I}} \sigma_x^2 dV + \int_{\text{vol.II}} \sigma_x^2 dV + \int_{\text{vol.III}} \sigma_x^2 dV\right).$$

Note that for constant stress over any volume element, the volume integral becomes

$$\int_{\text{vol}} dV = \iiint dx\,dy\,dz = \iint_{\text{area}} dA\,dx = A\int dx$$

for constant area A.

Now computing the total strain energy is straightforward:

$$U = \frac{1}{2E}\left(A_1\sigma_1^2\int_0^{5''} dx + A_2\sigma_2^2\int_{5''}^{10''} dx + A_3\sigma_3^2\int_{10''}^{22''} dx\right)$$

$$= \frac{\pi/4}{2(30\times10^6\,\text{psi})}\left[(0.375\,\text{in})^2(5\,\text{in})(16{,}300\,\text{psi})^2\right.$$

$$\left. +(0.375\,\text{in})^2(5\,\text{in})\,(9{,}050\,\text{psi})^2 + (0.25\,\text{in})^2\,(12\,\text{in})(20{,}400\,\text{psi})^2\right]$$

$$= \underline{10.5\,\text{lb}\cdot\text{in}}$$

3.7 Analysis of Simple Tensile Structures 'Y'

3.7.1 Elastic Stiffness

Every elastic body has associated with it a stiffness that relates the loads on the body to its displacement response. This is a fundamental concept in structural mechanics! If the body is linearly elastic, then the stiffness is constant (i.e., the stiffness is a constant of proportionality between load and response).

One of the simplest structural members we could imagine is the *translational spring* (Figure 3-27).

If we fix (restrain) node 1 from motion, that is require $u_1 = 0$, the response of the spring is given by $P_2 = ku_2$. The *spring stiffness* k (also sometimes called the *spring rate*) is a function of material properties and structural element geometry and has dimensions of force/length. For example, k for a helical coil spring can be shown equal to the following relation (Shigley and Mischke, 2001):

$$k = \frac{d^4 G}{8D^3 N} \tag{3.35}$$

where d and D are the wire and mean coil diameters, respectively, G is the shear modulus of the material, and N is the number of turns or coils. Note also that if the spring has a nonlinear response, the spring stiffness k cannot be called a spring "constant."

Noting that for equilibrium, $P_1 = -P_2 =$ the internal spring force (consider for yourself a free-body diagram to verify this), the general response of the spring is:

$P_1 = k(u_1 - u_2) = -P_2$ (P_1 as shown is positive, implying a compressive force if $u_1 > u_2$)
$P_2 = k(u_2 - u_1)$ (P_2 as shown is positive, implying a tensile force if $u_2 > u_1$)

These two nodal equations could be written as the single matrix equation

$$\left\{ \begin{array}{c} P_1 \\ P_2 \end{array} \right\} = \left[\begin{array}{cc} k & -k \\ -k & k \end{array} \right] \left\{ \begin{array}{c} u_1 \\ u_2 \end{array} \right\}$$

or

$$\{\mathbf{P}\} = [\mathbf{k}]\{\mathbf{u}\}, \tag{3.36}$$

where $\{\mathbf{P}\}$ and $\{\mathbf{u}\}$ are the local or nodal force and displacement vectors, respectively, and $[\mathbf{k}]$ is the *local stiffness matrix*.

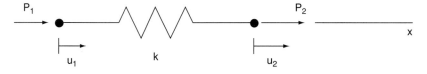

FIGURE 3-27 *Translational spring definition, where locations 1 and 2 are the nodal points, P_1 and P_2 are forces (loads) applied, u_1 and u_2 are the displacement responses at the nodes 1 and 2, respectively, and k is the elastic spring stiffness. Note that forces and displacements are defined as acting in the positive coordinate direction.*

The stiffness matrix by nature is rank deficient and hence singular. (Physically, the spring is unconstrained and can exhibit rigid body motion.) Also note of the stiffness matrix that:

- It is symmetric
- Rows and columns add to zero
- The determinate is zero (and hence the matrix singular)

Although we've drawn the "spring" to look like a helical spring of our common experience (say a coil spring from an automobile suspension or a ballpoint pen), again we emphasize that any elastic member is a "spring" (or can be modeled as a spring). The spring stiffness of a rod fixed also at node 1 is easily arrived at from the uniaxial constitutive relation

$$\sigma = E\varepsilon$$
$$\frac{P_2}{A} = E\frac{u_2}{L} \tag{3.37}$$
$$P_2 = \frac{EA}{L}u_2 = ku_2$$

Hence the stiffness of an axial structure is

$$\boxed{k = EA/L.} \tag{3.38}$$

Equation (3.38), and others like it that we will develop in later chapters, is an extremely important and fundamental relationship in structural mechanics.

EXAMPLE 3-5:

Consider an axial load of 10 N applied to the free end of a rod fixed at the other end (Figure E3-5). The length of the rod is 100 mm with cross-sectional area of 10 mm². The elastic modulus is E = 1 GPa. Determine the stiffness of the rod, the total deflection under load, as well as the axial stress.

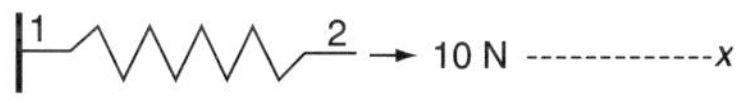

FIGURE E3-5

The rod stiffness is from Eq. (3.38):

$$k = (10^9 \text{ N/m}^2)(10\,\text{mm}^2)(10^{-6}\,\text{m}^2/\text{mm}^2)/(0.1\,\text{m}) = \underline{10^5\,\text{N/m}} = \underline{100\,\text{N/mm}}.$$

The total displacement is from the third of equations (3.37):

$$u_2 = 10\,\text{N}/(100\,\text{N/mm}) = \underline{0.1\,\text{mm}}.$$

The stress is from the second of equations (3.37):

$$\sigma = (10^9\,\text{N/m}^2)(0.1\,\text{mm})/(100\,\text{mm}) = \underline{1\,\text{MPa}}.$$

3.7.2 Direct Stiffness Method I

As explained in Chapters 1 and 2, the structural analysis based on displacement equilibrium, or the direct stiffness method, is a powerful analysis method that can be applied to any structural analysis problem. The key element of this method is that we write the displacement equilibrium equations at nodes, and assume a displacement field within the region of the body bounded by the nodes. This allows any body, no matter how complicated, to be discretized into smaller finite elements, with a very consistent procedure performed to achieve a solution. This leads to the ubiquitous *finite element method (FEM)*, which is the modern standard for structural analysis.

A brief outline of the method goes as follows (a more detailed outline will be given later):

1. Assemble known global stiffness matrix [k] and nodal load vector {P} from local stiffness and loads
2. Solve the system of equations $\{P\} = [k]\{u\}$ for the unknown nodal displacement vector $\{u\}$
3. Determine displacement field from assumed displacement field
4. Find strains from strain–displacement relations
5. Find stress from constitutive relations

The method is now demonstrated by a simple example.

EXAMPLE 3-6:

In the previous example, we considered an axial load P_2 of 10 N on a rod fixed at node 1. The stiffness of the rod was 100 N/mm. Again we let the rod be 100 mm in length, with a cross-sectional area of 10 mm^2 and modulus $E = 1$ GPa. We assume the displacement field $u(x)$ to be linear in x:

$$u(x) = ax + b.$$

The nodal displacement equilibrium equations are

$$\begin{Bmatrix} P_1 \\ P_2 \end{Bmatrix} = \begin{bmatrix} k & -k \\ -k & k \end{bmatrix} \begin{Bmatrix} u_1 \\ u_2 \end{Bmatrix}.$$

We set the boundary conditions $u_1(0) = 0$ and nodal load $P_2 = 10$ N,

$$\begin{Bmatrix} P_1 \\ 10 \text{ N} \end{Bmatrix} = \begin{bmatrix} k & -k \\ -k & k \end{bmatrix} \begin{Bmatrix} 0 \\ u_2 \end{Bmatrix}.$$

We then solve the second equation for u_2:

$$10 \text{ N} = (100 \text{ N/mm})(u_2)$$

or $u_2 = 0.1$ mm. Solving the first equation for P_1:

$$P_1 = (-100 \text{ N/mm})(0.1 \text{ mm}) = -10 \text{ N}.$$

EXAMPLE 3-6: *Cont'd*

(Note that the boundary condition on the first degree of freedom $u_1(0) = 0$ allows us in effect to zero out the associated first row and column of the stiffness matrix, leaving a reduced stiffness matrix that is no longer rank deficient and is much simpler to solve. We will take advantage of this as we proceed to more complicated problems.)

The displacement field can now be found (from the assumed field). Constants in the linear displacement relation are found by applying the boundary conditions (nodal displacements):

$$u(0) = u_1 = 0 = b$$

$$u(L) = u_2 = 0.1\,\text{mm} = aL \Rightarrow a = 0.1\,\text{mm}/100\,\text{mm} = 0.001.$$

Finally, then,

$$u(x) = 0.001x.$$

Now the strain is found from the appropriate strain–displacement relation:

$$\varepsilon_x = \frac{du}{dx} = 0.001$$

We would always have a constant strain in this case if we assume a linear displacement field. The stress is found from the uniaxial constitutive relation:

$$\sigma_x = E\varepsilon_x = (1\,\text{GPa})(0.001) = \underline{1\,\text{MPa}}.$$

Note that this is the same, for this simple case, as if we had calculated the stress from P/A, which shows that the solution was exact, and the assumption on the displacement field must have been correct.

The obvious question to ask is: "How did you know what form of the displacement field to assume, and what difference do different assumptions make?" The short answer to the first part is that there are formal ways to make sure that the displacement field is always compatible with the structure stiffness, so that we get the correct strain. Let's investigate by example what happens if we make a different assumption. Say, for example, that we assume a quadratic dependence for displacement:

$$u(x) = ax^2 + b.$$

Noting that the constant b still equals zero, but

$$u(L) = u_2 = 0.1\,\text{mm} = aL^2 \Rightarrow a = 0.1\,\text{mm}/(100\,\text{mm})^2 = 10^{-5}.$$

The displacement field now looks like:

$$u(x) = 10^{-5}x^2.$$

Continued

EXAMPLE 3-6: *Cont'd*

Calculating the strain in the same way yields

$$\varepsilon_x = 2 \times 10^{-5} x$$

and the stress is

$$\sigma_x = \underline{2 \times 10^4 x} \text{ Pa}$$

This clearly does not make sense, for example, that $\sigma_x(0) = 0$ when $P(0) = P_1 = -10\,\text{N}$. The problem is that the assumed displacement field is not compatible with the element stiffness. Again, there are formal methods to ensure the compatibility, but these are outside the scope of our discussion.

3.7.3 Direct Stiffness Method II

We have yet to discuss the idea of assembly of the global stiffness matrix. The power of the displacement method arises from the ability to decompose or discretize a structure into a number of small parts or *finite elements*. The response of each element is assumed to be known. The process of assembly then synthesizes the global structural model from each of the discrete parts.

In the previous example, with only one element, the local stiffness and global stiffness are the same. Now, let's consider two spring elements connected in series, as shown in Figure 3-28.

Spring 1 is defined by nodes 1 and 2, and by stiffness k_1; likewise, spring 2 is defined by nodes 2 and 3 (which were originally element or local nodes 1 and 2), and by stiffness k_2. Axial loads P_1, P_2 (not shown for clarity), and P_3 are applied at the respective nodes, and each node has associated with it a degree of freedom u_1, u_2, and u_3, respectively. These are global descriptions.

To see how these relate to the local or element descriptions, consider that $u_1 = u_1^{(1)}$, where the superscript notation (1) refers to element 1. By continuity, the displacement of node 2 of element 1 must equal the displacement of node 1 of element 2, or

$$u_2^{(1)} = u_1^{(2)} = u_2.$$

The complete relationship (called a *mapping*) of local and global nodal variables is given in Table 3-1.

Now we write down the nodal displacement equilibrium equations in local form:

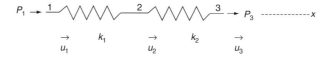

FIGURE 3-28 *Two spring elements connected in series.*

TABLE 3-1 Relationship of Local and Global Nodal Variables for Two Springs in Series

Local	$\Leftrightarrow$	Global
$u_1^{(1)}$	=	u_1
$u_2^{(1)}$	=	u_2
$u_1^{(2)}$	=	u_2
$u_2^{(2)}$	=	u_3
$P_1^{(1)}$	=	P_1
$P_2^{(1)} + P_1^{(2)}$	=	P_2
$P_2^{(2)}$	=	P_3

Note that equilibrium demands that $P_2 = P_2^{(1)} + P_1^{(2)}$.

• Spring element 1:

$$P_1^{(1)} = k_1[u_1^{(1)} - u_2^{(1)}]$$
$$P_2^{(1)} = k_1[u_2^{(1)} - u_1^{(1)}]$$

• Spring element 2:

$$P_1^{(2)} = k_2[u_1^{(2)} - u_2^{(2)}]$$
$$P_2^{(2)} = k_2[u_2^{(2)} - u_1^{(2)}].$$

Substituting into the first and last equations the global nodal variables for the local ones, as given in Table 3-1, we get

$$P_1 = k_1[u_1 - u_2] \ *$$
$$P_3 = k_2[u_3 - u_2] \ *.$$

Substituting the second and third equations into the force equilibrium equation at node 2 gives

$$P_2 = P_2^{(1)} + P_1^{(2)} = k_1[u_2^{(1)} - u_1^{(1)}] + k_2[u_1^{(2)} - u_2^{(2)}].$$

Now, substituting global variables for local variables, we get

$$P_2 = k_1[u_2 - u_1] + k_2[u_2 - u_3] \ *.$$

The three starred equations are the desired global displacement equilibrium equations, three equations now since we have three unknown displacements. In matrix form these become

$$\begin{Bmatrix} P_1 \\ P_2 \\ P_3 \end{Bmatrix} = \begin{bmatrix} k_1 & -k_1 & 0 \\ -k_1 & k_1 + k_2 & -k_2 \\ 0 & -k_2 & k_2 \end{bmatrix} \begin{Bmatrix} u_1 \\ u_2 \\ u_3 \end{Bmatrix} \tag{3.39}$$

where the stiffness matrix is the *assembled global stiffness matrix*, and the load and displacement vectors are global as well.

We have assembled the stiffness matrix by first writing out the equilibrium equations in local form, then converting them to global form. One can imagine that this becomes a tiresome process if the number of nodes increases! Yet the ability to have many elements and many nodes is the power of the displacement method, so we need an algorithm that lets us assemble the stiffness directly. It turns out that this is really very simple and is sometimes called the *assembly rule*:

1. Determine the structure or global "size" of the final assembled equations. N displacement degrees of freedom (i.e., N unknowns) requires an N by N system of equations.
2. Expand the local or element stiffness matrices to structure or global size.
3. Add the expanded element matrices together to form the global stiffness matrix.

Let's illustrate this process on the previous example:

1. Three displacement degrees of freedom = three unknown displacements = 3 by 3 system
2. We now expand the element stiffness matrices to structure size, recalling the relationship between local and global nodal variables:

$$[k^{(1)}] = \begin{matrix} 1 \\ 2 \\ 3 \end{matrix} \begin{bmatrix} k_1 & -k_1 & 0 \\ -k_1 & k_1 & 0 \\ 0 & 0 & 0 \end{bmatrix} \qquad [k^{(2)}] = \begin{matrix} 1 \\ 2 \\ 3 \end{matrix} \begin{bmatrix} 0 & 0 & 0 \\ 0 & k_2 & -k_2 \\ 0 & -k_2 & k_2 \end{bmatrix}$$
$$\phantom{[k^{(1)}] = } \begin{matrix} 1 & 2 & 3 \end{matrix} \qquad\qquad\qquad \begin{matrix} 1 & 2 & 3 \end{matrix}$$

The numbers on the outside of the brackets represent the global degrees of freedom. It is readily observed that we have kept the correct local–global mappings: DOF 1 and 2 of element 1 are associated with global DOF 1 and 2, respectively; element DOF 1 and 2 of element 2 are associated with global DOF 2 and 3, respectively.

3. We add together the expanded element stiffness matrix to get the assembled global stiffness matrix:

$$[k] = [k^{(1)}] + [k^{(2)}]$$

$$[k] = \begin{matrix} 1 \\ 2 \\ 3 \end{matrix} \begin{bmatrix} k_1 & -k_1 & 0 \\ -k_1 & k_1 + k_2 & -k_2 \\ 0 & -k_2 & k_2 \end{bmatrix}.$$
$$ \begin{matrix} 1 & 2 & 3 \end{matrix}$$

This is the same matrix arrived at in the previous example. The global force and displacement vectors are obviously

$$\{P\} = \begin{matrix} 1 \\ 2 \\ 3 \end{matrix} \begin{Bmatrix} P_1 \\ P_2 \\ P_3 \end{Bmatrix} \qquad \{u\} = \begin{matrix} 1 \\ 2 \\ 3 \end{matrix} \begin{Bmatrix} u_1 \\ u_2 \\ u_3 \end{Bmatrix}$$

We can now summarize the displacement or direct stiffness method for *any* number of elements:

1. Note element connectivity and how they map local nodal variables to global nodal variables
2. Expand each element stiffness matrix $[k^{(i)}]$ to structure size, keeping the local/global mappings consistent
3. Assemble known global stiffness matrix $[k] = \sum_{i=1}^{I} [k^{(i)}]$.
4. Assemble the nodal load vector $\{P\}$ from local loads
5. Enter displacement boundary conditions, eliminating rows and columns of the stiffness matrix that correspond to zero boundary displacements. (The load vector needs to be modified for any nonzero displacement boundary conditions, but this will not be discussed further here.)
6. Solve the (reduced) system of equations $\{P\} = [k]\{u\}$ for the unknown nodal displacement vector $\{u\}$
7. Determine displacement field from assumed displacement field
8. Find strains from strain–displacement relations
9. Find stress from constitutive relations

The method will now be demonstrated by an example.

EXAMPLE 3-7:

Consider two springs in series, with the left end fixed and a 100 N load placed on the right end as shown in Figure E3-7.

Given that $k_1 = 10{,}000\,\text{N/m}$ and $k_2 = 20{,}000\,\text{N/m}$, determine the displacements of the springs.

1. Local/global mapping due to element connectivity is given by

Local	$\Leftrightarrow$	Global
$u_1^{(1)}$	$=$	$u_1 = 0$
$u_2^{(1)}$	$=$	u_2
$u_1^{(2)}$	$=$	u_2
$u_2^{(2)}$	$=$	u_3
$P_1^{(1)}$	$=$	P_1
$P_2^{(2)}$	$=$	$P_3 = 100\,\text{N}$

Note that in this example, $P_2 = 0$ by definition.

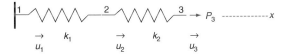

FIGURE E3-7

Continued

EXAMPLE 3-7: *Cont'd*

2. Expand each element stiffness matrix:

$$[k^{(1)}] = \begin{array}{c} 1 \\ 2 \\ 3 \end{array} \begin{bmatrix} 10 & -10 & 0 \\ -10 & 10 & 0 \\ 0 & 0 & 0 \end{bmatrix} k \text{ N/m} \qquad [k^{(2)}] = \begin{array}{c} 1 \\ 2 \\ 3 \end{array} \begin{bmatrix} 0 & 0 & 0 \\ 0 & 20 & -20 \\ 0 & -20 & 20 \end{bmatrix} k \text{ N/m.}$$
$$\phantom{[k^{(1)}] = } \begin{array}{ccc} 1 & 2 & 3 \end{array} \qquad\qquad\qquad\qquad \begin{array}{ccc} 1 & 2 & 3 \end{array}$$

3. Assemble the global stiffness matrix:

$$[k] = \begin{array}{c} 1 \\ 2 \\ 3 \end{array} \begin{bmatrix} 10 & -10 & 0 \\ -10 & 30 & -20 \\ 0 & -20 & 20 \end{bmatrix} k \text{ N/m.}$$
$$ \begin{array}{ccc} 1 & 2 & 3 \end{array}$$

4. Assemble the global load vector:
   Noting that the external nodal load at node 2 is zero,

$$\{P\} = \begin{array}{c} 1 \\ 2 \\ 3 \end{array} \left\{ \begin{array}{c} P_1 \\ 0 \\ 100 \text{ N} \end{array} \right\}$$

5. Enter boundary conditions and reduce equations accordingly:
   Since

$$\{u\} = \begin{array}{c} 1 \\ 2 \\ 3 \end{array} \left\{ \begin{array}{c} 0 \\ u_2 \\ u_3 \end{array} \right\},$$

   the first row and column of the stiffness matrix may be eliminated. This leaves the following reduced system to solve:

$$\left\{ \begin{array}{c} 0 \\ 100 \text{ N} \end{array} \right\} = \begin{bmatrix} 30 & -20 \\ -20 & 20 \end{bmatrix} \frac{\text{kN}}{\text{m}} \left\{ \begin{array}{c} u_2 \\ u_3 \end{array} \right\}$$

6. The solution of this system of equations gives:

$$\underline{u_2 = 10 \text{ mm}}$$
$$\underline{u_3 = 15 \text{ mm.}}$$

   (The student should check this solution. Does it make intuitive sense?)

3.8 Material Selection in Axial Structures

Material selection in general has previously been discussed in Chapter 2 (Section 2.6). The goal of material selection is to achieve a structure–material system that meets the design requirements in an optimal way. The approach we use to accomplish this goal was developed by Professor Ashby and colleagues at Cambridge University as previously discussed. It hinges on two important features:

- A collection of material properties in graphical form
- The definition of *material indices* (MI) to access the graphical information

For axial structures, common material indices for light–strong or light–stiff rods are particularly simple. Consider the case of a light–strong rod of given length L under a given axial load P, and materials are to be selected subject to the following constraints:

$$\text{Lightness constraint:} \, w \leq w_0$$

$$\text{Strength constraint:} \, \sigma \leq S_f/FS$$

where w_0 is a certain specified weight that should not be exceeded, and S_f is the failure strength of the material.

We note that both the weight [Eq. (3.40)] and stress [Eq. (3.41)] are functions of the cross-sectional area (A), which is a DDOF:

$$w = mg = \rho V g = \rho A l g \tag{3.40}$$

$$\sigma = P/A. \tag{3.41}$$

We now eliminate that DDOF from Eqs. (3.40) and (3.41), considering the equality condition on each constraint, to obtain:

$$A = \frac{w_0}{\rho L g} = \frac{P}{S_f/FS}$$

Setting up the light–strong MI gives

$$\frac{S_f}{\rho} = \frac{P L g (FS)}{w_0} \tag{3.42}$$

Note that as the material index S_f/ρ is maximized the weight w_0 is minimized as all other remaining quantities in Eq. (3.42) are fixed. (The student should verify that the MI equation is dimensionally consistent.)

In a similar manner, it can be shown (this is left as an exercise for the student) that the MI for a light–stiff rod is given by

$$\frac{E}{\rho} = \frac{P L^2 g}{\Delta_0 w_0} \tag{3.43}$$

where Δ_0 is the deflection constraint.

Other material indices can also be found. Graphical use of the MI is shown by an example in the following section.

3.9 Design of Tensile Axial Structures 'Y'🕸💻

We now illustrate the method by a design example:

It is necessary to deliver a force F to a bellcrank B in order to operate a gate valve on a large HVAC system (essentially lift a load at low duty cycle), as shown in Figure 3-29. However, because of space constraints, the force (the "actuator") is only available at 0.5 m distance away, and in the x-direction, from the bellcrank, as shown in the figure. The bellcrank is attached to the gate valve (which represents a 20,000 N load) and pivots on an approximately frictionless bearing. The bellcrank is made of mild (low carbon) steel with legs of 10-cm and 5-cm lengths, as shown.

Design a structure (link) to connect the force F to the bellcrank B, with a FS of 1.5 and total deflection less than 1 mm. The structure should be low cost, in keeping with the rest of the bellcrank system.

We now apply the design template:

I. *Problem Definition*

Performance requirements:
- Factor of safety = 1.5; this a specification on strength and, as such, is of primary importance. Then $\sigma \leq S_f/1.5$, where S_f is some as yet unspecified failure strength.
- Deflection $\Delta \leq \Delta_0 = 1$ mm; this is a specification on stiffness and is likely of secondary importance (unless some interference condition exists).

No other performance requirements are given.

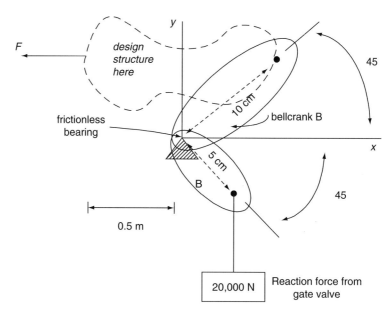

FIGURE 3-29 *Bellcrank design definition sketch.*

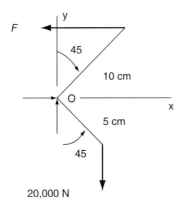

FIGURE 3-30 *Moment summation diagram.*

Service environment:
- Loads—are assumed to act statically. A summation of moments about the pivot O, shown graphically in Figure 3-30, shows that

$$\sum M_o = 0: \ F \cdot 10\,\text{cm} \cdot \cos 45 - 20{,}000\,\text{N} \cdot 5\,\text{cm} \cdot \cos 45 = 0$$

$$F = \frac{5}{10} \cdot 20{,}000\,\text{N} = 10{,}000\,\text{N}$$

(Note that we obtain a "mechanical advantage" through the unequal moment arms.)
- No other service environment specifications are given.

Project constraints:
- Low cost (implies generally available material, simplicity of design, ease of manufacture and assembly, reasonably low weight, etc.)
- No other constraints are given.

II. *Preliminary Design*

- Identify DDOFs:
  1. Geometry ⇒ minimize volume to keep weight low, area A
  2. Density ⇒ keep weight low, ρ
  3. Strength ⇒ design for yielding for low cost, S_Y
  4. Stiffness ⇒ modulus E
  5. Boundary conditions ⇒ pinned BC can keep link axial, that is, without moment reactions (more details about the connections will not be considered at this stage)
- Trade study: The desire to keep costs low (and as always to keep things as simple as possible!) suggests a simple axial structure or bar for the link. A circular cross-section of diameter d will keep the surface area to cross-sectional area of the link to a minimum, thus minimizing the volume of material; the circular cross-section also minimizes stress concentrations (Figure 3-31). Material considerations suggest simple, readily available materials to keep costs low, but analysis needs to confirm the specific choice.

FIGURE 3-31 *Link detail.*

- FMEA: Failure expected to be from simple global static yielding of the link. (A more detailed analysis would need to consider other failure modes, such as local yielding due to higher stress at contact points and stress concentrations, or perhaps corrosion failure.)

III. *Detailed Design*

- Determine stress in order to satisfy strength specification:

$$\sigma_x = F/A = 10{,}000\,\text{N}/A.$$

- For circular cross-section:

$$A = \pi d^2/4.$$

(Note that as a DDOF, geometry or A is unknown.)
- Determine required strength (again, note the DDOF S_Y is originally unknown):

$$S_Y \geq FS(\sigma_x) = 1.5\sigma_x = 1.5\,F/A = 15{,}000\,\text{N}/A$$

or

$$(F/A) \leq (S_Y/1.5) \quad \text{(I)}.$$

- Determine required stiffness:
The link deflection $\varDelta \leq \varDelta_0 = F/k$, where k is the link stiffness EA/L (note that DDOF E is unknown here). Then solving for k,

$$k \leq F/\varDelta_0$$

or

$$(EA/L) \leq (F/\varDelta_0) \quad \text{(II)}.$$

- Select candidate material:
The choice of a material depends on the application and the mechanical properties of the link called for by the application. Some desirable mechanical properties may be:

1. Elastic modulus, $E \Leftrightarrow$ stiffness
2. Yield strength, $\sigma_y \Leftrightarrow$ yielding by plastic flow of material
3. Durability $\Leftrightarrow$ fatigue life
4. Fracture toughness $\Leftrightarrow$ resistance to failure by crack-like defects

Depending upon the actual application, the link could function under different loading and environmental conditions. For example, such a link could be used to rotate the blades on a gas turbine or adjust the opening of the guide vanes in the water inlet system to a hydroelectric turbine. Thus some other desirable properties of the link may include:

- Minimum thermal distortion (for gas turbine)
- Durability under impact loading (for hydroelectric turbine)
- Resistance to thermal shock (for gas turbine)

For the link currently under consideration, we set our primary design goals as good strength ($\Rightarrow$ good resistance to failure by yielding S_Y), stiffness (E), and low cost ($\Rightarrow$ reasonably low weight or density ρ). To use the Material Selection charts, we form a material index S_Y/E by taking the equality case of Eqs. (I) and (II), solving each for A, and then equating to eliminate A:

$$A = \frac{1.5 F}{S_Y} = \frac{FL}{E \Delta_0}.$$

Then the material index becomes

$$\frac{S_Y}{E} = \frac{1.5 \Delta_0}{L} = \frac{(1.5)(1\,\text{mm})}{0.5 \times 10^3\,\text{mm}} = 3 \times 10^{-3}.$$

Now we can look up the E versus S_Y material selection chart and determine which materials satisfy the requirement that

$$\frac{S_Y}{E} \leq 3 \times 10^{-3}.$$

To make use of the numerical value of the MI, we first invert the MI so we may plot a straight line on the "Young's modulus versus strength" Material Selection chart. We do this by plotting a line between coordinates, say, ($S_Y = 0.1\,\text{MPa}$, $E = S_Y \times MI$) and (E/MI, $E = 1000\,\text{GPa}$) or (0.1 MPa, 0.333 GPa) and (3000 MPa, 1000 GPa), as seen in Figure 3-32. All materials above this line satisfy the light–strong criteria as specified.

Also, we want to obtain the best material in terms of strength and stiffness at the lowest possible cost. The cost constraint can be optimized by looking up the Material Selection Chart showing Young's modulus E versus relative cost per unit volume $C_R \rho$ (Figure 3-33), and the strength S_Y versus relative cost per unit volume $C_R \rho$ chart (see Appendix A3). We demonstrate the qualitative method used in ranking the modulus versus relative cost per unit volume in Figure 3-33 (the verification of the strength versus relative cost per unit volume is left as an exercise for the student).

The results from these charts are presented in Table 3-2. The material rankings are based on 1 being the best and 5 being the worst. Note that the unit cost for strength is valued (weighted) more highly than the unit cost for stiffness.

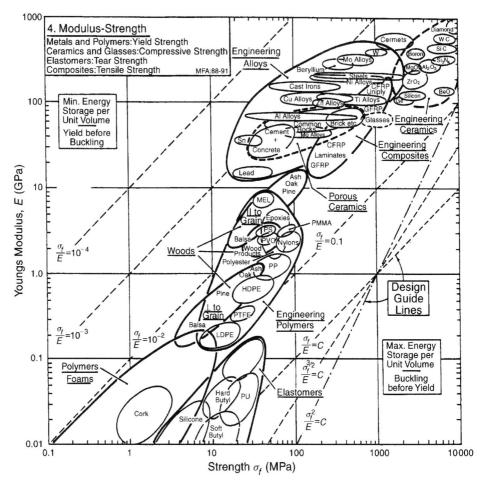

FIGURE 3-32 *Material index plotted (heavy line) on Ashby Material Selection Chart (with permission from Prof. M. Ashby).*

Hence the most suitable materials for this application based on the given constraints are steel and cast iron. In other words, the desired strength and stiffness at the lowest cost can be obtained with steel or cast iron. Note that we have not included factors such as manufacturability and market price fluctuations (for example, the price of aluminum may have changed since the printing of the Ashby charts we are using). If we want to include other factors, such as susceptibility to fracture, the methodology developed in this section can be further extended. This design example will be revisited in Chapter 7 on fracture. (We will see later that if we include fracture toughness as an additional requirement, cast iron is no longer a very viable alternative.) Hence the final choice for the material for the link is steel.

Proceeding with a low-cost steel, say AISI 1036 with $S_Y \approx 250\,\text{MPa}$ (36 ksi), then from Eq. (I):

$$A = \frac{\pi d^2}{4} \geq \frac{15,000\,\text{N}}{S_Y} = \frac{15,000\,\text{N}}{250 \times 10^6\,\text{N/m}^2} = 60\,\text{mm}^2$$

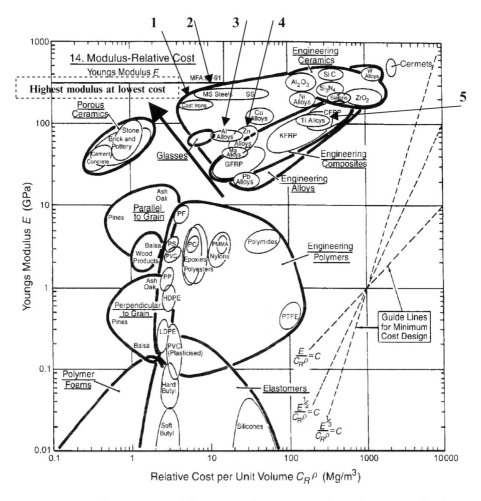

FIGURE 3-33 *Ashby Young's modulus versus relative cost per unit volume Material Selection Chart, showing relative ranking of the five materials listed in Table 3-2. A ranking of 1 means the material satisfied the criteria best, with 5 being the worst. Note that the stiffest materials at least cost per unit volume are located in the upper left-hand corner of the chart.*

(S_Y and A are now determined.)

Solving for the diameter d:

$$d \geq \sqrt{\frac{4}{\pi} 60\,\text{mm}^2} = 8.7\,\text{mm} \ (0.34\,\text{in.})$$

(Select common 10 mm or 3/8 in diameter stock)

- Check stiffness (calculate the deflection Δ):

$$F = k\Delta = \frac{EA}{L}\Delta.$$

TABLE 3-2 Materials Selection Chart for the Design Example

			Material ranking from the Ashby Charts *(Raw rankings and weighted rankings are given)*			
Possible materials from E versus S_Y chart	*E versus S_Y rank (weight = 0.25)*	*E versus $C_R \rho$ rank*	*Weighted rank (weight = 0.25)*	*S_Y versus $C_R \rho$ rank*	*Weighted rank (weight = 0.5)*	*Weighted subtotal*
Steel	0.25	2	0.5	1	0.5	1.25
Cast iron	0.5	1	0.25	2	1.0	1.75
Titanium alloy	0.75	5	1.25	5	2.5	4.5
Zinc alloy	1.0	4	1.0	4	2.0	4
Aluminum	1.25	3	0.75	3	1.5	3.5

(Link is essentially a spring.)

$$\Delta = \frac{FL}{EA} = \frac{10{,}000\,\text{N} \times 0.5\,\text{m}}{210 \times 10^9\,\text{N/m}^2 \times \frac{\pi \times 10^2}{4} \times 10^{-6}\,\text{m}^2}$$

$$= 0.30\,\text{mm}\ (0.008\,\text{in.})$$

(Meets the stiffness spec. of ≤ 1 mm.)

(Note that the strain $\varepsilon = \Delta/L = 0.30\,\text{mm}/500\,\text{mm} = 0.0006$ or 0.06% is within the linear range of steel.)

- Final link specification (Figure 3-34):

10 mm round stock × 0.5 m long A36 steel
(3/8 in round stock × 20 in long)

3.10   Analysis of Compact Compressive Structures Ⴤ

3.10.1   Direct Stiffness Method I

Our representation of an axial tensile structural element as a translational spring in Section 3.6 applies equally well to the compact compressive case as well. In fact, our generic "spring element" (Figure 3-35) shows forces applied that are both tensile and compressive.

The nodal equations are identical to Section 3.6:

$$\left\{ \begin{array}{c} P_1 \\ P_2 \end{array} \right\} = \left[\begin{array}{cc} k & -k \\ -k & k \end{array} \right] \left\{ \begin{array}{c} u_1 \\ u_2 \end{array} \right\}$$

or

$$\{P\} = [k]\{u\}$$

where again $\{P\}$ and $\{u\}$ are the local or nodal force and displacement vectors, respectively, and

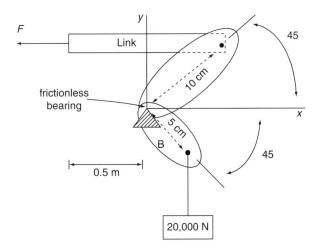

FIGURE 3-34 *Final bellcrank design.*

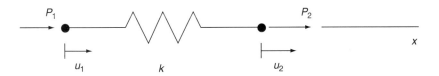

FIGURE 3-35 *Generic spring element.*

[**k**] is the local stiffness matrix. Let's rework the previous Example 3-6 from Section 3.7, but now with a compressive force.

EXAMPLE 3-8:

Consider a compressive axial load of 10 N applied to the free end of a rod fixed at the other end (Figure E3-8). The length of the rod is 100 mm with cross-sectional area of 10 mm^2. The elastic modulus is $E = 1$ GPa. Determine the stiffness of the rod and the total deflection under load, as well as the axial stress.

Note that the force is acting now to compress the spring. The rod stiffness is $k = EA/L$ or

$$k = (10^9 \text{ N/m}^2)(10 \text{ mm}^2)(10^{-6} \text{ m}^2/\text{mm}^2)/(0.1 \text{ m})$$
$$= 10^5 \text{ N/m} = 100 \text{ N/mm (same as before)}.$$

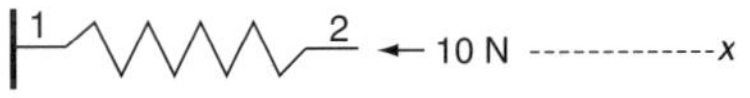

FIGURE E3-8 *Generic spring element.*

Continued

EXAMPLE 3-8: *Cont'd*

The total displacement is

$$u_2 = -10\,\text{N}/(100\,\text{N}/\text{mm}) = \underline{-0.1\,\text{mm}}.$$

(Note the negative sign indicating compressive response.) The stress is

$$\sigma = (10^9\,\text{N}/\text{m}^2)(-0.1\,\text{mm})/(100\,\text{mm}) = \underline{-1\,\text{MPa}}\ \text{(compressive!)}.$$

Thus the results are the same as before, except that we have a negative displacement, strain, and stress.

3.10.2 Direct Stiffness Method II

Based on the preceding discussion, it should be obvious that the direct stiffness method applies to compact axial compressive members just as it does for axial tensile members. It is left as an exercise to rework Example 3-7 from Section 3.7, but now with a compressive force.

3.11 Design of Compact Compressive Axial Structures Ⓨ📖💻

The design of compact compressive axial structures is illustrated by the following design example.

EXAMPLE 3-9:

A large manufacturing facility has 10 identical heavy machine tools, each weighing 90 kN (about 10 tons), that need to be elevated above a low-strength concrete floor by about 15 cm for cleaning purposes. (It is desired to hose down the floor with an alkaline solvent.) The machines currently have a "foot" on each corner, consisting of a cylindrical pad of 13 mm radius by 7 mm depth made from A-36 steel (see Figure E3-9A).

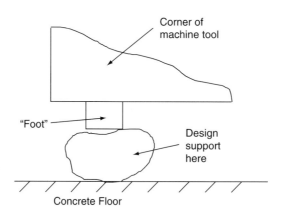

FIGURE E3-9A *Design problem definition sketch.*

EXAMPLE 3-9: *Cont'd*

Design a set of four inexpensive supports per machine (40 total for all 10 machines) to accomplish the elevation, with a factor of safety of 2. The existing "foot" (Figure 3-9A) should sit atop the support in each case.

I. *Problem Definition*

Performance requirements:
- Factor of safety $= 2$; this a specification on strength, and as such is of primary importance. Then $\sigma \leq S_f/2$, where S_f is some as yet unspecified failure strength.
- No other performance requirements are given.

Service environment:
- Loads are assumed to act statically. It can be assumed that the load per support is one-fourth the weight of a machine, or

$$P = 1/4(90\,\text{kN}) = 22.5\,\text{kN}.$$

(The assumption of static loads is made to keep the problem simple enough for educational purposes. In the real situation, the loads would clearly be dynamic.)
- There is a concern about the support being exposed to solvent.
- No other service environment specifications are given.

Project constraints:
- Low cost (implies generally available material, simplicity of design, ease of manufacture and assembly, reasonably low weight, etc.), particularly because of the large number of parts required.
- Existing strength of the concrete floor is low.
- No other constraints are given.

II. *Preliminary Design*

- Identify DDOFs:
  1. Geometry $\Rightarrow$ minimize volume to keep weight low, keep slenderness ratio < 10, keep contact stress low $\Rightarrow$ area A and length L as DDOFs
  2. Density $\Rightarrow$ keep weight low (ρ)
  3. Strength $\Rightarrow$ design for yielding where possible for low cost (S_Y); concrete strength is S_U
  4. Stiffness $\Rightarrow$ modulus (E)
  5. Boundary conditions $\Rightarrow$ pinned boundary conditions can keep link axial, that is, without moment reactions (more details about the connections will not be considered at this stage)
- Trade study: The desire to keep costs low (and as always to keep things as simple as possible!) suggests a simple axial structure for the support. A circular cross-section of diameter D will keep the surface area to cross-sectional area of

Continued

EXAMPLE 3-9: *Cont'd*

the link to a minimum, thus minimizing the volume of material. Material considerations suggest simple, readily available materials to keep costs low, but analysis needs to confirm the specific choice.

- FMEA: Failure expected to be from (1) simple static fracture of the concrete; (2) yielding due to higher stress at bearing between foot and support; and (3) corrosion failure due to use of alkaline solvent/wash.

III. *Detailed Design*

- Keep slenderness ratio of support < 10 to avoid failure by buckling:

$$\frac{L}{\sqrt{I/A}} < 10, \ I = \frac{\pi D^4}{64}, \ A = \frac{\pi D^2}{4}$$

$$\frac{L}{\sqrt{\dfrac{\pi D^4}{64}\dfrac{4}{\pi D^2}}} = \frac{L}{D/4} = \frac{4L}{D} < 10$$

or

$$L/D < 2.5.$$

- Keep bearing stress in concrete $< S_U/FS$:

Bearing stress in the concrete $= P/A \leq S_U/FS = 14\,\text{MPa}/2 = 7\,\text{MPa}$

Then solving for A:

$$A \geq (22.5\,\text{kN}/7 \times 10^6\,\text{N/m}^2) = 0.00321\,\text{m}^2.$$

Since $A = \pi D^2/4$, the minimum diameter D is then found from the equality case as

$$D = 4A/\pi = 4(0.00321\,\text{m}^2)/\pi = 64\,\text{mm}\ (2.52\,\text{in}).$$

Now the length of the support can be specified from the L/D ratio:

$$L \leq 2.5D = 2.5(6.4\,\text{cm}) = 16\,\text{cm}\ (6.3\,\text{in})$$

Note that that the 15 cm elevation desired is within this range.

At this point, we have defined the minimum geometry for the support, but have yet to specify the material. Two considerations that apply here are corrosion resistance and bearing stress.

EXAMPLE 3-9: *Cont'd*

- Review possible material choices based on corrosion resistance:

  Using the Ashby chart for comparative ranking of the resistance of materials to attack by different environments (Ashby, 1999), we find that the best possible material choices for resistance against alkali attack are steel, cast iron, titanium, polytetrafluoroethylene (PTFE), polypropylene (PP), alumina (Al_2O_3, a ceramic), and high-density polyethylene (HDPE). Assuming in advance that we will need relatively high strength to withstand the foot bearing stress, we further look to obtain the best possible combination of strength and low cost (this assumption can be checked later). Here we use the strength to relative cost per unit volume Ashby chart and find that, of the above materials, cast iron, steel, and alumina fare the best (as they fall on roughly the same guideline and are closest to the upper left-hand corner).

  However, looking at the Ashby chart it should be noticed that among these materials cast iron is the cheapest.

- Keep bearing stress at foot/support interface $< S_Y/FS$:

  This will require calculating the maximum bearing stress between the machine foot and the support. As before, we first calculate the bearing area A_f:

  $$A_f = \pi D_f^2/4 = \pi (25 \times 10^{-3}\,\text{m})^2/4 = 491 \times 10^{-6}\,\text{m}^2.$$

  The bearing stress of the foot on the support is then

  $$\sigma_z = P/A_f = 22.5\,\text{kN}/491 \times 10^{-6}\,\text{m}^2 = 45.8\,\text{MPa}.$$

  The required yield strength is then

  $$S_Y = \sigma_z FS \approx 92\,\text{MPa}$$

  Any of the three candidate materials will easily satisfy this requirement. So we choose the lowest cost material or cast iron.

  A few other notes about the design are in order:

- To keep the problem simple, we are assuming here that the machine will be globally stable when sitting upon the supports. In a real design, this would need to be carefully considered, especially since dynamic loads are involved.
- We haven't worried here about esthetics, for example, whether or not we're going to get rust stains on the concrete from the cast iron. This might be important in certain situations.

  The final design is shown in Figure E3-9B.

Continued

EXAMPLE 3-9: *Cont'd*

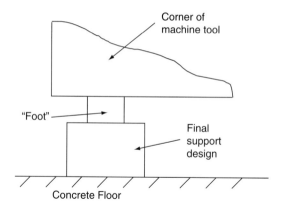

FIGURE E3-9B   *Final design sketch.*

- Final support specification:

64 mm (2.5 in) round bar × 15 cm (6 in) long cast gray iron

(Note: since 40 of these need to be made, one might order, say, 20 12-in lengths. These are readily available from suppliers.)

Key Points to Remember

For the elastic design of axial structures, the following design rules can be formulated:

- Nominal member stress is directly related to member force divided by member cross-sectional area. However, stress concentration effects need to be taken into account.
- Member strain is directly related to material elastic modulus for a given stress.
- Choice of cross-sectional shape may be based on a variety of considerations:
  1. Circular cross-sections give least volume for a given surface area
  2. Availability and cost of various shapes
  3. Fabrication and assembly issues associated with various shapes
- Failure by buckling can be avoided by keeping the compression member relatively compact, such as a slenderness ratio less than about 10.
- Boundary conditions should be such that moments are not developed in the axial member, such as pinned boundary. For compression members, boundary conditions must lead to configurational stability, such as pinned-roller boundaries.
- Material selection should be guided by use of Ashby's method for obtaining appropriate material indices, then use of a weighted decision matrix down-selection approach. In most cases, strength, weight, and cost are the most important considerations.
- For compression members, local crushing (bearing, contact) must be considered.
- Macroscopic mechanical behavior in tension and compression is governed by the response of the atomic structure to the applied loads.

References

Anonymous (1980). *Manual of Steel Construction*. American Institute of Steel Construction.

Ashby, M. F. (1999). *Materials Selection in Mechanical Design*, 2nd ed. Butterworth-Heinemann.

Eisenstadt, M. M. (1971). *Introduction to Mechanical Properties of Materials*. Macmillan, New York, pp. 164-169.

Freudenthal, A. M. (1950). *The Inelastic Behavior of Engineering Materials and Structures*. Wiley, New York.

Shigley, J. E., and Mischke, C. R. (2001). *Mechanical Engineering Design*. McGraw-Hill, New York.

Problems

3-1. A round bar of steel, 9 in long with a diameter of 0.5 in, is subjected to a tensile force of 10,000 lb. What are the dimensions of the bar after the stress is applied? Provide an answer in both the U.S. system of units and the SI system. Elastic modulus of steel is 30×10^6 psi and Poisson's ratio is 0.3.

3-2. A bridge is 1 km long and is made of steel. The temperature of the bridge varies from -20 to $40°C$. How much thermal expansion results? If the joints in the bridge cannot accommodate the expansion, what is the stress in the bridge at the lowest and maximum temperature if the bridge was erected at a temperature of $10°C$? The thermal expansion coefficient of steel is 6×10^{-6} mm/mm/$°C$ over this temperature range.

3-3. Three rods are hung vertically from a ceiling, each suspended from its top. Each rod is 10 m in length, with one made from steel, another from aluminum, and a third from wood. Compute the total elongation of each rod due to its own weight, and compare one to another. Why are the results independent of the cross-sectional area?

Material	Density ρ (kg/m^3)	Modulus E (GPa)
Steel	7850	200
Aluminum	2700	70
Wood	500	12

3-4. A helicopter needs to lift a 20-ton tank that is resting on the ground. A 300-foot titanium cable of 1 in diameter is lowered and connected to the tank. Assuming the tank lifts uniformly, how much elevation must the copter gain before the tank is lifted free from the ground?

3-5. A bar of length L, elastic modulus E, and density μ is to be hung vertically from the ceiling. A tensile force P is to be applied axially at the lower end. Derive an expression for the total elongation of the bar under the action of the applied force plus the self-weight of the bar. Under what conditions can the self-weight of the bar be neglected?

3-6. Axial loads are applied to a rod at locations A, B, C, and D as shown in Figure P3-6. The rod is made from steel with a cross-sectional area of 3 square inches. Using standard methods of statics, determine the axial stress in the bar:
   (a) on a section 10 in to the right of A
   (b) on a section 60 in to the right of A

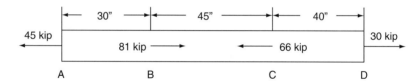

A   B   C   D

FIGURE P3-6

3-7. In Figure P3-7, A is a steel bar of rectangular cross-section 10×20 mm and B is a brass bar of rectangular cross-section 10×30 mm. Bar C may be considered rigid. All connections are pin joints. An 18-kN load is hung from C in such a place that the bar remains horizontal. Determine:
   (a) the axial stress in each bar A and B
   (b) the vertical displacement of bar C
   (c) the position of the line of action of the 18-kN force

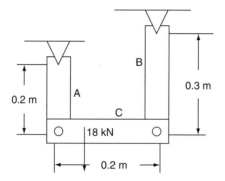

FIGURE P3-7

3-8. When a material is elastically deformed under axial loading if its volume remains constant, what is the Poisson's ratio?

3-9. A round bar of aluminum, which is 150 mm long and 12.5 mm in diameter, is subjected to a tensile load of 5000 N. What are the dimensions of the bar after the application of the load? The elastic modulus of aluminum is 72 GPa and the Poisson's ratio is 0.33.

3-10. Determine the stress in a body made of a material whose Poisson's ratio is equal to 0.5 if the three normal strains ε_{xx}, ε_{yy}, and $\varepsilon_{zz} = -500\,\mu\varepsilon$. Explain the results.

3-11. The potential energy of interaction between two atoms is given by $\phi(r) = -\frac{a}{r^m} + \frac{b}{r^n}$. Show that
   (a) The minimum potential energy is obtained when $\phi(r)$ is a minimum.
   (b) The condition $m > n$ must be satisfied for the following equation to describe an equilibrium state, or a state in which the crystal is stable:

$$\phi(r) = -ar^{-n}[1 - \frac{n}{m}(\frac{r}{r_o})^{-m}].$$

3-12. The Young's modulus of NaCl in the [100] direction has been experimentally determined to be 5.35×10^{10} N/m^2 and the Poisson's ratio as 0.16. Calculate E for NaCl and compare it to the experimentally measured value. Given that the ionic radius of Na is 0.97 Å, and for Cl is 1.81 Å, the value of n for NaCl is 8.

3-13. Fill in the blanks in the following statements:
   (a) The ratio of applied load to original cross-sectional area is defined as _____
   (b) The energy required to stretch a unit volume of a material to failure is defined as

   _____
   (c) Sudden localized reduction in the cross-sectional area of a specimen is defined as

   _____
   (d) The distance between the attachment points of an extensometer is defined as

   _____
   (e) The slope of the linear portion of the stress–strain curve is the graphic representation of _____
   (f) The type of fracture surface that is rough in texture and is perpendicular to the longitudinal axis of the specimen is characteristic of _____
   (g) After the lower yield point is reached more stress must be applied to increase the strain, the effect of this property is known as _____
   (h) The area under a stress–strain curve up to the proportional limit is the graphic equivalent of the _____
   (i) The area under a complete stress–strain diagram is the graphic equivalent of the

   _____
   (j) The ratio of negative lateral strain to longitudinal strain is defined as _____
   (k) The ratio of applied load to actual cross-sectional area is defined as _____
   (l) The slope of a graph of lateral strain versus longitudinal strain is constant in the elastic range and changes to a higher constant in the plastic range are graphic representation of _____
   (m) If pure shear stress is applied the elastic response of the material is characterized by

   _____

3-14. Are elongation and necking indicators of ductility or brittleness?
   (a) Ductility
   (b) Brittleness
   (c) Both
   (d) None

3-15. If one material has a higher elastic modulus than another, does it elongate more or less?
   (a) More
   (b) Less
   (c) None
   (d) No change

3-16. For three spring elements connected in series (Figure P3-16), show that the global stiffness matrix is given by:

$$[k] = \begin{bmatrix} k_1 & -k_1 & 0 & 0 \\ -k_1 & k_1 + k_2 & -k_2 & 0 \\ 0 & -k_2 & k_2 + k_3 & -k_3 \\ 0 & 0 & -k_3 & k_3 \end{bmatrix}$$

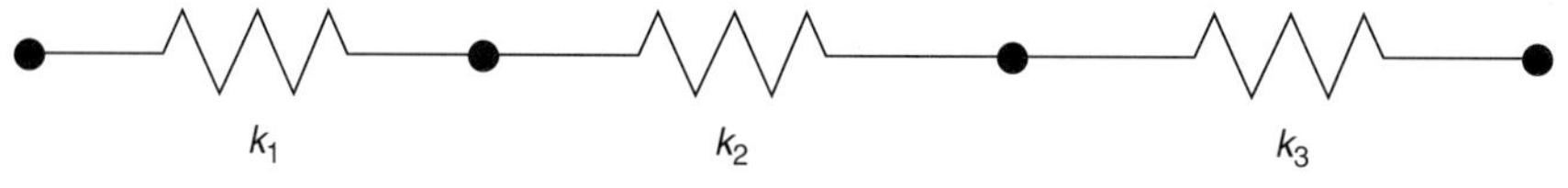

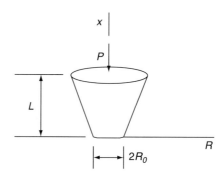

FIGURE P3-16

3-17. Rework Problem 3-6 and calculate the deflections along the rod using the direct stiffness method. Assume the rod has a fixed left boundary condition and roller right boundary condition. Assume a modulus of 10 million psi.

3-18. Rework Problem 3-1, except consider the force applied to be compressive. Assume that the rod will not buckle and it is stable.

3-19. A table leg has an "inverted taper" as shown in Figure P3-19.
(a) Show that the "taper" can be modeled by

$$R = R_0 \left(c \frac{x}{L} + 1 \right)$$

(b) Assuming small taper ($c << 1$), determine an expression for $\sigma_x(x)$.
(c) Assuming small taper ($c << 1$), determine an expression for the total axial deflection u_{total}.

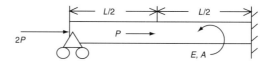

FIGURE P3-19

3-20. Use the direct stiffness method to determine the displacement and stress in the rod shown in Figure P3-20.

FIGURE P3-20

3-21. Rework Example 3-9 (two springs in series, with the left end fixed) but with a 100-N compressive load placed on the right end as shown in Figure P3-21.
Given that $k_1 = 10,000\,\text{N/m}$ and $k_2 = 20,000\,\text{N/m}$, determine the displacements of the springs.

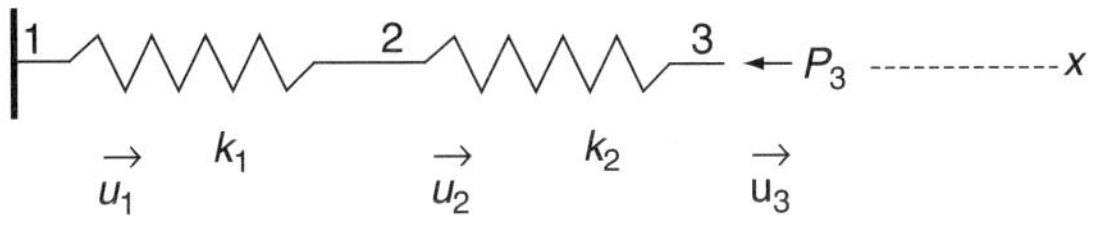

FIGURE P3-21

3-22. For a slenderness ratio less than 10, what is the limit on the length of a compression member having a square cross-section 1 in on a side?

3-23. Verify the ranking for strength versus relative cost per unit volume in Table 3-2.

Design Projects

3-24. Design a 1-m long prismatic rod that carries a 100-N load with a safety factor of 2 and weighs no more than 6.5 N. Make a list of possible materials. From this set of materials, choose the material that is the cheapest and least susceptible to fracture.

3-25. Design a 1-m long prismatic rod that carries a 100-N load with a safety factor of 2 and deflects no more than 1 mm. Make a list of possible materials. From this set of materials, choose the material that is the cheapest and least susceptible to fracture.

3-26. A heavy machine tool weighing 75 kN needs to be elevated above a rigid floor (infinitely strong) about 12 cm for cleaning purposes. It is desired to hose down the floor with an alkaline solvent. Design a set of four inexpensive supports to accomplish the elevation, with a factor of safety of 2.

4   Torsion Structures

C'mon let's twist, yeah just like this.

—Chubby Checker

Objective: This chapter will introduce several design degrees of freedom—J, G, S_s, β, ϕ— in the context of the design of torsion structures.

What the student will learn:

- Definition and configuration of torsion structures
- Equilibrium and deformation of torsion structures
- Elastic constitutive relation for shear
- Linkage of mechanics and materials concepts through the shear test
- Analysis methods for complex torsion structures
- Materials selection for torsion structures
- How to design static torsion structures

4.1   Introduction ⍦🏛💻

Thus far we have investigated axial structures, capable of carrying axial loads either in tension or compression. We have seen that axial structures can form elements of more complicated structural systems, e.g., trusses, capable of carrying limited transverse loads.

But what of twisting loads? If you put a wrench on a nut and twist it so as to tighten or loosen it, how is the load carried (Figure 4-1)? What is the internal response of the bolt? What are the DDOF that govern the bolt/nut design?

Many other examples of such torsion structures are common, such as the shafts on an automobile (Figure 4-2).

An infamous example of a torsion loading is the failure of the Tacoma Narrows Bridge in the state of Washington (e.g., see Billington, 1985). On November 7, 1940, just several months after it was opened, the bridge experienced severe torsion oscillations that ultimately led to its collapse. The peak amplitudes of the oscillations reached around 20 feet (Figure 4-3). Fortunately there was no loss of human life.

FIGURE 4-1 *Wrench being used to tighten a nut.*

Drive shaft   Axle shaft

FIGURE 4-2 *Axle and drive shafts on an SAE Mini-Indy racecar.*

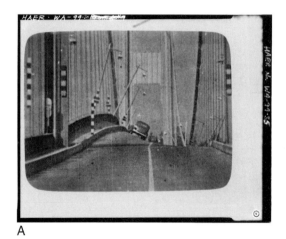

A

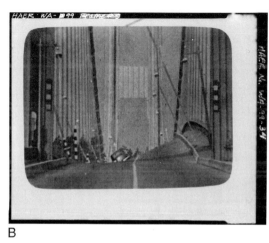

B

FIGURE 4-3 *Torsion oscillations of the Tacoma Narrows Bridge. (Courtesy of Library of Congress, Prints & Photographs Division, HAER, WASH, 27-TACO, 11-34 and 11-35).*

In practice, it is difficult to load any structural member purely in torsion. All of the examples above have transverse loads applied in addition to torques (students should be sure that they see this). In this chapter, however, we assume the loads to be purely torsional (we'll consider cases of combined loading in Chapter 6).

4.2 Equilibrium and Deformation ⵑ

4.2.1 Configuration

We can begin by considering a torsion structure of arbitrary cross-section (Figure 4-4). The cross-sections form a structural element that is straight and relatively narrow, that is, the dimension of any cross-section is small compared to the element length L. The cross-section is characterized by its area $A(x)$ and polar moment of inertia (polar second moment of area) $J(x)$ (this latter quantity will fall naturally out of the derivation later).

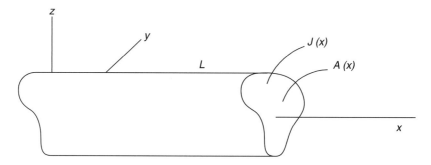

FIGURE 4-4 *Torsion structure of arbitrary cross-section.*

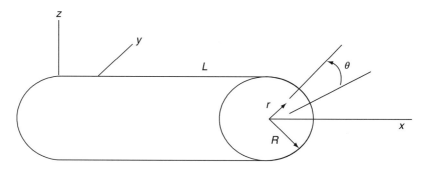

FIGURE 4-5 *Circular shaft of radius R. Note that a cylindrical-polar coordinate system (r, θ, x) is also referred to in this configuration.*

We first consider only cross-sections that are symmetric with respect to rotations about the long axis (*x*-axis in the figure), namely, *axisymmetric* cross-sections, and that are at the same time *prismatic*, that is, with $A(x) = $ constant. The simplest embodiment of an axisymmetric prismatic torsion structure is the solid circular shaft, whose cross-section is a circle of radius R (Figure 4-5).

In the first part of what follows, we will consider our torsion structural elements to be right circular cylinders (solid or hollow), but later other cross-sectional shapes will be considered. Any point on the circular cross-section $x = $ constant can be identified by the cylindrical-polar coordinates (r, θ, x).

As we did for axial structures in Chapter 3, it will be convenient to consider the locus of material particles parallel to the long (*x*) axis of the undeformed element. This collection of points represents a virtual material fiber in the element. The element may then be thought of as being composed of a large number of longitudinal fibers, one of which is colinear with the long axis of the element, that is, the *axial fiber*.

4.2.2 Equilibrium

The torsion structure is loaded by a *twisting moment* or *torque* **T**. In practice, this could arise from a couple of magnitude *Pd* applied to the shaft (Figure 4-6). The couple is self-equilibrating since no net force results. (Why?) Application of global statics would conclude that a moment $\mathbf{M} = -\mathbf{T}$ must be acting as a reaction for equilibrium to occur.

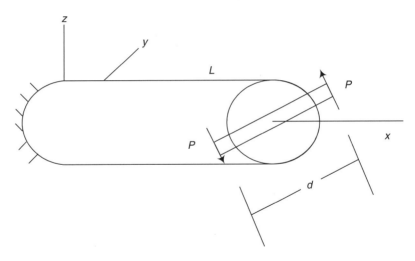

FIGURE 4-6 *Cylindrical shaft under the action of a couple of magnitude T = Pd. The end x = 0 is imagined to be clamped in the figure.*

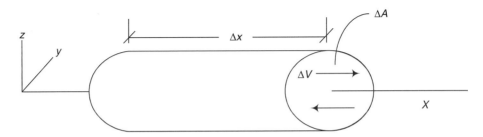

FIGURE 4-7 *Internal force system on a differential shaft element consisting of a shear force ΔV. (There is a shear force acting over every material particle in ΔA.)*

The applied couple is resisted internally by a shear force $\mathbf{V}$ of magnitude V. Figure 4-7 shows a cylindrical differential element of the shaft of length Δx and cross-section ΔA (radius $r < R$), which is coaxial with the original shaft of radius R. The associated direct shear stress on the element face is $\tau_{x\theta}$ given by

$$\tau_{x\theta} = \lim_{\Delta A \to 0} \frac{\Delta V}{\Delta A} \tag{4.1}$$

where ΔV is the incremental shear force. Because of symmetry, V or $\tau_{x\theta}$ can only be functions of r and x, but not θ.

Note that there is no net shear force V_y or V_z, since ΔV forms a self-equilibrating system of couples. (Why?)

Now the magnitude of the total internal moment $-\mathbf{M}$ resisting the applied load is given by the equilibrium equation $-\mathbf{M} = \mathbf{T}$, where

$$T = \int_A r\,dF = \int_A r\tau_{x\theta}\,dA. \tag{4.2}$$

4.2.3 Deformation Υ

We assume the pure torsion or twisting deformation to be smooth and regular. For a torsion element undergoing small twisting deformations, it is reasonable to further assume the following:

1. Cross-sections originally plane and normal to the element long axis remain plane and normal to the axis after deformation.
2. The originally symmetric cross-sections remain symmetric after deformation.
3. No displacement occurs in either the axial or radial directions.

For the circular shaft, these assumptions would imply that:

- A diameter $2R$ before deformation remains a diameter after deformation with the same value $2R$, that is, $\varepsilon_r = 0$.
- All diameters at any given section $A(x)$ rotate through the same *twist angle* $\phi(x)$, hence $\gamma_{r\theta} = \varepsilon_\theta = 0$.
- The length of the shaft remains unchanged, hence $\varepsilon_x = \gamma_{rx} = 0$.

Thus if a given cross-section $A(x)$ rotates by an angle $\phi(x)$, a nearby cross-section $A(x + \Delta x)$ rotates by an angle $\phi(x + \Delta x) = \phi(x) + \Delta\phi$ [and $A(x - \Delta x)$ rotates by an angle $\phi(x - \Delta x) = \phi(x) - \Delta\phi$]. Also, the only nonzero strain is $\gamma_{x\theta}$.

To understand the physical meaning of the shear strain $\gamma_{x\theta}$, consider the cylindrical differential element of radius r $(<R)$ and length Δx previously introduced in Figure 4-7 (shown now in Figure 4-8).

Imagine now a second differential element mapped onto the surface of the cylindrical differential element, with sides $r\Delta\theta$ and Δx. The curved sides of this element have (arc) length $r\Delta\theta$, but you can imagine that if the element is small enough ($\Delta\theta$ very small), the "curve" is essentially a straight line.

The original cylindrical shaft (and associated differential elements) is subjected to a twisting moment (torque) **T** such that the cross-sections at x and $x + \Delta x$ undergo rotations of ϕ and $\phi + \Delta\phi$, respectively (Figure 4-9a).

The *shearing strain* of the surface element is given by the small angular distortion α (recall that $\varepsilon_r = \varepsilon_\theta = 0$, so that the differential elements do not change size). From Figure 4-9 using the small angle approximation:

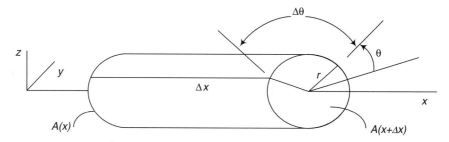

FIGURE 4-8 *Differential element of radius r coaxial with shaft of radius R with sector rΔθ shown.*

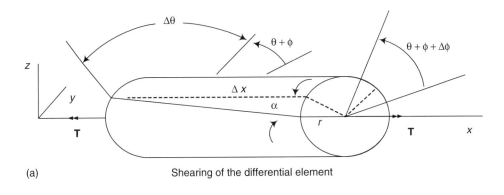

(a)   Shearing of the differential element

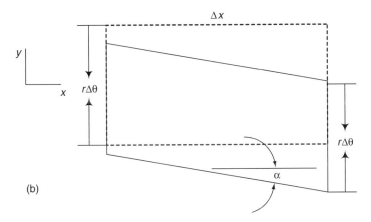

(b)

FIGURE 4-9 *Differential elements of the shaft under the action of a torque **T**. (a) Isometric view, (b) plane view. Note that the originally rectangular differential element rΔθ by Δx has been sheared into a parallelogram. The double-headed arrow is the right-hand rule representation of **T**.*

$$\alpha \approx \tan \alpha = \frac{r\Delta\phi}{\Delta x}. \tag{4.3}$$

Then the shear strain $\gamma_{x\theta}$ (the only nonzero strain) by definition is

$$\gamma_{x\theta} = \lim_{\Delta x \to 0} \alpha = \lim_{\Delta x \to 0} \frac{r\Delta\phi}{\Delta x} \tag{4.4a}$$

or

$$\boxed{\gamma_{x\theta} = r\frac{d\phi}{dx}} \tag{4.4b}$$

This is the *shear strain–(angular) displacement* relation. The shear strain is a function of radial position r and *twist gradient* $d\phi/dx$. The maximum shear strain thus exists at the surface (where r is maximum, i.e., $r = R$) for any twist gradient and for any radius wherever the twist gradient is maximum. (For a uniform torque on a shaft, the twist gradient is constant. Why?)

4.3 Constitution ⬠

In the theory of linear elastic homogeneous materials, shear or *deviatoric* effects (those relating to angular distortions of differential elements) are uncoupled (independent) from *dilatational* (volume changes in elements) effects associated with normal stresses and strains. In practice, this means that normal stresses are only related to normal strains, and shear stresses are only related to shear strains.

For the torsion of homogeneous linear elastic circular shafts, the only nonzero stress and strain components are related by

$$\tau_{x\theta} = G\gamma_{x\theta}$$

(4.5)

where G is the *shear modulus* of the material.

We can now substitute the constitutive relation (4.5) into the equilibrium relation (4.2), using the shear strain–displacement relation (4.4b):

$$T = \int_A r(G\gamma_{x\theta})dA = G\int_A r\left(r\frac{d\phi}{dx}\right)dA.$$

(4.6)

For constant twist gradient we have the *moment–twist (gradient)* relation

$$T = GJ\frac{d\phi}{dx}$$

(4.7)

where

$$J = \int_A r^2 dA$$

(4.8)

is the *polar moment of inertia* of the cross-section about the cylinder axis.

For a circular cross-section of diameter $D = 2R$, the integration indicated in (4.8) is easily performed since $dA = r\, dr\, d\theta$:

$$J = \int_0^{2\pi}\int_0^R r^2 rdrd\theta = 2\pi\frac{R^4}{4} = \pi\frac{R^4}{2} = \pi\frac{D^4}{32}.$$

(4.9)

We can recast the moment–twist relation (4.7) in terms of the twist gradient

$$\frac{d\phi}{dx} = \frac{T}{GJ}.$$

(4.10)

Then the total angle of twist, for a prismatic shaft of length L under constant torque of magnitude T, is found from integration of (4.10):

$$\phi_{max} = \int_0^L \frac{T}{GJ}dx = \frac{TL}{GJ}$$

(4.11)

---

where ϕ is in radians. [If the upper limit of integration was some intermediate point x, then the integration would supply the twist angle at x or $\phi(x)$.]

Recasting Equation (4.11) in terms of applied torque gives:

$$\boxed{T = K\phi} \tag{4.12}$$

Notice the similarity in form to the axial equation $F = ku$. The term K on the right-hand side of (4.12) is the *torsional stiffness* of the shaft

$$K = GJ/L \tag{4.13}$$

(Compare K to $k = EA/L$.) K gives the applied torque per unit twist angle [N-m/rad].

Finally, we can substitute the shear strain–displacement relation (4.4b) into the constitutive relation (4.5) to get the shear stress in terms of the applied torque:

$$\boxed{\tau_{x\theta} = Gr\frac{d\phi}{dx} = Gr\frac{T}{GJ} = \frac{Tr}{J}}. \tag{4.14}$$

We will see in a later chapter that there is an analogous expression for the flexural stress in a beam in terms of the applied bending moment. By inspection of (4.14):

$$\boxed{\gamma_{x\theta} = \frac{rT}{GJ} = \frac{rT}{KL}}. \tag{4.15}$$

Equation (4.15) shows that the shear strain is inversely proportional to the polar moment of inertia and torsional stiffness: as the stiffness decreases, the shear strain increases.

EXAMPLE 4-1:

A 2 ft long steel shaft of $1\frac{1}{4}$ in diameter is subject to a torque of 100 ft-lb. Determine the maximum shear stress in the shaft and the total angle of twist. The solution proceeds using Mathcad® (and as such the formatting may not always match precisely with that used in the text).

$$R := \frac{1.25}{2}\, \text{in} \qquad L := 24\ \text{in} \qquad G := 11.5 \cdot 10^6 \cdot \frac{\text{lb}}{\text{in}^2} \qquad T := 100\ \text{ft} \cdot \text{lb}$$

$$J := \frac{\pi \cdot R^4}{2}$$

$$J = 0.24\ \text{in}^4$$

$$\tau\,\text{max} := \frac{T \cdot R}{J}$$

$$\tau\,\text{max} = 3.129 \times 10^3\, \frac{\text{lb}}{\text{in}^2}$$

$$\phi\,\text{max} := \frac{T \cdot L}{G \cdot J}$$

$$\phi\,\text{max} = 0.01\ \text{rad} = 0.599°$$

EXAMPLE 4-2:

A solid tapered shaft has a diameter that varies linearly from D_0 at $x = 0$ to $2D_0$ at $x = L$ as shown in Figure E4-2A. It is subjected to an end torque T at $x = 0$ and is attached to a rigid wall at $x = L$. The shear modulus of the shaft is G.

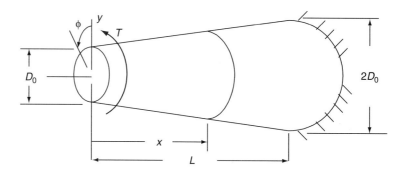

FIGURE E4-2A

(a) Determine an expression for the maximum (cross-sectional) shear stress in the tapered shaft as a function of the distance x from the origin of coordinates.
(b) Determine an expression for the total angle of twist of the shaft.

Consider a cutting plane coincident with the $z = 0$ plane. The outer fiber exposed by this plane has an equation:

$$y(x) = R(x) = R_0\left(1 + c\frac{x}{L}\right)$$

where $R_0 = D_0/2$ and $c = 1$ for the problem at hand. Then

$$J(x) = \frac{\pi R^4(x)}{2} = \frac{\pi}{2}R_0^4\left(1 + \frac{x}{L}\right)^4.$$

Now with $T(x) = T$,

$$\tau_{x\theta,\ max} = \frac{T(x)R(x)}{J(x)} = \frac{2T}{\pi R_0^3\left(1 + \frac{x}{L}\right)^3}.$$

To understand what this result means, we plot it in Figure E4-2B.
We can see that the maximum shear stress exists at the small end ($x = 0$) and decays to an eighth of that value at the large end ($x = L$). This makes sense since the smallest torsion stiffness $K(x)$ exists at the smallest end, hence the strain, and thus stress, are highest there.
The total angle of twist is found from

EXAMPLE 4-2: *Cont'd*

Let q = x/L   q : = 0,0.1.. 1

Let tau(q) = tau(x)/tau$_{max}$

$$\text{tau (q) } := \frac{1}{(1+q)^3}$$

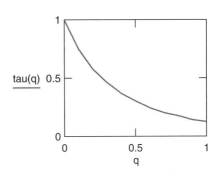

FIGURE E4-2B

$$\phi(L) = \int_0^L \frac{T}{GJ(x)} dx.$$

(Note that if the upper limit were x instead of L, we would then find the twist deflection as a function of x along the shaft.)

Substituting in for $J(x)$ and performing the indicated integration gives

$$\phi(L) = \frac{2T}{\pi R_0^4 G} \int_0^L \frac{dx}{\left(1+\frac{x}{L}\right)^4} = \ldots = \frac{-2TL}{3\pi R_0^4 G}\left[\frac{1}{\left(1+\frac{x}{L}\right)^3}\right]_0^L = \ldots = \underline{\underline{\frac{7TL}{8\pi R_0^4 G}}}.$$

(The intermediate steps missing above are left as an exercise for the student. Problems associated with this example are given at the end of the chapter.)

EXAMPLE 4-3:

A 1-in diameter 2024-0 aluminum shaft is built in at both ends as shown in Figure E4-3A. An intermediate torque is applied at point B of magnitude $T = 600 \text{ in} \cdot \text{lb}$. Determine the twist deflection at point B. $G = 3.67 \text{ M}_{si}$.

We first consider a free-body diagram of the shaft as shown in Figure E4-3B. A summation of moments gives:

$$\Sigma M = 0: M_A + M_C - T_B = 0$$

This is one equation in two unknowns, M_A and M_C and therefore the system is statically indeterminate (overconstrained—why?). To solve for the unknown reaction moments in

Continued

EXAMPLE 4-3: *Cont'd*

this formulation, we need to introduce another independent equation, and we will find one in a *constraint equation*. This is readily found by the compatibility of rotations at point B, namely that the total rotation from A to B must equal the total rotation from B to C, or:

$$\phi_{AB} = \phi_{BC}.$$

That is,

$$\frac{M_A L_{AB}}{GJ} = \frac{M_C L_{BC}}{GJ}$$

or

$$M_A L_{AB} = M_C L_{BC}$$

Now substituting this relation into the moment equilibrium equation gives

$$M_A + M_A \frac{L_{AB}}{L_{BC}} - 600 \, \text{in} \cdot \text{lb} = 0$$

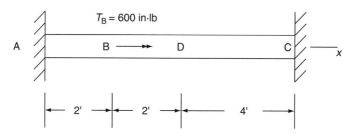

FIGURE E4-3A

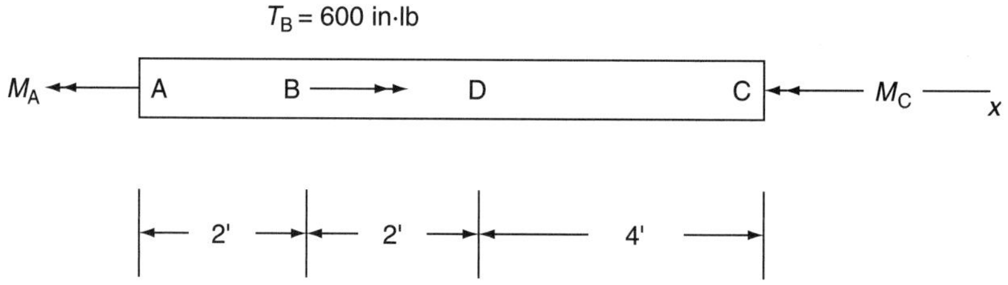

FIGURE E4-3B

EXAMPLE 4-3: *Cont'd*

Then

$$M_A = 450 \text{ in} \cdot \text{lb} \ (= 0.75T_B)$$

$$M_C = 150 \text{ in} \cdot \text{lb} \ (= 0.25T_B)$$

$$\phi_{AB} = [(450 \text{ in} \cdot \text{lb})(24 \text{ in})] \bigg/ \left[(3.67 \times 10^6 \text{ psi}) \frac{\pi}{2} (0.5 \text{ in})^4 \right] = 0.030 \text{ rad}$$

(We'll see later in the chapter how to solve this problem easily with the direct stiffness method.)

4.4 Mechanics ⇔ Materials Link: Shear Stress and the Shear Test ϓ🕸

When a tensile load is applied to a specimen consisting of a single crystal of a ductile material, it will permanently deform if stressed beyond its elastic limit. In other words, the material is said to have yielded. The stress at which the material just yields is called the yield stress. It has been experimentally observed that yielding occurs due to the process of *slip* that takes place in several narrow regions of the specimen. At the onset of yielding, *slip lines* appear on the surface of the specimen (see Tension Test virtual lab on-line for more details). Thus the deformation mechanism involves the slipping on selective planes relative to the adjacent planes, which suggests that shear forces are acting parallel to the slip planes and causing deformation. Hence, though macroscopically a tensile stress is being applied, the physical microscopic mechanisms that cause permanent deformation are due to a shearing phenomenon. Polycrystalline materials also exhibit a similar behavior, with the same deformation mechanism discussed earlier being present in each grain of the polycrystalline metal. It is for this reason that a more detailed discussion on the slip mechanisms is provided.

4.4.1 Directions and Planes in Crystals

At the atomistic level, strength, and hence yielding or plastic deformation in crystalline materials is related to *dislocations* (see Tension Test virtual lab on-line for more details). Dislocations are defects in what otherwise would be a regular crystal lattice. In one of the great experiments in mechanics and materials, Griffith (see Gordon, 1984) grew single-crystal "whiskers" of glass and showed that they were considerably stronger than bulk glass. The single-crystal whiskers were defect free and thus no dislocations were present to move, causing fully realized strength.

One important way that dislocations move is by *glide* or *slip*, that is, when atoms in a crystal plane "slip" over another adjacent plane, called a *slip plane* (again, see the Tension Test virtual lab on-line). A *critical shear stress* is required to move dislocations in glide. Hence the deformation of crystalline materials, particularly ductile materials, depends on shearing of specific crystal planes in specific directions. For analysis, it is important to be able to specify these planes and directions. *Miller indices* are used for this purpose.

Crystal Directions

Each point P in a crystal lattice can be identified by a vector from a chosen origin:

$$\mathbf{R} = u\hat{\mathbf{i}} + v\hat{\mathbf{j}} + w\hat{\mathbf{k}} \qquad (4.16)$$

The Miller indices are represented by [uvw], with the numbers u, v, w, being the smallest prime integers. Negative numbers are represented by an overbar, say $[1\ \bar{2}\ 3]$ for 1, -2, 3.

"Families" of directions are indicated by $<>$, e.g., the body diagonals [111], $[\bar{1}11]$, $[1\bar{1}1]$, and $[11\bar{1}]$ are the family $\langle 111 \rangle$, being all combinations of 1 and -1.

Crystal Planes

In a similar fashion, crystal planes can be represented by Miller indices ( ). In the material science literature, these are often found by:

1. Identifying the intercepts of the plane on the crystal axes $\frac{1}{h}$, $\frac{1}{k}$, $\frac{1}{l}$
2. Taking the reciprocal of the intercepts
3. Clearing the fraction if necessary
4. Representing the Miller indices as the smallest set (hkl)

Families of planes are denoted by { }.

There is another way for describing the orientation of crystal planes that would be more "intuitive" for engineers: describe the vector $\mathbf{n}$ that is the inward-seeking normal to the plane. This method can equally determine the Miller indices. Several examples will be shown to demonstrate the method.

EXAMPLE 4-4:

Determine the Miller indices for the directions shown within the cubic unit cell in Figure E4-4.

Let's take direction C as an example. To go from tail to head of a vector in the C-direction, we go -1 in the x-direction, $+1$ in the y-direction, and $+2/3$ in the z-direction, or

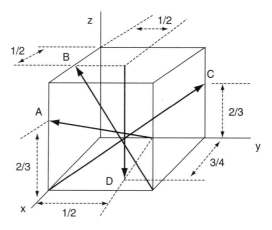

FIGURE E4-4 *Unit cube with various directions shown.*

EXAMPLE 4-4: *Cont'd*

$$\mathbf{n} = -\mathbf{i} + \mathbf{j} + 2/3\,\mathbf{k}.$$

Then the Miller indices would nearly be the vector components

$$[-1,\ 1,\ 2/3].$$

We need only multiply through by the least common denominator, 3, to get the Miller indices

$$[\bar{3}\quad 3\quad 2]$$

Table E4-4 gives values for the other directions as well (students should make sure they can complete the table for themselves).

TABLE E4.4

Direction	Vector *n*	Vector components	Miller indices
A	$1\,\mathbf{i} - 1/2\,\mathbf{j} + 2/3\,\mathbf{k}$	$[1, -1/2, 2/3]$	$[6\quad \bar{3}\quad 4]$
B	$-1/2\,\mathbf{i} - \mathbf{j} + \mathbf{k}$	$[-1/2, -1, 1]$	$[\bar{1}\quad \bar{2}\quad 2]$
C	$-\mathbf{i} + \mathbf{j} + 2/3\,\mathbf{k}$	$[-1, 1, 2/3]$	$[\bar{3}\quad 3\quad 2]$
D	$1/4\,\mathbf{i} - \mathbf{k}$	$[1/4, 0, -1]$	$[1\quad 0\quad \bar{4}]$

EXAMPLE 4-5:

Crystal planes are defined by the inward-seeking vector normal to the plane. The normal vector is easily found by forming the cross product of two adjacent *edge vectors* of the plane. Consider the following plane "B" in the unit cube shown in Figure E4-5A.

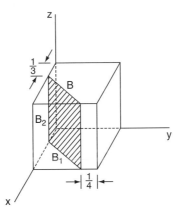

FIGURE E4-5A *A crystal plane in the unit cube.*

Continued

EXAMPLE 4-5: *Cont'd*

We define two edge vectors $\mathbf{B}_1$ and $\mathbf{B}_2$ as shown in Figure E4-5B (other edge vectors could also be chosen):

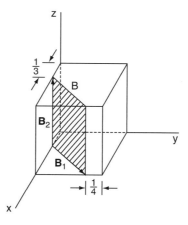

FIGURE E4-5B   *Edge vectors for the plane B.*

$$\mathbf{B}_1 = 2/3 \; \mathbf{i} + 3/4 \; \mathbf{j}, \; \mathbf{B}_2 = \mathbf{k}$$

We can form the inward-seeking normal vector by

$$\mathbf{n} = \mathbf{B}_2 \times \mathbf{B}_1 = \mathbf{k} \times (2/3 \; \mathbf{i} + 3/4 \; \mathbf{j}) = -3/4 \; \mathbf{i} + 2/3 \; \mathbf{j}$$

Now as before

$$[-3/4, \; 2/3, \; 0] \Rightarrow [\bar{9} \quad 8 \quad 0]$$

4.4.2   Inelastic Deformation

Inelastic or plastic deformation in crystalline materials, such as metals, occurs primarily due to dislocation motion. When an external stress is applied to the material, the deviatoric stress component causes dislocations to move. The movement of dislocations results in the slipping of one atomic plane relative to another in a certain preferred direction. In this type of motion, the lattice dimensions and the density of the crystals remain practically unchanged. The slip direction coincides with a crystallographic direction of maximum linear atomic density (that is, maximum number of lattice points per unit length). The slip planes are generally parallel to those crystallographic planes containing the slip directions in which the atomic density (that is, the maximum number of lattice points per unit area) exceeds or is equal to that of any other crystallographic plane containing the same slip direction. For example, in an FCC lattice slip occurs on {1 1 1} planes and in the direction [1 1 0]. In FCC materials there are four {1 1 1} planes—(1 1 1), ($\bar{1}$ 1 1), (1 $\bar{1}$ 1), ($\bar{1}$ $\bar{1}$ 1)—and three slip directions in each plane—for example, [$\bar{1}$ 1 0], [$\bar{1}$ 0 1], [0 $\bar{1}$ 1], resulting in 12 slip systems (the student should sketch out the plane normal vectors and the slip direction vectors). In BCC materials there are 12 slip planes and

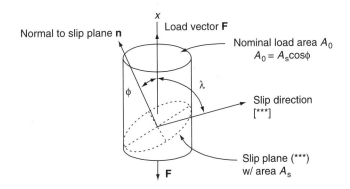

FIGURE 4-10 *Slip system generalized definition sketch.*

four slip directions in each plane, resulting in 48 slip systems. (Review Module 1 gives a summary of the slip systems for common crystal structures.)

Among the slip systems in a crystal, the system that actually becomes active is determined by the *critical resolved shear stress* (CRSS) in the slip plane and in the slip direction. For simplicity, consider a tension (or compression) specimen that consists of only one slip system as shown in Figure 4-10.

Note that the projection of the load vector onto the slip plane is given by

$$\frac{F\cos\lambda}{A_s} = \frac{F\cos\lambda}{A_0/\cos\phi} = \frac{F}{A_0}\cos\lambda\cos\phi. \qquad (4.17)$$

The component of the applied stress F/A_0 in the slip direction is the *resolved shear stress* (RSS) on the slip plane, and is given by

$$\tau_{RSS} = \frac{F}{A_0}\cos\phi\cos\lambda. \qquad (4.18)$$

When τ_{RSS} becomes equal to or greater than a certain critical value characteristic of the material, plastic flow initiates. The CRSS is then designated by τ_{CRSS}, and is related to the tensile yield strength S_Y by

$$S_Y = \tau_{CRSS}/(\cos\phi\,\cos\lambda). \qquad (4.19)$$

The τ_{CRSS} is practically a constant for a given material (however, BCC materials exhibit some exceptions). Also, temperature, strain rate, initial work hardening, and so forth can affect the value of τ_{CRSS}. The difference in the value of τ_{CRSS} for different materials is due to the difference in the magnitude of the interatomic forces.

To recap, in order to analyze the shear stress failure in a material, we need to know:

- Load magnitude and direction
- Load plane area A_0
- Slip plane orientation **n**
- Slip direction within the slip plane
- Slip plane area A_s

EXAMPLE 4-6:

A 10-mm diameter rod made of single-crystal copper (Cu) is placed in axial tension. The axis of the rod is parallel to the $\lfloor \bar{1}\,0\,1 \rfloor$ crystal direction. When an applied load of $3\,kN$ is applied, slip is observed on the $(\bar{1}\,\bar{1}\,1)$ plane. Determine the slip system(s) activated and the critical resolved shear stress (see Figure E4-6).

From Appendix A1 we find that copper has an FCC crystal structure, with a slip system $< 110 >$. We can identify the slip directions by realizing that the slip takes place in the slip plane defined by the slip direction $\mathbf{n}$, that is,

$$\mathbf{n} \cdot (\text{unknown slip direction}) = 0.$$

In other words, since the slip direction is perpendicular to the slip plane vector $\mathbf{n}$, the inner or scalar product of the two associated vectors must be zero. Based on the slip plane given in the problem statement, we write

$$(-\mathbf{i} - \mathbf{j} + \mathbf{k}) \cdot (x\mathbf{i} + y\mathbf{j} + z\mathbf{k}) = 0$$

or

$$-x - y + z = 0$$

must be enforced. Now, given the $< 110 >$ system, the possible values for x, y, and z are shown in Table E4-6.

To determine the activated slip system(s), we recall that for the critically resolved shear stress

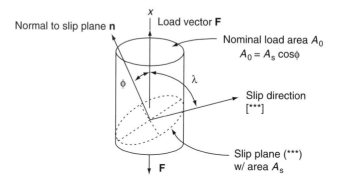

FIGURE E4-6

TABLE E4-6

If	then	Resulting slip direction
$x = 0$	$y = 1$, $z = 1$ or $y = -1$, $z = -1$	$[0\ 1\ 1]$
$y = 0$	$x = 1$, $z = 1$ or $x = -1$, $z = -1$	$[1\ 0\ 1]$
$z = 0$	$x = -1$, $y = 1$ or $x = 1$, $y = -1$	$[\bar{1}\ 1\ 0]$

EXAMPLE 4-6: *Cont'd*

$$\tau_{RSS} = \frac{F}{A_0} \cos \phi \cos \lambda$$

to be nonzero, $\cos \lambda$ must be nonzero. We determine $\cos \lambda$ by forming the unit scalar product of the load direction and the possible slip directions:

$$\cos \lambda_1 = \frac{[\bar{1}01] \cdot [011]}{\sqrt{2}\sqrt{2}} = \frac{1}{2}$$

$$\cos \lambda_2 = \frac{[\bar{1}01] \cdot [101]}{\sqrt{2}\sqrt{2}} = 0$$

$$\cos \lambda_3 = \frac{[\bar{1}01] \cdot [\bar{1}10]}{\sqrt{2}\sqrt{2}} = \frac{1}{2}$$

Hence the slip systems activated are $(\bar{1} \ \bar{1} \ 1) \ [0 \ 1 \ 1]$ and $(\bar{1} \ \bar{1} \ 1) \ [\bar{1} \ 1 \ 0]$. The CRSS is straightforward to calculate after determining

$$\cos \phi = \frac{[\bar{1}01] \cdot [\bar{1}\bar{1}1]}{\sqrt{2}\sqrt{3}} = \frac{2}{\sqrt{6}}.$$

Then

$$\tau_{CRSS} = \frac{3000 \, \text{N}}{\pi/4 \, (0.01 \, \text{m})^2} \frac{1}{2} \frac{2}{\sqrt{6}} = \underline{\underline{15.6 \, \text{MPa}}}$$

4.4.3 The Shear Test

The most appropriate test to determine inelastic or plastic behavior is one that subjects the material to a state of pure shear. (This is because no "hydrostatic stress" component is present, a concept to be discussed later in Chapter 6.) This is the most appropriate test for studying plastic behavior as it applies a state of pure deviatoric stresses to the test specimen. The test is typically conducted by applying an increasing torque to one end of a fixed thin-walled cylinder. The use of a thin-walled specimen also eliminates the effect of nonuniform stress distribution along the diameter of the cylinder.

The torsion test is conducted at a constant strain rate (mm/mm/s) or at a constant stress rate (Pa/s). This test is more suited for studying plastic behavior than the tension test, as the absence of the hydrostatic stress component eliminates volumetric expansion and the phenomena of necking. Hence the range of plastic strain that can be investigated with a torsion test greatly exceeds that produced in a tension test prior to fracture. This difference increases with increasing brittleness of the material. For example, a highly cold-worked material will show more ductile behavior in a torsion test than in a tension test.

4.4.4 Failure in Shear

Although shear tests can be performed, they are more complicated and hence more expensive than tensile tests. Thus it is common to estimate shear properties from tensile data. For

TABLE 4-1 Shear Stress in Failure of Ductile and Brittle Materials

	$\tau_{\max}$	$\tau_{\text{allowable}}$
Ductile materials ($\varepsilon_{\text{failure}} \gtrsim 0.05$)	$= 0.577 S_{\text{Yield, Tension}} \approx 0.5 S_{\text{Yield, Tension}}$ ("distortion energy theory")	$\approx \dfrac{S_{\text{Y,T}}}{2(FS)}$
Brittle Materials ($\varepsilon_{\text{failure}} \lesssim 0.002$)	$= S_{\text{Ultimate, Shear}} \approx S_{\text{Ultimate, Tension}}$ ("Maximum Normal Stress Theory")	$\approx \dfrac{S_{\text{U,T}}}{FS}$

example, it would be convenient and economical to deduce shear strength from tensile test results. Essentially, brittle materials (strain at failure may be less than about 0.02) fail by fracture, which relates to crack propagation. Cracks propagate readily under tension. Ductile materials (strain at failure may be greater than 0.05), on the other hand, fail by yielding, which relates to dislocations motion along shear planes as described earlier. More details on this subject are given in Chapter 6, but we summarize the results in Table 4-1.

4.5 Energetics ϒ

Previously we considered the energy stored in an axial structure during deformation. We saw that the stored energy of a rod can be calculated in terms of stress and strain through use of the constitutive relations. For the one-dimensional Hooke's law model, we get an axial strain energy per unit volume of a prismatic rod as follows:

$$U = \frac{1}{2}ku^2 = \frac{1}{2}Pu = \frac{1}{2}\sigma_x A \varepsilon_x L \tag{4.20a}$$

or

$$U = \frac{1}{2}\sigma_x \varepsilon_x (AL). \tag{4.20b}$$

In the analogous situation for static torsion structures (recall that in isotropic elastic structures the shear deformation is decoupled from the normal deformation), the torsional strain energy per unit volume is

$$U = \frac{1}{2}\tau_{x\theta}\gamma_{x\theta}(AL). \tag{4.21}$$

Hence the strain energy per unit volume is given by $1/2\tau_{x\theta}\gamma_{x\theta} = 1/2G\gamma_{x\theta^2} = 1/2\tau_{x\theta^2}/G$. In general, for the one-dimensional case:

$$U = \frac{1}{2}\int_{\text{volume}} \tau_{x\theta}\gamma_{x\theta}dV. \tag{4.22}$$

Two- and three-dimensional forms of the strain energy will be discussed later.

EXAMPLE 4-7:

In Example 4-1, a 2 ft long steel shaft of $1\frac{1}{4}$ in diameter was subjected to a torque of 100 ft-lb. Determine the total strain energy stored in the shaft.

Equation (4.22),

$$U = \frac{1}{2} \int_{\text{volume}} \tau_{x\theta} \gamma_{x\theta} dV,$$

can be recast using $\frac{1}{2}\tau_{x\theta}\gamma_{x\theta} = \frac{1}{2}\tau_{x\theta}{}^2/G$ as:

$$U = \frac{1}{2} \int_{\text{volume}} \frac{\tau_{x\theta}{}^2}{G} dV$$

Since $\tau_{x\theta} = Tr/J$ and $dV = dA\ dx$,

$$U = \frac{1}{2} \int_L \frac{T^2}{J^2 G} dx \int_A r^2 dA.$$

The last integral is simply the definition of J; hence

$$U = \frac{1}{2} \int_L \frac{T^2}{JG} dx = \frac{1}{2}\frac{T^2 L}{JG}.$$

Using data and results from Example 4.1 we get:

$$J := 0.24\,\text{in}^4 \quad G := 11.5 \cdot 10^6\,\frac{\text{lb}}{\text{in}^2} \quad T := 100\,\text{ft} \cdot \text{lb}$$

$$U := \frac{1}{2}\frac{T^2 \cdot L}{G \cdot J} \quad U = 0.522\,\text{lb ft}$$

4.6   Analysis of Static Torsion Structures ⅄

We have so far considered the torsion of shafts of circular cross-section, be they solid or hollow, prismatic or nonprismatic. We found that the twist angle varies linearly along the shaft length and varies inversely with the polar moment of inertia. We also found that the shear stress varies linearly from the center of the shaft, so that only the outer fibers are fully stressed. It makes sense then that high-performance structures, such as aerospace vehicles that demand high torsion resistance at minimum weight, would rely on thin-walled shafts.

In this section we will examine the analysis of thin-walled torsion elements with both open and closed cross-sections. We will then introduce a measure to compare the torsional efficiency of various shapes. Finally, we will see how to analyze torsional structures using the direct stiffness method.

4.6.1   Shear Flow

Consider a thin-walled prismatic tube of arbitrary cross-section under pure torsion loading as shown in Figure 4-11. Figure 4-12 shows details of the elements in Figure 4-11.

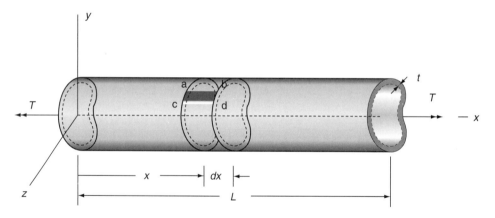

FIGURE 4-11 *Thin-walled prismatic tube of arbitrary cross-section under pure torsion loading (following Gere and Timoshenko, 1984).*

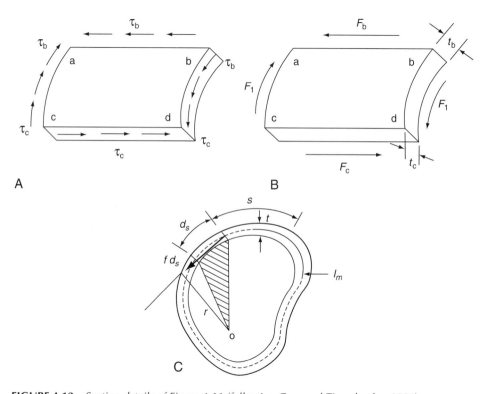

FIGURE 4-12 *Section details of Figure 4-11 (following Gere and Timoshenko, 1984).*

We assume that, since the thickness t is so small, the shear stress is approximately constant across the wall thickness. In Figure 4-12, for the arbitrary element abcd to be in equilibrium, the force $F_b(= \tau_b t_b dx)$ must be equal in magnitude to $F_c(= \tau_b t_b dx)$. Then

$$\tau_b t_b = \tau_c t_c \qquad (4.23)$$

or

$$\tau t = \text{constant}. \tag{4.24}$$

The quantity τt is called the *shear flow*. Note that this implies that the largest shear stress occurs where the thickness is smallest. (Hereafter we drop the subscript on τ and t for convenience, since no confusion should arise.)

We now need to relate the shear flow to the applied torque T. Looking at Figure 4-12c, we see that an increment of torque dT is applied by the shear flow as:

$$dT = r\tau t \, ds \tag{4.25}$$

where ds is an increment of arc length along the mean circumference l_m. Integrating along the entire mean length l_m gives

$$T = \tau t \int_0^{l_m} r \, ds = 2\tau t A_m \tag{4.26}$$

where A_m is the mean area enclosed by the mean circumference. Note that the integral

$$\int_0^{l_m} r \, ds = 2A_m \tag{4.27}$$

since from Figure 4-12C, $r \, ds$ represents twice the area of the shaded triangle. Finally, we then have a useful relationship for the shear stress on a thin-walled shaft of arbitrary cross-section:

$$\tau = \frac{T}{2tA_m}. \tag{4.28}$$

4.6.2 Thin-Walled Shafts of Closed Cross-Section

Consider a shaft of thin circular closed cross-section as shown in Figure 4-13.

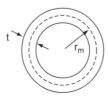

FIGURE 4-13 *Shaft of thin circular closed cross-section. The mean radius is given by r_m and t is the wall thickness.*

The polar moment of inertia J is calculated in the usual way as before:

$$J = \frac{\pi}{2} \left[\left(r_m + \frac{t}{2} \right)^4 - \left(r_m - \frac{t}{2} \right)^4 \right]. \tag{4.29}$$

After expansion and some simplification this becomes

$$J = \frac{\pi r_m t}{2} \left(4 r_m^2 + t^2\right) = 2\pi r_m^3 t \left[1 + \left(\frac{t}{2 r_m}\right)^2\right]. \tag{4.30}$$

Now for a thin wall, where $t/r_m \ll 1$ and $(t/2r_m)^2 \ll 1$, Eq. (4.28) simplifies to

$$J \approx 2\pi r_m^3 t = 2 A_m r_m t \tag{4.31}$$

where A_m is the area enclosed by the mean radius.

The shear stress is then

$$\tau = \frac{T r_m}{J} \approx \frac{T r_m}{2\pi r_m^3 t} = \frac{T}{2\pi r_m^2} = \frac{T}{2 A_m t} \tag{4.32}$$

For values of $t/r_m = 0.2, 0.1$, and 0.05, the approximate shear stress is $92\%, 95\%$, and 98%, respectively, of the exact shear stress. Note also that since τ is assumed not to vary across the thickness, the stress found here is the maximum shear stress.

The total twist angle is given as before:

$$\phi_{\text{total}} = \frac{TL}{GJ} = \frac{TL}{2 G A_m r_m t}. \tag{4.33}$$

Of course, for the thin-walled closed circular cross-section, the approximation to J provides us little advantage. However, we can apply this same approximation technique to more complex thin-walled shapes, of either open or closed section, where calculating the exact polar moments of inertia would be difficult. (In these cases of noncircular cross-section, the factor J is no longer the polar moment of inertia but is called more generally the *torsion constant*, and is in general less than the polar moment of inertia.)

A formal method for finding the torsion constant can be developed from strain energy considerations. Recall that the strain energy expression developed previously was

$$U = \frac{1}{2} \int_{\text{volume}} \frac{\tau^2}{G} dV. \tag{4.34}$$

For the thin-walled section, $dV = dA\, dx = t\, ds\, dx$, where ds is an increment of wall length measured along the mean circumference. Upon substitution and some manipulation we get

$$U = \frac{\tau^2 t^2}{2G} \int_0^{l_m} \left(\int_0^L dx\right) \frac{ds}{t} = \frac{(\tau t)^2 L}{2G} \int_0^{l_m} \frac{ds}{t} = \frac{T^2 L}{8 G A_m^2} \int_0^{l_m} \frac{ds}{t} = \frac{T^2 L}{2GJ} \tag{4.35}$$

where the final term has the same form as the strain energy for a circular shaft, except that here J is the torsion constant found from

$$J = \frac{4 A_m^2}{\int_0^{l_m} \frac{ds}{t}}. \tag{4.36}$$

EXAMPLE 4-8:

Consider for example the thin square closed cross-section shown in Figure E4-8.

The foregoing formula for J can be applied here. The value of the integral in the denominator is simply $4b/t$, and A_m is b^2, giving $J = b^3 t$. Hence the average shear stress (maximum shear stress must take into account the stress concentrations in the corners) is

$$\tau = \frac{T}{2tA_m} = \frac{T}{2tb^2},$$

and the total twist angle is

$$\phi_{total} = \frac{TL}{Gb^3 t}.$$

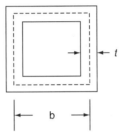

FIGURE E4-8 *Shaft of thin square closed cross-section. Each side has mean length given by* b *and a wall thickness* t.

EXAMPLE 4-9:

Compare the shear stress and total twist angle in two thin-walled shafts having the same length L and net cross-sectional area, which are subjected to the same torque T, except that one shaft is of circular cross-section while the other has a square cross-section.

For the circular tube:

$$A_c = 2\pi r_m t, \quad A_{mc} = \pi r_m^2, \quad J_c = 2\pi r_m^3 t.$$

For the square tube:

$$A_s = 4bt = A_c = 2\pi r_m t \Rightarrow b = \pi r_m/2 \quad A_{ms} = b^2, \quad J_s = b^3 t = \pi^3 r_m^3 t/8.$$

Now

$$\frac{\tau_s}{\tau_c} = \frac{T/2tA_{ms}}{T/2tA_{mc}} = \frac{A_{mc}}{A_{ms}}\frac{\pi r_m^2}{\pi^2 r_m^2/4} = \frac{4}{\pi} = 1.27, \qquad \frac{\phi_s}{\phi_c} = \frac{J_c}{J_s} = \frac{2\pi r_m^3 t}{\pi^3 r_m^3 t/8} = \frac{16}{\pi^2} = 1.62.$$

Hence the shear stress and total twist angle for the square tube are 27% and 62% greater than those for the circular tube. The square shape is clearly less efficient in torsion.

4.6.3 Torsional Shape Factor

We can now begin to see how we might characterize the shape efficiency of a given cross-sectional configuration. We define a *shape factor* β that compares the relative efficiency of a given cross-section to a reference cross-section. Following Ashby, we arbitrarily take a solid circular shaft as the reference. This reference shaft has length L_0, cross-sectional area $A_0 = \pi R_0^2$, shear modulus G_0, and torsion constant

$$J_0 = \frac{\pi}{2} R_0^4 = \frac{1}{2} A_0 R_0^2. \tag{4.37}$$

Under the action of a torque T, the reference shaft has a maximum shear stress and total twist angle given as before:

$$\tau_0 = \frac{T R_0}{J_0} = \frac{2T}{A_0 R_0} \tag{4.38}$$

$$\phi_0 = \frac{T L_0}{G_0 J_0} = \frac{2 T L_0}{G_0 A_0 R_0^2} \tag{4.39}$$

Then we use β to compare the efficiency of other shaft configurations having shear modulus G, but equivalent cross-sectional area A_0 and length $L = L_0$ (this keeps the shaft weights the same), under the action of the same torque T. That is:

$$\beta_\tau = \frac{\tau}{\tau_0} = \frac{\tau}{2T/A_0 R_0} \tag{4.40}$$

$$\beta_\phi = \frac{\phi}{\phi_0} = \frac{TL/GJ}{TL_0/G_0 J_0} = \frac{G_0 J_0}{GJ} = \frac{G_0}{G} \frac{A_0 R_0^2}{2J}. \tag{4.41}$$

A value of $\beta < 1$ means that the shaft under consideration is more efficient than the reference solid circular shaft, that is, for the same load the stress and/or stiffness are less than for the reference shaft.

EXAMPLE 4-10:

Determine β for the thin-walled circular shaft of Example 4-9.

$$A_c = 2\pi r_m t = A_0 = \pi R_0^2 \Rightarrow R_0 = \sqrt{2 r_m t}$$

$$A_{mc} = \pi r_m^2$$

$$\beta_\tau = \frac{\tau_c}{\tau_0} = \frac{T/2tA_{mc}}{2T/A_0 R_0} = \frac{A_0 R_0}{4 t A_{mc}} = \frac{(2\pi r_m t)\sqrt{2 r_m t}}{4 t \pi r_m^2} = \cdots = \sqrt{\frac{t}{2 r_m}} = 0.707 \sqrt{\frac{t}{r_m}}$$

$$\beta_\phi = \frac{J_0}{J_c} = \frac{\frac{1}{2} A_0 R_0^2}{2\pi r_m^3 t} = \frac{1}{4\pi} \frac{(2\pi r_m t)(2 r_m t)}{r_m^3 t} = \frac{t}{r_m}.$$

EXAMPLE 4-10: *Cont'd*

Table E4-10 compares both the strength and stiffness efficiency of the thin-walled circular shaft for various wall thicknesses (compared to the solid circular shaft of same cross-sectional area). As shown, as the wall thickness becomes thinner, the shaft becomes more efficient. (Note that this trend cannot go on indefinitely, since the shaft is getting larger to keep the area constant and equal to A_0. As the wall thickness gets thinner, other failure modes will start to take place, such as wall buckling.)

TABLE E4-10

t/r_m	β_τ	β_ϕ
1/10	0.224	0.1
1/100	0.0707	0.01
1/1000	0.0224	0.001

4.6.4 Thin-Walled Shafts of Open Cross-Section

Slicing a thin-walled shaft of closed circular cross-section longitudinally results in an *open cross-section* (Figure 4-14).

However, this cut "releases" the shear stress depicted in Figure 4-12c. This significantly increases flexibility of the structure in torsion, and hence thin-walled open sections are considerably less efficient in torsion than comparable closed cross-sections.

The exact form of the torsion constant for a thin solid rectangular section h by t ($h > t$) (Figure 4-15) can be shown to be (Boresi et al., 1993).

$$J = \frac{ht^3}{16}\left[\frac{16}{3} - 3.36\frac{t}{h}\left(1 - \frac{t^4}{12h^4}\right)\right]. \tag{4.42}$$

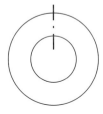

FIGURE 4-14 *A longitudinal cut transforms a closed cross-section into an open cross-section.*

FIGURE 4-15 *Cross-section of a thin solid rectangular shaft.*

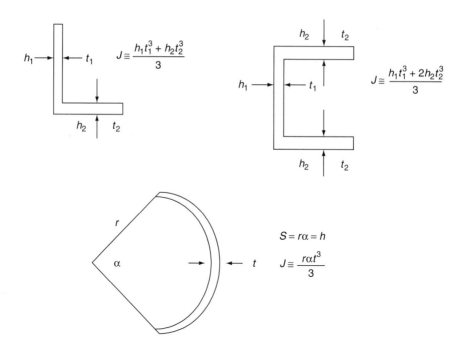

FIGURE 4-16 *Approximate torsion constants for some thin-walled open sections.*

For $t/h \ll 1$, $J \approx ht^3/3$. Thus the torsion constant for open sections can be approximated as a sum of thin rectangular sections. Figure 4-16 provides several examples.

The (average) shear stress (away from sharp corners) and the total twist angle are found as

$$\tau_{\text{avg}} = \frac{Tt}{J}, \quad \phi = \frac{TL}{GJ}. \qquad (4.43\text{a, b})$$

EXAMPLE 4-11:

Compare the torsion efficiency under the load T of two thin-walled tubes of circular section and equal cross-sectional area, one being closed and the other formed by a thin longitudinal cut in the wall:

$$\frac{\tau_{\text{open}}}{\tau_{\text{closed}}} = \frac{Tt/J_{\text{open}}}{Tr_m/J_{\text{closed}}} = \frac{t}{r_m} \frac{2\pi r_m^2 t}{2\pi \frac{r_m}{3} t^2} = \underline{3 \frac{r_m}{t}}$$

$$\frac{\phi_{\text{open}}}{\phi_{\text{closed}}} = \frac{J_{\text{closed}}}{J_{\text{open}}} = \frac{2\pi r_m^3 t}{2\pi \frac{r_m}{3} t^3} = \underline{3\left(\frac{r_m}{t}\right)^2}.$$

As can be seen, the open cross-section carries much more shear stress and rotates significantly more than that of the equivalent closed cross-section.

4.6.5 Direct Stiffness Method for Static Torsion Structures

Consider the generic *torsion spring element* shown in Figure 4-17.

If we fix (restrain) node 1 from displacement (that is, require $\phi_1 = 0$), the response of the spring is given by:

$$\phi_2 = \frac{T_2 L}{GJ} = \frac{T}{K}, \text{ where } K = \frac{GJ}{L} \tag{4.44}$$

and G, J, and L are the characteristics of the element as previously defined.

Since the form of this relation is identical to that for the axial case, that is, $T = K\phi \Leftrightarrow F = ku$, we can use the method previously derived for the axial case in Chapter 3 and now apply it to the torsion case, with only a simple substitution of variables:

$$T \leftarrow F$$
$$K \leftarrow k$$
$$\phi \leftarrow u$$

Then the nodal equations for the torsion element can be written in matrix form as

$$\begin{Bmatrix} T_1 \\ T_2 \end{Bmatrix} = \begin{bmatrix} K & -K \\ -K & K \end{bmatrix} \begin{Bmatrix} \phi_1 \\ \phi_2 \end{Bmatrix} = \frac{GJ}{L} \begin{bmatrix} 1 & -1 \\ -1 & 1 \end{bmatrix} \begin{Bmatrix} \phi_1 \\ \phi_2 \end{Bmatrix} \tag{4.45a}$$

or

$$\{T\} = [K]\{\phi\}. \tag{4.45b}$$

where $\{T\}$ and $\{\phi\}$ are the *local* or *nodal torque* and *(angular) displacement* vectors, respectively, and $[K]$ is the *local (torsion) stiffness matrix*.

FIGURE 4-17 *Generic torsion spring definition, where locations 1 and 2 are the nodal points, T_1 and T_2 are torques (loads) applied, ϕ_1 and ϕ_2 are the rotational displacement responses at the nodes 1 and 2, respectively, and K is the elastic torsion spring constant. Note that loads and displacements are defined as acting in the positive coordinate direction.*

EXAMPLE 4-12:

Do Example 4-3, this time using the direct stiffness method and a two-element approximation.

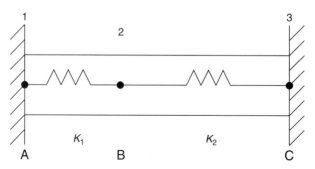

FIGURE E4-12A

$$K_1 = \frac{GJ}{L_1}, \ K_2 = \frac{GJ}{L_2}$$

Writing the matrix equation for the system, including boundary conditions, gives

$$\left\{ \begin{array}{c} T_1 \\ T_2 = 600 \\ T_3 \end{array} \ \text{lb} \cdot \text{in} \right\} = \left[\begin{array}{ccc} K_1 & -K_1 & 0 \\ -K_1 & K_1 + K_2 & -K_2 \\ 0 & -K_2 & K_2 \end{array} \right] \left\{ \begin{array}{c} \phi_1 = 0 \\ \phi_2 \\ \phi_3 = 0 \end{array} \right\}.$$

After striking rows and columns associated with the constrained degrees of freedom, we have

$$600 \, \text{in} \cdot \text{lb} = (K_1 + K_2)\phi_2$$

or

$$\phi_2 = \frac{600 \, \text{in} \cdot \text{lb}}{GJ\left(\frac{1}{L_1} + \frac{1}{L_2}\right)} = \frac{600 \, \text{in} \cdot \text{lb}}{\left(3.67 \times 10^6 \, \frac{\text{lb}}{\text{in}^2}\right)\frac{\pi}{2}(0.5 \, \text{in})^4\left(\frac{1}{24 \, \text{in}} + \frac{1}{72 \, \text{in}}\right)} = \underline{\underline{0.03 \, \text{rad.}}}$$

The deflection at "D" can be found from the assumed constant strain element

$$\alpha = r\frac{d\phi}{dx}, \ \text{i.e.,} \ \phi = \frac{\alpha}{r}x + c \ \text{or} \ \phi \sim x:$$

$$\frac{0.03}{6} = \frac{\phi_{oc}}{4} \Rightarrow \phi_{oc} = \frac{2}{3}(0.03) = \underline{\underline{0.02 \, \text{rad}}}$$

Finally

$$T_1 = -K_1\phi_2 = -\frac{3.67 \times 10^6 \, \frac{\text{lb}}{\text{in}^2}\left(\frac{\pi}{2}\right)(0.5 \, \text{in})^4}{24 \, \text{in}}(0.0239 \, \text{rad}) = \underline{\underline{-450 \, \text{lb} \cdot \text{in}}}$$

and

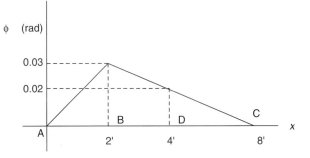

EXAMPLE 4-12: *Cont'd*

FIGURE E4-12B

$$T_2 = -k_2\phi_2 = -\frac{(\quad)}{72\text{ in}} = \underline{\underline{-150\text{ lb}\cdot\text{in}}}$$

4.7 Material Selection in Static Torsion Structures

We can include shape selection into the material selection process by assuming that the shape modification translates into "effective materials properties". This may be best illustrated by example.

EXAMPLE 4-13:

Torsion of a light, strong shaft-material and shape:

Solid circular cross-section A_0

$$J_0 = \frac{\pi}{2}R_0^4$$

$$\tau_0 = \frac{TR_0}{J_0} = \frac{2}{\pi}\frac{T}{R_0^3} \leq S_f \text{ \underline{strong}}$$

$$\rightarrow R_0 = \sqrt[3]{\frac{2T}{\pi S_f}} \qquad\qquad (E1)$$

$$mg = \rho A_0 Lg = \rho\pi R_0^2 Lg \leq w_f \text{ \underline{light}}$$

$$\rightarrow R_0^2 = \frac{w_f}{\rho\pi Lg} \qquad\qquad (E2)$$

Equating $(E1)^2$ and (E2):

$$R_0^2 = \left(\frac{2}{\pi}\frac{T}{S_f}\right)^{2/3} = \frac{w_f}{\rho\pi Lg}$$

Continued

EXAMPLE 4-13: *Cont'd*

$$\frac{S_f^{2/3}}{\rho} = \frac{\left(\frac{2}{\pi}\right)^{2/3} \pi L g (T)^{2/3}}{w_f}$$

$$= \underline{\left(2T\sqrt{\pi}\right)^{2/3} \frac{Lg}{w_f}}$$

This is the Material Index sought.

Any other thin-walled cross-section with $A = A_0$

$$\beta_\tau = \frac{A_0 R_0}{4t A_m} \text{(see Example 4-10)}$$

$$\tau = \beta_\tau \tau_0 = \beta_\tau \frac{2}{\pi} \frac{T}{R_0^3} \le S_f \ \underline{\text{strong}}$$

$$R_0 = \sqrt[3]{\frac{2T\beta_\tau}{\pi S_f}} \quad \text{(E3)}$$

$$mg = \rho A_0 L g = \rho \pi R_0^2 L g \le w_f \ \underline{\text{light}}$$

$$\to R_0^2 = \frac{w_f}{\rho \pi L g} \quad \text{(E4)}$$

Equating $(E3)^2$ and (E4):

$$R_0^2 = \left(\frac{2T}{\pi} \frac{\beta_\tau}{S_f}\right)^{2/3} = \frac{w_f}{\rho \pi L g}$$

$$\left(\frac{S_f/\beta_\tau}{\rho}\right)^{2/3} = \left(2T\sqrt{\pi}\right)^{2/3} \frac{Lg}{w_f}$$

$$\frac{(S_f)^{2/3}}{\rho} = \underline{(\beta_\tau)^{2/3} \left(2T\sqrt{\pi}\right)^{2/3} \frac{Lg}{w_f}}$$

This is the Material Index sought.

4.8 Design of Static Torsion Structures ᛉ⌂🖳

4.8.1 Design Example
We now provide an example design of a torsion structure.

EXAMPLE 4-14:

A 1-m long shaft is required to transmit 211 kW of power at 3600 rpm. The maximum shaft weight is limited to 10 N and the shaft should not fail by yielding with a factor of safety of 2.

Assume that the shaft is well supported such that no bending or axial loads exist.

Consider two different cross-sectional shapes and provide a set of design specifications for a low-cost solution.

EXAMPLE 4-14: *Cont'd*

I. *Problem Definition*

Performance requirements:
- Weight $\leq 10\,\text{N}$
- Factor of safety $= 2$; this is a specification on *strength*, and as such is of primary importance. Then $\tau_{max} \leq S_f/2$, where S_f is specified as yield failure strength (in torsion in this case).
  No other performance requirements are given.

Service environment:
- Loads are assumed to act statically. The load is:

$$T = \text{power}/(2\pi \times \text{frequency})$$
$$= (211 \times 10^3 \text{ Nm/s})/(2\pi \times 3600 \text{ rev/min} \times 1/60 \text{ min/s}) = 560 \text{ Nm}.$$

(The assumption of static loads is made to keep the problem simple enough for educational purposes. In the real situation, the loads would clearly be dynamic.)
  We can assume that the shaft is only loaded in torsion due to proper boundary conditions.
- The shaft length is required to be 1 meter.
  No other service environment specifications are given.

Project constraints:
- Low cost (implies generally available material, simplicity of design, ease of manufacture and assembly, reasonably low weight, etc.), particularly due to the large number of parts required.
- We are to trade off two different cross-sectional configurations.
  No other constraints are given.

II. *Preliminary Design*

- Identify DDOFs:
  1. Geometry $\Rightarrow$ minimize volume to keep weight low $\Rightarrow$ area (A) and cross-section shape (J)
  2. Density $\Rightarrow$ keep weight low (ρ)
  3. Strength $\Rightarrow$ design for yielding (S_Y)
  4. Stiffness $\Rightarrow$ modulus (G)
  5. Boundary conditions $\Rightarrow$ roller BC keeps shaft in pure torsion, that is, without moment reactions (more details about the connections will not be considered at this stage)

Continued

EXAMPLE 4-14: *Cont'd*

- Trade study: The desire to keep costs low (and as always to keep things as simple as possible!) suggests a simple torsion structure for the support. A circular cross-section of diameter D will keep minimized the volume of material and stress concentrations. Both solid and hollow cross-sections should be considered.

    Material considerations suggest simple, readily available materials to keep costs low, but analysis needs to confirm the specific choice.
- FMEA: Failure expected to be from simple static yielding of the shaft material.

III. *Detailed Design*

From Example 4-13, the material index for a light, strong solid circular shaft is

$$
\begin{aligned}
\frac{(S_f)^{2/3}}{\rho} &= (2\sqrt{\pi})^{2/3} g T^{2/3} \frac{L}{w_f} \\
&= \left(22.82 \frac{N}{kg}\right)(560\,N \cdot m)^{2/3} \frac{(1\,m)}{10\,N} \\
&= 155 \frac{(N \cdot m)^{2/3} \left(\frac{10^6\,m^3}{10^6\,m^3}\right)^{2/3}}{kg/m} \\
&= 155 \frac{(MPa)^{2/3}}{10^4\ kg/m^3} \\
&= \underline{\underline{15.5 \frac{(MPa)^{2/3}}{Mg/m^3}}}
\end{aligned}
$$

We can plot this material index on the Ashby strength versus density chart as follows:

$$
15.5 \frac{(MPa)^{2/3}}{Mg/m^3} = \frac{\sigma_f^{2/3}}{30} \Rightarrow \sigma_f = 10{,}027 \approx 10{,}000.
$$

Thus we plot a straight line between the coordinates (1, 15.5) and (30, 10,000). All materials above this line satisfy the light–strong criteria as specified. Note, however, that steel is barely represented in this group. Now let us reanalyze for a thin-walled circular cross-section. From Example 4-10:

$$
\beta_\tau = 0.707 \sqrt{\frac{t}{r_m}} \quad \text{thin-walled circular crosssection}
$$

But what are representative "commercial off-the-shelf" (COTS) options for t and r_m? If we consult a typical parts catalog (e.g., McMaster Carr®), we might find the following representative values:

EXAMPLE 4-14: *Cont'd*

OD (in)	ID (in)	t (in)	r_m (in)	β_τ
1.5	0.89	0.305	0.5975	0.505
2.0	1.25	0.375	0.813	0.48

Hence using a representative value of $\beta_\tau = 0.5$ we have from Example 4-13:

$$\frac{(S_f)^{2/3}}{\rho} = (0.5)^{2/3}15.5 = 9.76 \; \frac{(\text{MPa})^{2/3}}{\text{Mg/m}^3}.$$

We plot this material index as before, now as a straight line between coordinates $(1, 9.76)$ and $(30, 5000)$. This new line is shifted down from the first line for the solid shaft, and in particular steel has now a larger presence in the cleared group of materials. Choosing steel to keep cost low, a COTS ground hollow steel shaft might have $S_Y = 700\,\text{MPa}$; then

$$t_{\text{allow}} = \frac{S_Y/2}{2} = 175\,\text{MPa} = \begin{cases} \dfrac{TR}{J} = \dfrac{2T}{\pi R_0^3} & \text{solid} \\[3mm] \dfrac{T}{2tA_m} & \text{hollow} \end{cases}$$

Solving first for the solid shaft outside diameter:

$$\pi R_0^3 = \frac{2T}{\tau_{\text{allow}}} = \frac{2(560\,\text{N}\cdot\text{m})}{175 \times 10^6 \,\text{N/m}^2} = 6.4 \times 10^{-6}\,\text{m}^3$$

$$R_0 = 0.0127\,\text{m} = 12.7\,\text{mm} \Rightarrow D_0 = \underline{25.4\,\text{mm}}$$

Thus a 25.4-mm (1.0 in) diameter solid steel shaft satisfies the design requirements. The dimensions of the hollow shaft are readily found by equating $A_m = A_0$ and solving for the wall thickness t from the "hollow" equation:

$$t = \frac{T}{2A_m \tau_{\text{allow}}} = \frac{(560\,\text{N}\cdot\text{m})}{2\pi(12.7 \times 10^{-3}\,\text{m})^2(175 \times 10^6\,\text{N/m}^2)} = 3.16 \times 10^{-3}\,\text{m}.$$

Now using $\beta_\tau = 0.707\sqrt{\dfrac{t}{r_m}} = 0.5 \Rightarrow r_m = 2t = 2(3.16\,\text{mm}) = 6.32\,\text{mm}$. Then the outside diameter of the hollow shaft is $2r_m + t = 2(6.32\,\text{mm}) + 3.16\,\text{mm} =$ 15.8 mm.

We can now provide the shaft design specifications:

- COTS steel shaft, 1 m (3.28 ft) long
- Solid shaft: 25.4 mm (1.0 in) diameter
- Hollow shaft: 16 mm (0.625 in) outside diameter by 3 mm (0.125 in.) wall thickness

4.8.2 Summary: Design of Static Torsion Structures

For the static elastic design of torsion structures, the following design rules can be formulated:

- Nominal member stress is directly related to member moment divided by member torsion constant. However, stress concentration effects need to be taken into account for detailed and accurate stress prediction.

Solid section

$$4J = \int_A r^2 dA$$

$$\tau = \frac{Tr}{J}$$

Closed thin-walled section

$$J \approx \frac{4A_m^2}{\int_0^{L_m} \frac{ds}{t}}$$

$$\tau = \frac{Tr_m}{J} \approx \frac{T}{2tA_m}$$

Open thin-walled section

$$J \approx ht^3/3$$

$$\tau_{avg} \approx \frac{Tt}{J}$$

- Member strain is directly related to material elastic shear modulus for a given stress.
- Choice of cross-sectional shape may be based on a variety of considerations:
  - Circular cross-sections are generally more efficient than other shapes.
  - Closed cross-sections are always more efficient than comparable open cross-sections.
  - Fabrication and assembly issues associated with various shapes may differ.
- Failure by buckling in torsion is an issue for thin-walled sections, especially if they are open. Further details are not discussed here.
- Material selection should be guided by use of appropriate performance indices, then use of a weighted decision matrix down-selection approach. In most cases, strength, weight, and cost are the most important considerations. Shape can be directly included in the material selection by use of the torsion shape factor.

Key Points to Remember

- Pure twisting loads result in *shear* stresses and strains.
- Shear strain relates to the distortion of a volume element of material. The shear strain is related geometrically to the gradient of the *twist angle*.
- The shear stress is the internal resistance to the applied torque.
- The shear stress and strain are related through a constitutive model, such as linear elasticity. In that case, the shear stress and strain are decoupled or unrelated to the extensional stress and strain, and are related by the *shear modulus*.
- The *torsion constant*, which for circular cross-sections is the *polar area moment of inertia*, is a measure of the distribution of matter in a plane about the twist axis. The torsion constant is one variable in the *torsion stiffness* of a structure, the others being shear modulus and twist axis length.
- Shear deformation at the atomistic level is related to *dislocation glide*, which is an important factor in strength of ductile materials. The *critical shear stress* must be exceeded at the crystal level to initiate glide and yielding.
- The concept of *shear flow* provides a convenient method for analyzing shear in thin-walled structures, particularly of open-walled cross-section.
- Unlike in axial structures, the cross-sectional shape does affect the load-carrying resistance of a structure. The *torsion shape factor* is used to quantify the effect of shape and can be incorporated directly into the material selection process.

References

Billington, D. P. (1985). *The Tower and the Bridge*. Princeton University Press, Princeton, NJ.

Boresi, A. P., Schmidt, R. J., and Sidebottom, O. M. (1993). *Advanced Mechanics of Materials*. Wiley, New York.

Gere, J. M., and Timoshenko, S. P. (1984). *Mechanics of Materials*, 2nd ed. PWS Engineering, Boston, Massachusetts.

Gordon, J. E. (1984). *The New Science of Strong Materials*. Princeton University Press, Princeton, NJ.

Mangonon, P. L. (1999). *The Principles of Materials Selection for Engineering Design*. Prentice-Hall, Upper Saddle River, NJ.

Problems

4-1. A hollow prismatic circular shaft has an outside diameter $D_0 = 60$ mm and an inside diameter $D_i = 40$ mm. When subjected to an end torque T, the maximum shear stress in the shaft is 90 MPa.

   (a) Sketch the shear stress distribution on an arbitrary cross-section, indicating the values of the maximum and minimum shear stresses.

   (b) By integrating over the cross-section, determine the torque T.

   (c) Use the torsion formula $\tau = TR/J$ to determine the torque T and compare to your answer in (b).

4-2. A 32-in long by 1.6-in diameter solid prismatic circular shaft is made of an aluminum alloy that has an allowable shear stress (strength) of 10 ksi, and a shear modulus of elasticity $G = 3800$ ksi. If the allowable angle of twist over the 32 in length of the shaft is 0.10 radian, what is the value of the allowable end torque? (Hint: you must investigate from both strength and stiffness perspectives, then compare.)

4-3. A 2-m long solid prismatic circular shaft is made of brass that has an allowable shear stress (strength) of 120 MPa and a shear modulus of elasticity $G = 39$ GPa. Over the shaft length, the angle of twist allowed is 0.10 radian. If the shaft is subjected to a maximum end torque of 25 kNm, what is the required minimum diameter of the shaft? (Hint: you must investigate from both strength and stiffness perspectives, then compare.)

4-4. A hollow tapered shaft has an outer diameter that varies linearly from D_0 at $x = 0$ to $2D_0$ at $x = L$ (Figure P4-4). The inner diameter is a constant $D_i = D_0/2$. It is subjected to an end torque T_0 at $x = 0$ and is attached to a rigid wall at $x = L$. The shear modulus of the shaft is G.
  (a) Determine an expression for the maximum (cross-sectional) shear stress in the tapered shaft as a function of the distance x from the origin of coordinates.
  (b) Determine an expression for the total angle of twist of the shaft.

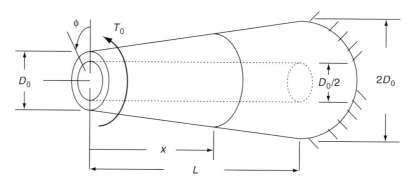

FIGURE P4-4

4-5. A solid tapered shaft has a diameter that varies linearly from $2D_0$ at $x = 0$ to D_0 at $x = L$. It is subjected to an end torque T_0 at $x = 0$ and is attached to a rigid wall at $x = L$. The shear modulus of the shaft is G. Determine an expression for the maximum (cross-sectional) shear stress in the tapered shaft as a function of the distance x from the origin of coordinates. Determine an expression for the total angle of twist of the shafts. Compare your answers with Example 4-2. (This problem is Example 4-2 "turned around.")

4-6. Two solid shafts, one with a diameter of D_0 and the other $2D_0$, are each subjected to an end torque T_0 at $x = 0$ and are each attached to a rigid wall at $x = L$. The shear modulus of the shafts is G. Determine an expression for the maximum (cross-sectional) shear stress in each shaft as a function of the distance x from the origin of coordinates. Determine an expression for the total angle of twist of the shafts. Compare your answers with those from Example 4-2 and discuss.

4-7. Calculate the maximum shear stress in Example 4-2 from the relation $\tau_{x\theta} = Gr\frac{d\phi}{dx}$ and compare to the result in the example.

4-8. Rework Example 4-5 by choosing another set of edge vectors.

4-9. A 10-mm diameter single crystal rod of an FCC (face-centered cubic) metal was pulled in tension along the [001]. Slip was observed when the load was 49 N. When the slip plane is the $(1\bar{1}1)$, determine the possible slip directions on this plane and the critical resolved shear stress.

4-10. An iron single crystal with critical resolved shear stress of 35 MPa was pulled in tension along the [100]. Determine which of the following slip systems will be activated: (a) $(110)[\bar{1}11]$, (b) $(101)[\bar{1}11]$, and (c) $(01\bar{1})[\bar{1}11]$. What will be the applied tensile force if the diameter of the single crystal rod is 25 mm?

4-11. Show that the maximum tensile yield strength of single crystals is twice the critical resolved shear stress.

4-12. What is the theoretical shear strength of material? Why is the strength not exhibited by single crystals?

4-13. Calculate the total strain energy for Example 4-2.

4-14. Calculate the total strain energy for Example 4-3.

4-15. Calculate the torsion constant for a thin-walled shaft of rectangular cross-section (Figure P4-15).

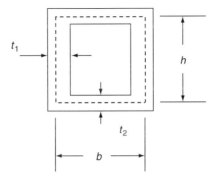

FIGURE P4-15

4-16. Compute the shape factor β for the thin-walled square shaft of Example 4-8.

4-17. Compute an approximate torsion constant for the "I" section in Figure P4-17.

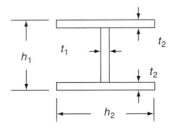

FIGURE P4-17

4-18. A stepped steel shaft AC ($G = 12 \times 10^6$ psi) is subjected to an external torque T_B at B and is fixed to rigid supports at ends A and C, as shown in Figure P4-18. (a) Using the direct stiffness method, determine ϕ_B, the angle of rotation of the shaft at joint B.

(b) Determine T_1 and T_2, the internal torques in segments AB and BC, respectively.
(c) Finally, determine the maximum shear stress in each segment.

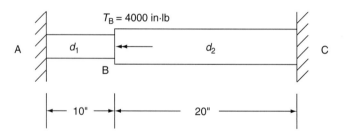

$T_B = 4000$ lb·in, $d_1 = 1.0$ in, $L_1 = 10.0$ in, $d_2 = 2.0$ in, $L_2 = 20.0$ in

FIGURE P4-18

4-19. Do Example 4-3 using the direct stiffness method and a three-element approximation.

4-20. Following Example 4-13, show that for a light, stiff shaft of length L subjected to a torque T:

- The performance index for a solid circular cross-sectional area A_0 is

$$\frac{\sqrt{G}}{\rho} = \frac{g\sqrt{2\pi L^3}\sqrt{T/\phi_f}}{w_f}$$

- The performance index for any other thin-walled cross-section of area $A = A_0$ is

$$\frac{\sqrt{G}}{\rho} = \sqrt{\beta_\phi}\,\frac{g\sqrt{2\pi L^3}\sqrt{T/\phi_f}}{w_f}$$

4-21. Using the fundamental definition of the polar (area) moment of inertia, $J = \int_A r^2 dA$, derive J as given in the text for a thin-walled circular shaft:

$$J = \frac{\pi}{2}\left[\left(r_m + \frac{t}{2}\right)^4 - \left(r_m - \frac{t}{2}\right)^4\right],$$

which was shown to be approximately equal to $2A_m r_m t$. Next show that the approximate relation

$$J = \frac{4A_m^2}{\displaystyle\int_0^{L_m} \frac{ds}{t}}$$

gives the same approximate result $2A_m r_m t$ for the thin-walled circular shaft.

Design Projects

4-22. A prismatic shaft of length L, fixed at both ends, carries an intermediate torque T as shown in Figure P4-22. Specify a light, stiff shaft design that satisfies the specifications in the table. For one or more of the shafts in Table P4-22, investigate at least two different cross-sectional shapes, one of which is not closed. Investigate at least three different materials.

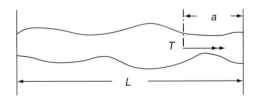

FIGURE P4-22

TABLE P4-22

Shaft	Maximum weight (lb)	Torque T (lb-in)	L (in)	a (in)	ϕ_{max} (degree)
1	2.0	800	24	6	1.5
2	2.5	850	24	6	1.5
3	3.0	900	24	6	1.5
4	3.5	950	24	8	1.5
5	4.0	1000	24	8	1.5

5 Flexural Structures

A large portion of success is derived from flexibility.

—Alice Foote MacDougall (1928)

Objective: This chapter will introduce three design degrees of freedom—I, S_F, E_F—in the context of the design of flexural structures.

What the student will learn:

- Definition of flexural structures
- Equilibrium and deformation of flexural structures
- Linkage of mechanics and materials concepts through the flexural test
- Analysis of flexural structures using singularity functions and the direct stiffness method
- Materials selection for flexural structures
- How to design a simple flexural structure

5.1 Introduction ♈☫💻

Flexural structures carry transverse loads primarily through bending (flexing), which is accompanied by a stress state of combined tension and compression. The tension and compression state creates an internal bending moment. Common flexural structures, called *beams*, are typically straight, narrow members that are longer than they are deep (but short and stubby, as well as curved, beam configurations are also used). A plate or shell-like structure can also be used as a flexural structure. Whether or not a given structure responds, or can be represented to respond, as a flexural structure depends on the nature of the loading, and on the structure's boundary conditions. A structural element will be a flexural structure if its boundaries and loading are such that an internal bending moment can be developed.

Simplifying assumptions are often made in analyzing flexural structures. For example, we will ignore the fact that all real structures deform simultaneously in three dimensions under the action of any load. We can get away with this neglect, for now, because we will only consider the structure's response to transverse loads. Later, when we consider combined loading, we will be required to look at multidimensional response. In this text we will also disregard the effects of a small amount of shear that may develop in flexural deformation.

Beams are one of the most common of structural types. In nature, tree limbs are a most familiar example (Figure 5-1). The cantilever beam in Figure 5-2, from Galileo's *The Two New Sciences* (1638), may have indeed been motivated by the tree limb.

Beams resting on columns are fundamental building blocks of civil structures as seen in Figure 5-3.

FIGURE 5-1 *Tree limbs are cantilever beams. Note that the limbs (and trunks) are always thicker near their attached ends. Why?*

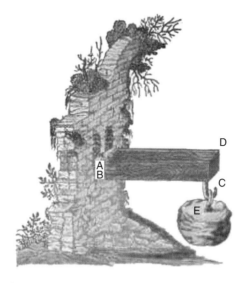

FIGURE 5-2 *Galileo's famous flexural experiment from* The Two New Sciences *(1638).*

FIGURE 5-3 *The Parthenon (built around 440 BC), an early beam/column structure. Short, stubby horizontal beams are seen spanning the vertical columns.*

The wings of a modern airplane are highly evolved and complex beams (Figure 5-4). Even the teeth of gears or the arms of a mechanical link are in fact beams (Figure 5-5).

5.2   Equilibrium and Deformation: Flexural Loading Ⱡ

5.2.1   Configuration

The beams we consider here have cross-sections that are relatively narrow and compact. We will for now consider that the cross-sections are symmetrical with respect to the plane of bending (say the x–y plane; see Figure 5-6). (This makes for simplification in the analysis of flexural structures. The added complications from nonsymmetrical cross-sections will not be discussed here; see, e.g., Crandall et al., 1972.)

The cross-section is characterized by its area $A(x)$ and *area moment of inertia* (second moment of area) $I(x)$, where in general A and I can be functions of position x along the beam. The area moment of inertia is given by

$$I(x) = \int_A y(x)dA \tag{5.1}$$

For now, however, we will only consider *prismatic* beams, that is, beams with constant cross-sectional properties, where A and I are constants along the length of the beam. In this case, we can represent the beam as a one-dimensional structural element of length L, with cross-sectional properties A and I (Figure 5-7).

FIGURE 5-4 *Wing of NASA's KC-135 "vomit-comet" is a cantilever beam.*

FIGURE 5-5 *Variable stator linkage on a GE aircraft turbine engine (courtesy of Mr. Andrew Johnson, GE Aircraft Engines).*

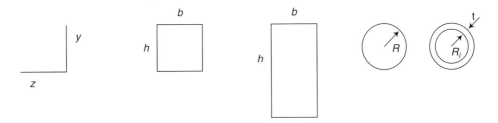

FIGURE 5-6 *Simple beam cross-sections symmetrical with the bending plane (arbitrarily selected as the x−y plane).*

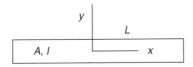

FIGURE 5-7 *Prismatic beam as a one-dimensional structural element.*

It is convenient to think about the beam as being comprised of longitudinal "fibers," running the length of the beam and parallel to the long axis (x-axis in the Figure 5-7). (The "fiber" imagined here is just a collection or locus of material points along a given $y = $ constant line and not a real fiber, although real fibers do exist, e.g., in fiber-reinforced composite materials.) In a rectangular cross-section, the outer fibers would be located on the surfaces $\pm h/2$. Another important fiber is colocated with the neutral axis of the beam and will be described in more detail later. Unless otherwise specified, we will assume the x-axis to be coincident with the neutral axis of the beam.

Of course, in order to carry loads, the beam must be supported in space. Unlike the axial and torsion structures considered earlier, there is considerable variety of boundary conditions available for beam support, since the loading and subsequent reactions can be more complex. The supports can occur at any x location along the length of the beam, although the location may limit the type of boundary conditions applicable.

A beam has three *coordinate degrees of freedom* (CDOF) associated with its response: axial translation u, transverse translation v, and rotation in the plane of bending θ (Figure 5-8). One or more of these CDOF may be constrained at the boundary, defining the boundary condition, depending on the nature of response desired in the beam. (Because the beam is modeled as one-dimensional, we automatically assume that all other translations and rotations are constrained.)

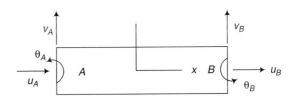

FIGURE 5-8 *Coordinate degrees of freedom for a beam.*

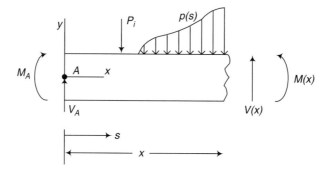

FIGURE 5-9 *Free-body diagram of beam under general loading. The global axial coordinate for the beam is* x, *while the local coordinate along the free body is* s.

We now outline the principal beam boundary conditions (refer to Chapter 2 for further details):

- Pin or simple support ($u = 0$, $v = 0$, $\theta \neq 0$)
- Roller support ($u = 0$, $v \neq 0$, $\theta \neq 0$) or ($u \neq 0$, $v = 0$, $\theta \neq 0$)
- Guide support ($u = 0$, $v \neq 0$, $\theta = 0$) or ($u \neq 0$, $v = 0$, $\theta = 0$)
- Fixed or clamped support ($u = 0$, $v = 0$, $\theta = 0$)

5.2.2 Equilibrium

Consider the free-body diagram of a beam subjected to transverse concentrated forces P_i, distributed force $p(x)$ (force/length), and concentrated moments M_i (force × length) as shown in Figure 5-9.

The internal reaction system required to enforce equilibrium under any general system of applied loads includes an axial force N, shear force V, and moment M. (We will not consider here any loads that generate a net axial force N in the beam. Formally, such a structure would be called a *beam-column*.) In order to relate the internal forces and moments to the externally applied loads, we sum forces and moments in the usual way from the left-side FBD for convenience (the same results arise from a summation over the right-side FBD, and this is left as an exercise for the student):

$$-V_A + P_i + \int_0^x p(s)ds - V(x) = 0$$

$$-M_A - V_A x + M_i + P_i(x - s_i) + \int_0^x (x - s)p(s)ds + M(x) = 0 \qquad (5.2a,b)$$

Note that:

1. The signs and magnitudes of the reaction forces and moments (V_A and M_A in this case) would be determined from global statics in the usual way.
2. We have summed the internal moments about the cut at x, in order not to have to include the moment due to $V(x)$, which at the present is unknown.
3. If there are m forces P_i and n moments M_i, then one would replace P_i and M_i everywhere in the equations with $\Sigma P_i (i = 1 \ldots m)$ and $\Sigma M_i (i = 1 \ldots n)$.
4. Equations (5.2) can now be solved for $V(x)$ and $M(x)$ in terms of the applied loads.

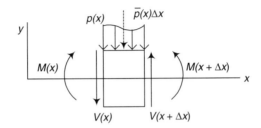

FIGURE 5-10 *Incremental element of a beam subjected to transverse distributed load* p(x).

The internal stresses σ_x and τ_{xy} on the cross-section can be related to the internal shear and moment, and hence to the applied loads, by

$$V(x) = \int_A \tau_{xy}(x)dA \tag{5.3}$$

$$M(x) = \int_A y\sigma_x(x)dA \tag{5.4}$$

To relate the internal forces and moments to the applied loads, let's first consider an incremental element (slice) of a beam Δx long, under a distributed load $p(x)$ only (Figure 5-10).

In Figure 5-10, $\bar{p}(x)\Delta x$ represents an average force over the element. Satisfying force equilibrium in the y-direction gives

$$\Sigma F_y = 0: \quad -V(x) - \bar{p}(x)\Delta x + V(x+\Delta x) = 0 \Rightarrow \bar{p}(x) = \frac{V(x+\Delta x) - V(x)}{\Delta x}. \tag{5.5}$$

In the limit, as $\Delta x \to 0$ (remember stress and strain are point quantities), $\bar{p}(x) \to p(x)$ and

$$\boxed{p(x) = \lim_{\Delta x \to 0} \frac{V(x+\Delta x) - V(x)}{\Delta x} = \frac{dV(x)}{dx}}. \tag{5.6}$$

Satisfying moment equilibrium at x gives

$$\Sigma M_z = 0: \quad -M(x) - \bar{p}(x)\Delta x\frac{\Delta x}{2} + V(x+\Delta x)\Delta x + M(x+\Delta x) = 0 \tag{5.7}$$

or

$$\frac{M(x+\Delta x) - M(x)}{\Delta x} = \bar{p}(x)\frac{\Delta x}{2} - V(x+\Delta x). \tag{5.8}$$

Again, in the limit as $\Delta x \to 0$ and using Eq. (5.5)

$$\boxed{\frac{dM(x)}{dx} = -V(x)}. \tag{5.9}$$

Equations (5.6) and (5.9) are fundamental differential equations relating beam loads. [The sign in Eq. (5.9) has no immediate practical significance, but is helpful when plotting the moment distribution, as we will later see.] These two equations are also useful in beam design since they allow us to locate the position along the beam length of the maximum internal shear force and bending moment, the later particularly driving the design. Note that by Eq. (5.9), the location of the maximum internal moment is found by setting the internal shear function $V(x) = 0$ and solving for the roots (one of which gives the location of the global maxima). This is a critical concept as it allows us to find the maximum bending stress when doing a strength design, and we will come back to this point later.

From Eq. (5.6) we see that the internal shear force can be found by an integration of the distributed load:

$$\int V(x)dx = \int p(x)dx.$$

(5.10)

Similarly, from Eq. (5.9) we see that the internal moment can be found from an integration of the shear force:

$$\int M(x)dx = -\int V(x)dx.$$

(5.11)

Equations (5.10) and (5.11) are the integral equation versions of Eqs. (5.8) and (5.9). They give rise to what are called *shear* and *moment diagrams*, respectively. For simple loading $p(x)$, the shear and moment diagrams can be manually constructed by simple graphical methods (see, e.g., Crandall *et al.*, 1972). For more complicated loadings, however, the modern engineer usually relies directly on Eqs. (5.10) and (5.11).

EXAMPLE 5-1:

A simply supported beam (pinned–roller) of length L has a distributed load as shown in Figure E5-1A.

Here, the loading function $p(x) = p_0 x / L$, where p_0 is the intensity of the load at the right beam end in dimensions of [force/length]. Let's find the internal shear force and moment expressions and the maximum moment.

First, it is always a good idea to apply the equations of statics and solve for the reaction forces and moments (which, among others things, satisfies us as to whether the problem is statically determinate or not). For our example:

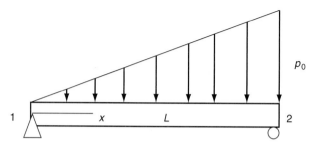

FIGURE E5-1A

EXAMPLE 5-1: *Cont'd*

$$\Sigma F_y = 0: \ R_1 - \frac{p_0 L}{2} + R_2 = 0$$

$$\Sigma M_1 = 0: \ -\frac{p_0 L}{2}\frac{2L}{3} + R_2 L = 0\text{-}$$

Solving these two equations for the unknown reactions gives $R_2 = p_0 L/3$ and $R_1 = p_0 L/6$.

From Eq. (5.10),

$$\int_0^x V(\zeta)d\zeta = \int_0^x \frac{p_0}{L}\zeta d\zeta$$

$$V(x) - V_1 = \frac{p_0}{2L}x^2$$

where we have used $V_1 = V(0)$. From Eq. (5.11),

$$\int_0^x M(\zeta)d\zeta = -\int_0^x \left[V_1 + \frac{p_0}{2L}\zeta^2 \right]d\zeta$$

$$M(x) = -V_1 x - \frac{p_0}{6L}x^3$$

where we have taken that $M(0) = 0$ from the condition at boundary 1. Using the other moment boundary condition $M(L) = 0$, we solve for $V_1 = p_0 L/6 = R_1$, which should come as no surprise and makes a good check. Finally then our internal shear force and moment distributions are

$$V(x) = -\frac{p_0 L}{6} + \frac{p_0}{2L}x^2$$

$$M(x) = \frac{p_0 L}{6}x - \frac{p_0}{6L}x^3$$

Then we can also check $V(L) = V_2 = p_0 L/3 = R_2$. Setting $V(x_{max}) = 0$ gives

$$V(x_{max}) = 0 = -\frac{p_0 L}{6} + \frac{p_0}{2L}x_{max}^2 \Rightarrow x_{max} = \frac{L}{\sqrt{3}}.$$

Then M_{max} is given by

$$M_{max}(x_{max}) = \frac{p_0 L}{6}\left(\frac{L}{\sqrt{3}}\right) - \frac{p_0}{6L}\left(\frac{L}{\sqrt{3}}\right)^3 = \ldots = \frac{p_0 L^2}{9\sqrt{3}}.$$

Let's plot the internal shear force and moment distributions in Figure E5-1B and see what they look like.

Continued

EXAMPLE 5-1: *Cont'd*

Note that the shear force is maximum at boundary 2, and that the moment is maximum when $V(x) = 0$ at $x = x_{max} = 0.577L$ as predicted.

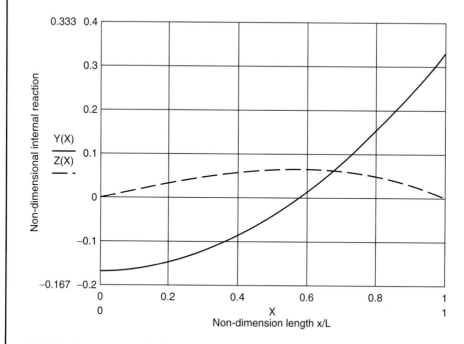

FIGURE E5-1B *Internal shear force and moment reactions along the beam.* $Y(x) = V(x)/p_0L$ and $Z(x) = M(x)/p_0L^2$.

Last, we need to deal with concentrated forces and moments, as well as distributed loads that do not extend continuously across the beam. The previous differential relations were developed based upon an assumption of smooth and continuous loading function $p(x)$. To proceed in a similar manner with a discontinuous (i.e., concentrated) load, we will find it convenient to use *singularity functions* (aka *discontinuity* functions) to represent these loads. In fact, what we will find most powerful is to have a single expression each for shear and moment distribution that include every type of load applied along the beam, be it distributed or concentrated. We will take up this subject in Section 5.6.

Deformation

For a beam undergoing small deflection, it is convenient and reasonable to make the following simplifying assumptions:

- The beam bends symmetrically about a so-called *neutral axis*.
- The neutral axis is colocated with a unique fiber, the *neutral fiber*, which remains *inextensional* during deformation, that is, the axial strain along the fiber is zero. Because of symmetry and zero neutral fiber strain, it is convenient to associate the origin of coordinates along the neutral axis.

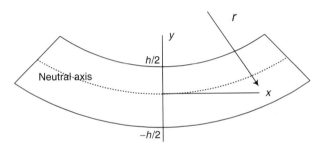

FIGURE 5-11 *Deformation of a beam. The beam deflections shown are greatly exaggerated given the Euler–Bernoulli assumptions. (Why?) The radius of curvature of an axial fiber is r.*

- Cross-sectional planes initially perpendicular to the neutral axis remain plane and perpendicular to the neutral axis during deformation.

(These assumptions are known as the *Euler–Bernoulli assumptions*.)

Consider a beam under the action of transverse forces and/or moments applied to the surface (Figure 5-11).

Note that the fiber colocated at $y = -h/2$ is extended, while the one at $y = +h/2$ is contracted. Let's now examine a slice of the beam of width dx, taken arbitrarily along the beam at x, before and after the deformation. We consider an arbitrary fiber of initial length $ds = dx$ located at a position y in the slice (Figure 5-12).

Note that the outer fibers are shown extended and contracted as previously discussed (e.g., $ds \rightarrow ds^*$), and the neutral fiber does not change length (equal to dx before and after deformation). The cutting planes perpendicular to the neutral axis remain plane and perpendicular after deformation. The initially straight neutral fiber is bent into a curved line, with radius of curvature $r_0(x)$, which is a large number since the deflections are assumed very, very small.

In the deformed slice, the neutral axis has length $dx^* = r_0 d\theta = dx$, which is equal to dx because of the inextensional condition. The new length of the arbitrary fiber segment ds is ds^*, with length:

$$ds^* = (r_0 - y)d\theta.$$

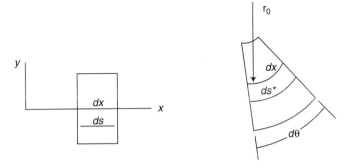

FIGURE 5-12 *Elemental slice through the beam, before and after deformation.*

Equilibrium and Deformation: Flexural Loading 161

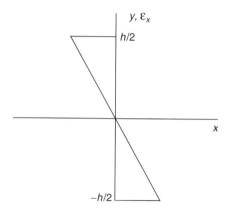

FIGURE 5-13 *Extensional fiber strain distribution at a cross-section.*

From the definition of extensional strain:

$$\varepsilon_x(y) = \frac{ds^* - ds}{ds} = \frac{(r_0 - y)d\theta - r_0 d\theta}{r_0 d\theta} = -\frac{y}{r_0}. \tag{5.12}$$

(We have assumed here that the y-location of ds is the same before and after deformation. For the small deflections assumed here, this is quite reasonable.)

We see that the extensional strain of a fiber ε_x is a linear function of its distance y from the neutral axis. The strain distribution must then look something like Figure 5-13.

Along any given $y = $ constant fiber, the strain increases with increasing curvature, as would be expected. From elementary calculus, the curvature $1/r_0$ (say) of a line is related to the first and second derivatives of the function describing the line by

$$\frac{1}{r_0} = \frac{d^2u/dx^2}{\sqrt[3]{1 + (du/dx)^2}}. \tag{5.13}$$

For small deflections, $(du/dx)^2 \ll 1$, and $1/r_0 \approx d^2u/dx^2$. Then the strain in Eq. (5.12) is given by

$$\boxed{\varepsilon_x = -\frac{y}{r_0} \approx -y\frac{d^2u}{dx^2} = -yu''(x)}. \tag{5.14}$$

This is the *strain–curvature relationship* for the beam.

EXAMPLE 5-2:

A cantilever beam of length L and depth h, such that $L/(h/2) = 30$, is loaded by a pure moment of magnitude M_0 as shown in Figure E5-2A.

(a) Determine an expression for the normalized radius of curvature $r_0/(h/2)$ if the bottom fiber (at $y = -h/2$) is at the tensile yield strain of the material, $\varepsilon_{Y,T}$.

EXAMPLE 5-2: *Cont'd*

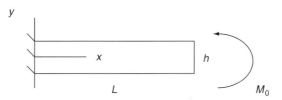

FIGURE E5-2A

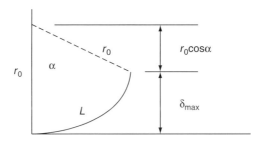

FIGURE E5-2B *The deflected beam is represented in the figure only by its neutral axis. The pure moment loading results in the deformed shape representing a circular arc.*

$$\varepsilon_x = -\frac{y}{r_0} = -\frac{(-h/2)}{r_0} = \varepsilon_{Y,T}$$

$$\underline{\frac{r_0}{h/2} = \frac{1}{\varepsilon_{Y,T}}}$$

(b) Determine the normalized tip deflection $\delta/(h/2)$ for the given loading (see Figure E5-2B).

For small deflection, a pure moment loading results in a circular arc for a deflection curve (the proof is left as an exercise for the student). Then $L = r_0 \alpha$, and $\delta_{\max} = r_0 - r_0 \cos \alpha = r_0(1 - \cos \alpha)$.

$$\alpha = \frac{L}{r_0} = \frac{L}{h/2} \frac{h/2}{r_0} = 30\varepsilon_{Y,T}$$

$$\underline{\frac{\delta_{\max}}{h/2} = \frac{1}{\varepsilon_{Y,T}} \left[1 - \cos\left(30\varepsilon_{Y,T} \right) \right]}$$

(c) Determine r_0 and $\delta_{\max}$ if $L = 15$ ft and $\varepsilon_{Y,T} = 0.001$.

$$r_0 = \frac{L}{30\varepsilon_{Y,T}} = \frac{15\,\text{ft}}{30(0.001)} = \underline{\underline{500\,\text{ft}}}$$

$$\delta_{\max} = r_0[1 - \cos(30\varepsilon_{Y,T})] = (500\,\text{ft})[1 - \cos(0.03\,\text{rad})] = \underline{0.225\,\text{ft}} = \underline{2.70\,\text{in}}$$

One last issue remains to be discussed here, and that is the *shear deformation* in flexure. Looking back on Figure 5-11 (which, recall, is a greatly exaggerated view) and Figure 5-13, it is clear that there must be a relative "slip" or "sliding" between fibers. That is, the outermost tensile fiber extends just infinitesimally more than its contiguous neighbor, which extends just infinitesimally more than its contiguous neighbor, and so on and so forth. Thus there is a shearing that takes place through the cross-section of the beam. Flexing a deck of cards will give you a very quick intuition of this effect, imagining each card as a "fiber." The minimum relative slip exists at the outer fiber (the shear is zero on the free surface of the beam) and the maximum relative slip (maximum shear) takes place at the neutral axis (the difference between the maximum strain of the outer fiber and the zero strain of the neutral axis).

For us, the question is: "Is this important?" The answer is: "It depends." For a large class of flexural structures, those that are homogenous and respond within the assumptions of the linear theory (think of metallic beams used in commercial buildings, for example), the shear contribution is likely an order of magnitude or more smaller than the normal contribution. In such cases, the shear effects can likely be disregarded. We will discuss this further in Section 5.5 in regard to the flexural test.

On the other hand, there are many flexural structures where shear deformation is important, even within the linear theory. In particular are inhomogeneous beams, such as *laminated* beams, those that are built up by adhering several thin layers or "lamina" together (sometimes called "glue-lam beams"). For these beams, the interfacial shear strength, that is, the strength in shear of the adhesive between the lamina, may be the controlling design feature. Also, whenever the beam deformations become large, the shear contribution becomes increasingly large as well and may need to be investigated by the engineer. However, such topics are outside the scope of this text, and we refer the interested reader elsewhere for further information (Crandall et al., 1972).

5.3 Constitution 🕸

For a linear elastic isotropic homogeneous beam, Hooke's law provides an adequate model for the constitutive relations. In terms of strain (see Chapter 6 for a full discussion of three-dimensional Hooke's law):

$$\varepsilon_x = \frac{1}{E}\left[\sigma_x - v\left(\sigma_y + \sigma_z\right)\right]$$
$$\varepsilon_y = \frac{1}{E}\left[\sigma_y - v(\sigma_x + \sigma_z)\right] \tag{5.15}$$
$$\varepsilon_z = \frac{1}{E}\left[\sigma_z - v(\sigma_x + \sigma_y)\right].$$

Now for the beams we study here, there will only be transverse loads in the x–y plane (i.e., transverse to the long axis of the beam in the bending plane). Thus it is reasonable to assume that $\sigma_y = \sigma_z = 0$ (especially since the surfaces $x = 0$, L and $z = \pm h/2$ are traction free, i.e., unloaded). Then the constitutive relations for the beam simplify to

$$\varepsilon_x = \frac{\sigma_x}{E}$$
$$\varepsilon_y = \varepsilon_z = -v\frac{\sigma_x}{E} \tag{5.16}$$

Note that the negative sign in the latter equations implies that the extensional fibers ($\sigma_x > 0$) contract due to Poisson's effect, while the contractile fibers ($\sigma_x < 0$) expand! In most practical cases, however, we need only concern ourselves with the first of Eqs. (5.16).

Now we substitute the constitutive relation into the strain–curvature relation (5.14), which gives a *stress–curvature* relationship:

$$\sigma_x = -yEv''(x). \tag{5.17}$$

Computing the *stress resultant N* by integrating the stress through the symmetric cross-section gives

$$N = \int_A \sigma_x dA = -\int_{-b/2}^{b/2}\int_{-h/2}^{h/2} yEv''(x)dydz = -bEv''(x)\frac{y^2}{2}\Big|_{-h/2}^{h/2} = 0. \tag{5.18}$$

This is an important result that verifies the symmetry of the bending stress distribution about the neutral plane, that is, there is no net axial force on the beam as a result of the transverse loads.

A *stress couple* or moment can be calculated in a similar manner:

$$\underline{\underline{M = \int_A \sigma_x ydA = -\int_{-b/2}^{b/2}\int_{-h/2}^{h/2} y^2Ev''(x)dydz = -Ev''(x)\int_{-b/2}^{b/2}\int_{-h/2}^{h/2} y^2dydz = -EIv''(x)}}. \tag{5.19}$$

This is the important *moment–curvature* relation, where we have used the definition of moment of inertia I given earlier. Note that if the moment produced by the load is a known function of x (meaning the beam is statically determinate), two integrations of the moment–curvature relation equation (5.19) can be performed to find the resulting deflection $v(x)$. The first integration gives $v'(x)$, that is, the slope of the beam at x or $v'(x) = \theta(x)$,

$$v'(x) = -\frac{1}{EI}\int_0^x M(x)dx + C_1, \tag{5.20}$$

while the second gives the deflected shape or *elastic curve*:

$$v(x) = -\frac{1}{EI}\int_0^x \left[\int_0^x M(x)dx\right]dx + C_1 x + C_2. \tag{5.21}$$

The constants C_1 and C_2 are determined from the two boundary conditions at the supported end of the free body.

Returning to the beam boundary conditions, we now see that $v = 0$ implies $V \neq 0$ at the boundary and vice versa. Similarly, $\theta = 0$ at the boundary implies $M \neq 0$ at the boundary and vice versa. Hence the simple and roller supports provide a momentless boundary with a shear force, while the fixed and guided supports each provide a moment, but no shear force in the latter case.

Substituting the moment–curvature relation into the above stress–curvature relation gives

$$\boxed{\sigma_x(x) = \frac{M(x)y}{I}}, \tag{5.22}$$

which is useful in calculating the stress in a beam cross-section from a given moment-producing load.

EXAMPLE 5-3:

A cantilever beam of length L, modulus E, and area moment of inertia I is tip-loaded by a transverse concentrated force P as shown in Figure E5-3A. Determine the following quantities:

$$v(x) = ? \quad v'(x) = ? \quad \sigma_x(x) = ?$$

$$v_{max} = ? \quad v'_{max} = ? \quad \sigma_{x,\,max} = ?$$

The moment can easily be found from a section analysis (the FBD is shown in Figure E5-3B):

$$\Sigma M_c = 0: \ M(x) - P(L - x) = 0$$

$$M(x) = P(L - x)$$

$$\int_0^x M(x)dx = \int_0^x P(L - x)dx = PLx - \frac{P}{2}x^2$$

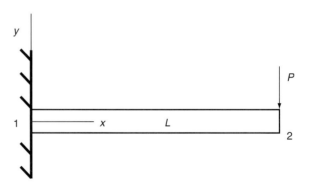

FIGURE E5-3A

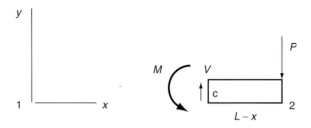

FIGURE E5-3B

EXAMPLE 5-3: *Cont'd*

$$v(x) = -\frac{P}{EI}\int_0^x \left(Lx - \frac{x^2}{2}\right)dx + C_1 x + C_2$$

$$= -\frac{P}{EI}\left(\frac{L}{2}x^2 - \frac{1}{6}x^3\right) + C_1 x + C_2$$

$$v'(x) = -\frac{P}{EI}\left(Lx - \frac{x^2}{2}\right) + C_1$$

$$v(0) = 0 \Rightarrow C_2 = 0$$

$$v'(0) = 0 \Rightarrow C_1 = 0$$

$$\underline{\underline{v(x) = -\frac{P}{EI}\left(\frac{L}{2}x^2 - \frac{x^3}{6}\right)}}$$

$$\underline{\underline{v'(x) = -\frac{P}{EI}\left(Lx - \frac{x^2}{2}\right)}}$$

$$\frac{dv'}{dx}\bigg|_{x=x_{max}} = -\frac{P}{EI}(L - x_{max}) = 0 \Rightarrow x_{max} = L$$

$$v_{max} = v(L) = -\frac{P}{EI}\left(\frac{L^3}{2} - \frac{L^3}{6}\right) = \underline{\underline{-\frac{PL^3}{3EI}}}$$

$$v'_{max} = v'(L) = -\frac{P}{EI}\left(L^2 - \frac{L^2}{2}\right) = \underline{\underline{-\frac{PL^2}{2EI}}}$$

$$\sigma_x(x) = \frac{M(x)y}{I} = \underline{\underline{\frac{P(L-x)y}{I}}}$$

$$\sigma_{x,\,max} = \sigma_x(0) = \underline{\underline{\frac{PLc}{I}}}.$$

EXAMPLE 5-4:

We wish to minimize the amount of material in a beam such that, under a specific loading condition, each cross-section will be at the maximum allowable stress (at the outer fiber). Applications are leaf springs, gear teeth, and bridge girders. Consider a cantilever beam of rectangular cross-section b by $h(x)$, with tip load F as shown in Figure E5-4.

Note that for this beam, $I = bh^3(x)/12$.

At every section, we vary $h(x)$ to maintain

$$\sigma_{allow} = \frac{6M}{bh^2(x)} = \frac{6Px}{bh^2(x)}.$$

Continued

EXAMPLE 5-4: *Cont'd*

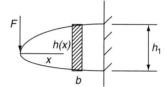

FIGURE E5-4

At the fixed end:

$$\sigma_{\text{allow}} = \frac{6PL}{bh_1^2} \Rightarrow h_1 = \sqrt{\frac{6PL}{b\sigma_{\text{allow}}}}.$$

Anywhere else:

$$h(x) = \sqrt{\frac{6Px}{b\sigma_{\text{allow}}}}\left(\sqrt{\frac{L}{L}}\right) = h_1 \sqrt{\frac{x}{L}}$$

or

$$x(h) = \frac{L}{h_1} h^2.$$

This is the equation of a parabola, and the beam is called a *constant-strength parabolic beam*.

At this point, it might be helpful to summarize a few key results for beams. Since the beams we consider here are assumed to have a one-dimensional stress–strain state only, we can make an interesting comparison to the axial structures of Chapter 3 (Table 5-1):

TABLE 5-1 Comparison of Some Key Relationships between Axial and Flexural Structures

	Flexural structures	*Axial structures*
Equilibrium	$\dfrac{dM}{dx}\big\vert_{x_{\max}} = \dfrac{d}{dx}\left(\dfrac{\sigma_{x,\max}I}{y}\right) = 0$	$\dfrac{d\sigma_x}{dx} = 0$
Kinematics	$\varepsilon_x = -\dfrac{y}{r_0} \approx -y\dfrac{d^2u}{dx^2} = -\dfrac{d}{dx}\left(y\dfrac{du}{dx}\right)$	$\varepsilon_x = \dfrac{du}{dx}$
Constitution	$\sigma_x = E\varepsilon_x$	$\sigma_x = E\varepsilon_x$

5.4 Energetics Ɣ

If we maintain that the stress–strain state in a flexural structure is one-dimensional, then the stored energy (internal or strain energy) under flexural loading is the same as it was for the axial structures we considered in Chapter 3. Recalling Eq. (3.37):

$$U = \frac{1}{2}\int_{vol} \sigma_x \varepsilon_x dV = \frac{1}{2}\int_{vol} \frac{\sigma_x^2}{E} dV, \quad dV = dxdydz. \tag{3.37}$$

If, on the other hand, we were to consider the contribution of shear to the stress–strain state, then the strain energy relation would include shear terms. Such multiaxial strain energy is discussed in Chapter 6. In this chapter, we will use the one-dimensional strain energy relation in the context of the flexural test (Section 5.5).

5.5 Mechanics ⇔ Materials Link: Flexural Test, Flexural Modulus, and Flexural Strength Ɣ🕸

There are several advantages of *flexural testing* over tensile testing. Flexural specimens are simpler in design and to manufacture than tensile specimens. For materials where the clamping force of the grips may cause problems, as in very brittle materials such as ceramics, flexural testing obviates those concerns. Even for small strains, the displacements under flexure can be considerably more than those for tensile loading, making measurements easier.

Several testing standards cover flexural testing. A common standard is ASTM 790-03, *Standard Test Methods for Flexural Properties of Unreinforced and Reinforced Plastics and Electrical Insulating Materials*. The standard configurations for flexural testing are either *three-point bending* or *four-point bending*. The three-point bend configuration is shown in Figure 5-14.

If the Young's modulus of a material is found from flexural testing (E_F), it should be very close to that found from tensile testing (E) if the compression modulus is the same as the tension modulus (which it is for many materials). Flexural strengths (S_F) may or may not correlate as well with tensile strengths (S_T).

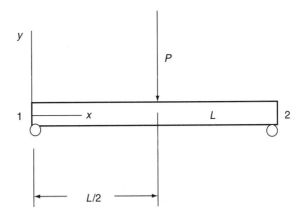

FIGURE 5-14 *Three-point bending configuration for flexural testing.*

EXAMPLE 5-5:

Let's determine the Young's modulus from a three-point bend test, using a beam of length L with rectangular cross-section beam of depth h and width b. We start with Eq. (3.37), and upon substituting in Eq. (5.22):

$$U = \frac{1}{2E_F} \int \left(\int \frac{M^2 y^2}{I^2} dA \right) dx.$$

The moment is easily found by the method of sections. First we find the reactions $R_1 = R_2 = P/2$ by global statics. Then an FBD shown in Figure E5-5 is used for finding the moment M.

Summing moments around c we find that $M = (P/2)x$. Noting that

$$\int y^2 dA = I$$

and that M is only a function of x, the strain energy equation just given becomes

$$U = \frac{1}{2E_F I} \int M^2 dx.$$

We note that the moment is discontinuous at $x = L/2$ (left as an exercise for the student), but symmetrical about $x = L/2$; we complete the integration by integrating from $x = 0$ to $x = L/2$ and multiply by 2:

$$U = \frac{2}{2E_F I} \int_0^{L/2} \frac{P^2 x^2}{4} dx = \frac{P^2 L^3}{96 E_F I}.$$

For the rectangular cross-section, $I = bh^3/12$. Recasting the strain energy in terms of the work done on the beam, $U = 1/2P\Delta$, where Δ is the deflection incurred under the load P (see Section 3.6):

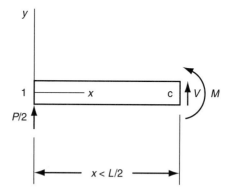

FIGURE E5-5 *Beam with discontinuous loads.*

EXAMPLE 5-5: *Cont'd*

$$U = \frac{P^2 L^3}{8E_F bh^3} = \frac{P}{2\Delta} \Rightarrow \quad E_F = \underline{\underline{\frac{PL^3}{4bh^3\Delta}}}.$$

If the beam is relatively slender (standard test methods recommend that $L/h > 15$), then the contribution of shear to the deflection Δ is greater than 10 times smaller than the normal deflection, and are thus usually ignored in the modulus calculation.

(As was stated previously, in many materials the elastic modulus in compression is not significantly different than in tension, and in those cases the flexural and tensile moduli should be the same. In what follows, we will assume this to be the case and drop the subscript F from the modulus symbol.)

5.6 Analysis of Flexural Structures 🏋

5.6.1 Singularity Functions

As we mentioned earlier, in order to use our developed beam relationships, it will be helpful to describe the loading, internal shear, and internal moment as a single function of position along the beam. This can be illustrated by a simple example. Consider the pinned–pinned beam with concentrated and distributed loads as shown in Figure 5-15.

If we wish to solve Eq. (5.10),

$$\int V(x)dx = \int p(x)dx,$$

for the internal shear along the beam, $p(x)$ must be continuous to give $V(x)$ valid for any x. However, each load M_0, P, and p_0 in Figure 5-15 is only "active" over a portion of the beam,

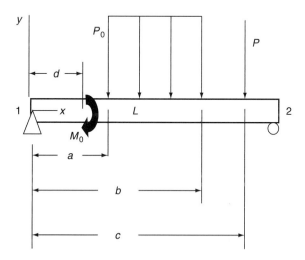

FIGURE 5-15 *Beam with discontinuous loads.*

and the load expression $p(x)$ has to reflect this. We could analyze the beam "section by section," that is, develop a free-body diagram for every section of the beam where we expect the moment to change and compute the moment expression for that FBD. This can be a tedious process when there is more than a load or two applied. (For simple loadings, as in Examples 5-3 and 5-4, this "method of sections" is fast and convenient.)

On the other hand, if we had some way to "turn on" and "turn off" the individual load functions at the correct locations in the "global" load function, then we could have a single statement for the moment at every point. This is the role *singularity functions* will play for us. They will act like "on-off switches" to describe the load function continuously across the beam.

Singularity functions are close relations of well-known discontinuous mathematical functions (sometimes called "pathological functions") such as the Dirac delta function and the Heaviside step function. Such functions do not exhibit the smoothness properties of continuity and differentiability but are in fact discontinuous. We do not provide here a rigorous treatment of these kinds of functions, but simply discuss the essential points for structural analysis and design (see Shames and Cozzarelli, 1992, for a more thorough treatment). We provide definition of three useful singularity functions, then give rules for their differentiation and integration, and finally provide examples of how they are used to describe discontinuous loading.

We will have use for three singularity functions $<x-a>^0$, $<x-a>$, and $<x-a>^2$. (The $<>$ brackets apparently date from Macaulay, 1919; see footnote in Crandall et al., p. 164.) By definition:

$$\langle x - a \rangle^n = \begin{cases} (x-a)^n, & \text{for } x \geq a \\ 0, & \text{for } x < a \end{cases} \tag{5.23}$$

and $n \geq 0$. We see in the first of Eqs. (5.23) that the angle brackets are replaced by regular parentheses whenever the quantity in the $<>$ brackets is nonnegative (zero or positive), and are replaced by zero whenever the quantity in the $<>$ brackets is negative. The three singularity functions are plotted in Figure 5-16.

As seen in Figure 5-16, the function $<x-a>^0$ looks like a "step" and is in fact referred to as a *step function*. Based on the definition in Eqs. (5.23) and that any quantity raised to the zero power is unity, our step singularity function goes as

$$\langle x - a \rangle^0 = \begin{cases} 1, & \text{for } x \geq a \\ 0, & \text{for } x < a \end{cases} \tag{5.24}$$

It is easy to see from Eqs. (5.24) how this singularity function can act as an "on–off" switch for concentrated forces and moments. Turning on and off discontinuous distributed loads will require a little more discussion, and we will do that by examples that follow.

The rules for integration and differentiation of singularity functions are straightforward:

$$\int \langle x - a \rangle^n dx = \frac{1}{n+1} \langle x - a \rangle^{n+1}, \ n \geq 0 \tag{5.25}$$

and

$$\frac{d}{dx} \langle x - a \rangle^n = n \langle x - a \rangle^{n-1}, \ n \geq 1. \tag{5.26}$$

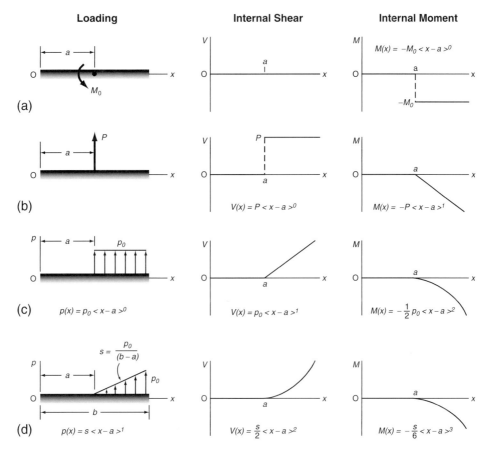

| Loading | Internal Shear | Internal Moment |

(a)

$M(x) = -M_0 < x - a >^0$

(b)

$V(x) = P < x - a >^0$

$M(x) = -P < x - a >^1$

(c)

$p(x) = p_0 < x - a >^0$

$V(x) = p_0 < x - a >^1$

$M(x) = -\frac{1}{2} p_0 < x - a >^2$

(d)

$s = \dfrac{p_0}{(b - a)}$

$p(x) = s < x - a >^1$

$V(x) = \dfrac{s}{2} < x - a >^2$

$M(x) = -\dfrac{s}{6} < x - a >^3$

FIGURE 5-16 *The first three singularity functions of Eqs. (5.23).*

We are now ready to use singularity functions in applications. The rules for assembling the loading, internal shear, and internal moment functions are given next, and several examples follow that illustrate the method.

Rule for Assembling the Loading, Shear, and Moment Functions

1. Assemble the distributed loading function $p(x)$. Use singularity functions to describe any discontinuous distributed loads.
2. To obtain the internal shear force function $V(x)$, integrate the loading function $p(x)$ and add to the result any concentrated forces (described using singularity functions), including any reaction force at $x = 0$.
3. To obtain the internal moment function $M(x)$, integrate the shear function $V(x)$ [observing the sign in Eq. (5.6)] and add to the result any concentrated moments (described using singularity functions), including any reaction moment at $x = 0$.
4. If the maximum moment is desired, set the shear function $V(x) = 0$, find the root $x_{\max}$, and then find $M_{\max}$ at $x_{\max}$.

EXAMPLE 5-6:

Consider the discontinuous uniform distributed load as shown in Figure E5-6A.

Turning on the load is easy with the singularity function $p_0(x) <x-a>^0$, which equals $p_0(x)(1)$ for $x \geq a$. But how do we turn it off at $x = b$? What we have to do is imagine the original load turned on at $x = a$ and remaining "on" along the entire length of the beam, but cancelled ("turned off") where we don't want it by applying an equal and opposite load in that region. A look at Figure E5-6B should clear this up.

Thus we turn on an equal and opposite load at $x = b$. The distributed loading function is then written as

$$p(x) = -p_0 <x-a>^0 + p_0 <x-b>^0 .$$

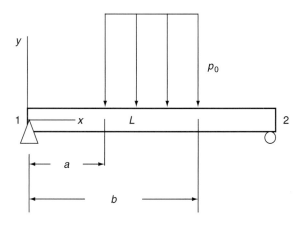

FIGURE E5-6A

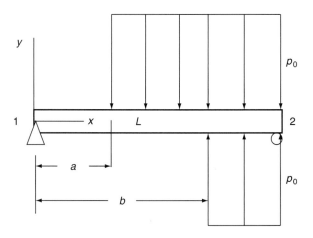

FIGURE E5-6B

EXAMPLE 5-6: *Cont'd*

Noting that the vertical reaction forces are R_1 and R_2, and there are no reaction moments, the shear and moment are readily given then as

$$V(x) = \int_0^x p(x)dx + R_1 = -p_0 \left(\frac{1}{0+1}\right) \langle x-a \rangle^{0+1} + p_0 \left(\frac{1}{0+1}\right) \langle x-b \rangle^{0+1} + R_1$$

or

$$V(x) = -p_0 \langle x-a \rangle^1 + p_0 \langle x-b \rangle^1 + R_1$$

Similarly,

$$M(x) = -\int_0^x V(x)dx = \frac{p_0}{2} \langle x-a \rangle^2 - \frac{p_0}{2} \langle x-b \rangle^2 - R_1 x.$$

For example, let's say the load is centrally located, running from $a = L/4$ to $b = 3L/4$. Then $R_1 = p_0(b-a)/2$, and the moment at two arbitrary locations is given by

$$M\left(\frac{L}{2}\right) = \frac{p_0}{2}\left(\frac{L}{2} - \frac{L}{4}\right)^2 - \frac{p_0}{2}\left(\frac{3L}{4} - \frac{L}{4}\right)\frac{L}{2} = \frac{-3p_0 L^2}{32}$$

$$M\left(\frac{7L}{8}\right) = \frac{p_0}{2}\left(\frac{7L}{8} - \frac{L}{4}\right)^2 - \frac{p_0}{2}\left(\frac{7L}{8} - \frac{3L}{4}\right)^2 - \frac{p_0}{2}\left(\frac{3L}{4} - \frac{L}{4}\right)\frac{7L}{8} = \frac{-p_0 L^2}{32}.$$

EXAMPLE 5-7:

Consider a discontinuous linearly increasing load on a cantilever beam as shown in Figure E5-7A.

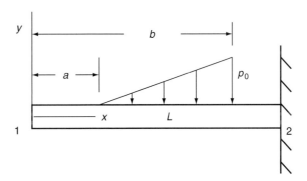

FIGURE E5-7A

Continued

EXAMPLE 5-7: *Cont'd*

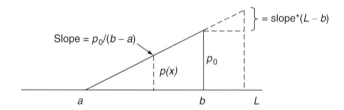

FIGURE E5-7B

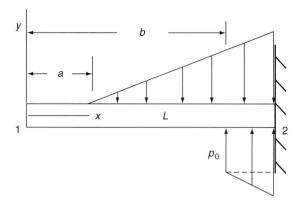

FIGURE E5-7C

Note in Figure E5-7B that, by similar triangles,

$$p(x) = \frac{p_0}{b-a}(x-a).$$

The load is turned off as shown in Figure E5-7C.

Note that the "turn off" load comprises a uniform load of intensity p_0 and a linearly increasing load of the slope $p_0/(b-a)$. The distributed loading function is then

$$p(x) = \frac{-p_0}{b-a}\langle x-a\rangle^1 + p_0\langle x-b\rangle^0 + p_0\frac{L-b}{b-a}\langle x-b\rangle^1.$$

The internal shear function is readily found (we omit the formal integration statement here):

$$V(x) = \frac{-p_0}{2(b-a)}\langle x-a\rangle^2 + p_0\langle x-b\rangle^1 + \frac{p_0}{2}\frac{L-b}{b-a}\langle x-b\rangle^2.$$

(Why was it convenient here to place the origin of coordinates at the free end of the beam?)

EXAMPLE 5-8:

Let's revisit Example 5-6, this time adding a concentrated force P acting at a distance c and a concentrated moment acting at a distance d, as seen in Figure E5-8.

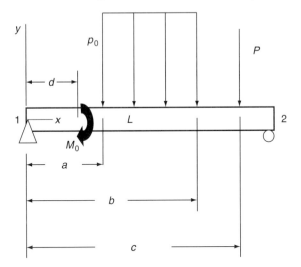

FIGURE E5-8

The distributed load function for this problem is unchanged from Example 5-6. We will need to add the concentrated force to the shear function and the concentrated moment to the moment function. First, we must recalculate the vertical support reaction R_1. The support reaction R_1 can be found from

$$\sum M_2 = 0: -R_1 L - M_0 + p_0(b-a)\left[(L-b) + \frac{b-a}{2}\right] + P(L-c) = 0.$$

Then

$$V(x) = -p_0 \langle x - a \rangle^1 + p_0 \langle x - b \rangle^1 + R_1 - P \langle x - c \rangle^0$$

and

$$M(x) = \frac{p_0}{2} \langle x - a \rangle^2 - \frac{p_0}{2} \langle x - b \rangle^2 - R_1 x + P \langle x - c \rangle^1 - M_0 \langle x - d \rangle^0.$$

Note that, from the moment summation equilibrium equation, the sign of the moment contribution of M_0 must be the same as from the support reaction R_1.

We can use the same singularity functions to aid us in determining the deflection and slope of beams. In that case, we rely on the moment–curvature relation Equation (5.19); the integrals of Eq. (5.19) give us the slope and deflection, and these were given in Eqs. (5.20) and (5.21), respectively. We illustrate the method next by example.

EXAMPLE 5-9:

Let us return to the symmetric discontinuous uniform load of Example 5-6. The rectangular cross-section steel beam now has length 10 m, depth $h = 0.1$ m, and width $B = 0.01$ m. The intensity of the uniform load is 1 kN/m. Determine the maximum stress and deflection in the beam.

It is easy to show that the maximum shear location occurs in the center of the beam:

$$V(x_{max}) = 0 = -p_0\left(x_{max} - \frac{L}{4}\right) + \frac{p_0}{2}\left(\frac{3L}{4} - \frac{L}{4}\right) \Rightarrow \underline{x_{max} = \frac{L}{2}}.$$

(Note that we have assumed that x_{max} must occur at $x < b$, but you should prove to yourself that if you assume x_{max} exists for $x \geq b$, then you have no meaningful solution.)

Then the maximum moment is as we've already discovered:

$$M\left(\frac{L}{2}\right) = \underline{\underline{\frac{-3p_0L^2}{32}}}.$$

Then the maximum stress is given by (the calculations here and elsewhere are performed using Mathcad®, and as such formatting may not match precisely with the text):

$$p_0 := 100 \, \frac{N}{m} \quad L := 10 \, m \quad h := 0.1 \, m \quad B := 0.01 \, m \quad E := 270 \cdot 10^9 \, \frac{N}{m^2}$$

$$M_{max} := \frac{3 \cdot p_0 \cdot L^2}{32} \qquad M_{max} = 937.5 \, J$$

$$I := \frac{B \cdot h^3}{12}$$

$$\sigma_{max} := \frac{M_{max} \cdot \frac{h}{2}}{I} \qquad \sigma_{max} = 5.625 \times 10^6 \, Pa$$

The maximum stress of about 5.6 MPa is certainly less than the 210 MPa of a typical structural steel.

Recall the first integral of the moment–curvature relation

$$v'(x) = -\frac{1}{EI}\int_0^x M(x)dx + C_1.$$

Upon substituting the moment expression of Example 5-6,

$$M(x) = -\int_0^x V(x)dx = \frac{p_0}{2}\langle x - a\rangle^2 - \frac{p_0}{2}\langle x - b\rangle^2 - R_1 x,$$

EXAMPLE 5-9: *Cont'd*

we have

$$v'(x) = -\frac{1}{EI}\left[\frac{p_0}{6}\langle x-a\rangle^3 - \frac{p_0}{6}\langle x-b\rangle^3 - \frac{R_1}{2}x^2\right] + C_1$$

Substituting this expression into the second of the moment–curvature integrals

$$v(x) = -\frac{1}{EI}\int_0^x\left[\int_0^x M(x)dx\right]dx + C_1x + C_2$$

gives

$$v(x) = -\frac{1}{EI}\left[\frac{p_0}{24}\langle x-a\rangle^4 - \frac{p_0}{24}\langle x-b\rangle^4 - \frac{R_1}{6}x^3\right] + C_1x + C_2$$

Applying the boundary condition $v(0) = 0$ to the preceding equation means that $C_2 = 0$, since every other term is zero (the arguments of all singularity terms are negative at $x = 0$ and hence the terms themselves are zero). Now applying $v(L) = 0$ gives

$$v(L) = 0 = -\frac{p_0}{24EI}\left[\left(L - \frac{L}{4}\right)^4 - \left(L - \frac{3L}{4}\right)^4 - L^4\right] + C_1L \Rightarrow C_1 = -\frac{11p_0L^3}{384EI}.$$

Now also by symmetry the maximum deflection must occur at $x = L/2$ (but we could find this by setting the slope equal to zero). With a little bit of algebra using $v(x_{max})$, we find the maximum deflection v_{max} as

$$v_{max} := \frac{-p_0 \cdot L^4}{E \cdot I} \cdot \left[\left(\frac{1}{24}\right) \cdot \left(\frac{1}{2} - \frac{1}{4}\right)^4 - \frac{1}{24} \cdot \left(\frac{1}{2}\right)^3 + \frac{11}{384} \cdot \left(\frac{1}{2}\right)\right]$$

$$v_{max} = -0.041\text{m}$$

The resulting maximum deflection is about half the beam depth, which is not out of keeping with our linear assumption.

5.6.2 Direct Stiffness Method for Beams

We have seen that in the case of statically determinate beams, the internal moment can be found from the differential equation (5.6) using singularity functions if necessary. Then the maximum moment can be found and the corresponding maximum flexural stress. In the case of deflections, these can be found from integration of the moment–curvature relationship, a second-order linear ordinary differential equation (ODE). For statically indeterminate cases, the following fourth-order ODE must be integrated in a similar fashion:

$$(EIv'')'' = p(x). \tag{5.23}$$

This equation results from substituting $dV/dx = p(x)$ into the derivative of $dM/dx = -V$, then substituting the resulting equation into two additional derivatives of the moment–curvature relation.

We don't derive the fourth-order Eq. (5.23) here (it is left as an exercise for the student), since there is a better and more powerful approach that is equally applicable to both statically determinate and indeterminate strength and deflection problems. In fact, for all but the simplest loading cases, the modern engineer will turn to the finite element method, which as we have previously discussed is based on the direct stiffness method.

As we found in previous chapters, the direct stiffness method has the advantage of handling both statically determinate and indeterminate structures, in a straightforward and consistent manner. We now consider the direct stiffness method for beams. A typical two-node, six-CDOF beam element is shown in Figure 5-17.

We will assume in what follows that local and global axes coincide. In that case, the element equilibrium equation (generalized force-displacement) would have the form:

$$\{N_1\ V_1\ M_1\ N_2\ V_2\ M_2\}^{\mathrm{T}} = [k]_{6\times6}\{u_1\ v_1\ \theta_1\ u_2\ v_2\ \theta_2\}^{\mathrm{T}}. \tag{5.24}$$

To get a qualitative sense of what the stiffness terms look like in the stiffness matrix $[k]$, let's recall the maximum deflection found for the cantilever beam under tip load. (Formal derivations of the beam stiffness matrix can be found in most any FEM book.) We had, upon rewriting,

$$P = V(L) = -\frac{3EI}{L^3}v(L) = -kv_{\max}. \tag{5.25}$$

In this form, it is apparent that the beam bending stiffness, relating force to deflection, has the form

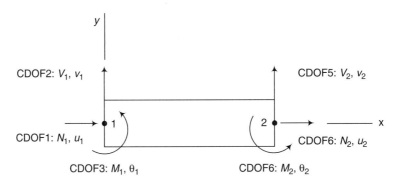

2-Node, 6CDOF beam element

FIGURE 5-17 *Two-node, six-CDOF beam element. N and V are the axial and shear forces, respectively, and M is the moment; u, v, and θ are the axial, transverse, and rotational displacements, respectively. (We include the axial force N here for completeness.)*

$$k \sim \frac{EI}{L^3} \text{[force/length]}. \tag{5.26}$$

The relation between moment and deflection for the same beam may be found from

$$V = -\frac{dM}{dx} \Rightarrow M = -\int V dx = -Px. \tag{5.27}$$

Then, for example,

$$M_{\text{max}} = M(0) = PL = \frac{3EI}{L^2} v_{\text{max}} \tag{5.28}$$

so the stiffness term here goes as EI/L^2.

The other CDOF is rotation, which is the local slope of the neutral axis (*elastic curve*). From the small deflection assumption, we have

$$\theta \approx \tan \theta \approx \frac{\Delta v}{\Delta x} \sim v'. \tag{5.29}$$

Recalling that

$$\begin{aligned} v'_{\text{max}} &= -\frac{PL^2}{2EI} \Rightarrow P = -\frac{2EI}{L^2} v'_{\text{max}} \\ M_{\text{max}} &= PL = -\frac{2EI}{L} v'_{\text{max}} \end{aligned} \tag{5.30}$$

then the rotation is related to the shear and moment by

$$V \sim \frac{EI}{L^2} \theta \qquad M \sim \frac{EI}{L} \theta \tag{5.31}$$

where the stiffness terms look like EI/L^2 and EI/L, respectively.

With only applied transverse loads and small transverse deflections, axial deflections are negligible, and Eq. (5.24) simplifies. Finally, the beam element equilibrium equation is given by

$$\begin{Bmatrix} V_1 \\ M_1 \\ V_2 \\ M_2 \end{Bmatrix} = \frac{EI}{L^3} \begin{bmatrix} 12 & 6L & -12 & 6L \\ 6L & 4L^2 & -6L & 2L^2 \\ -12 & -6L & 12 & -6L \\ 6L & 2L^2 & -6L & 4L^2 \end{bmatrix} \begin{Bmatrix} v_1 \\ \theta_1 \\ v_2 \\ \theta_2 \end{Bmatrix}. \tag{5.32}$$

EXAMPLE 5-9:

As a simple example, let's consider a one-element model of the cantilever beam under tip load previously analyzed in Example 5-3. Since v_1 and θ_1 both vanish, we eliminate the corresponding row and column in Eq. (5.32) to leave

$$\left\{ \begin{array}{c} V_2 \\ M_2 \end{array} \right\} = \frac{EI}{L^3} \left[\begin{array}{cc} 12 & -6L \\ -6L & 4L^2 \end{array} \right] \left\{ \begin{array}{c} v_2 \\ \theta_2 \end{array} \right\}.$$

Now

$$V_2 = 12\frac{EI}{L^3}v_2 - 6\frac{EI}{L^2}\theta_2 = -P$$
$$M_2 = -6\frac{EI}{L^2}v_2 + 4\frac{EI}{L}\theta_2 = 0$$

The second equation leads to

$$6\frac{v_2}{L} = 4\theta_2 \Rightarrow \theta_2 = \frac{3}{2}\frac{v_2}{L},$$

while the first now gives

$$-P = 12\frac{EI}{L^3}v_2 - 6\frac{EI}{L^2}\left(\frac{3}{2}\frac{v_2}{L}\right) = -3\frac{EI}{L^3}v_2$$

or

$$\boxed{v_2 = -\frac{PL^3}{3EI}}$$

This is just what we found before!

EXAMPLE 5-10:

A steel cantilever beam of length $L = 1\,\text{m}$ has Young's modulus $E = 210\,\text{GPa}$. The beam has a rectangular cross-section of $5\,\text{cm}$ by $20\,\text{cm}$. At mid-span, a concentrated transverse force of $3000\,\text{N}$ and moment of $1000\,\text{Nm}$ are applied.

Using a two-element approximation, determine the nodal displacements and the maximum stress in the beam.

We first generate the two-element stiffness matrices, then combine them into the global stiffness matrix:

$$E: = 210 \cdot 10^9\,\frac{\text{N}}{\text{m}^2} \qquad I: = 3.3 \cdot 10^{-5}\text{m}^4 \qquad L: = 1\,\text{m}$$

EXAMPLE 5-10: *Cont'd*

$$k1 := \begin{pmatrix} 12 & 6 & -12 & 6 & 0 & 0 \\ 6 & 4 & -6 & 2 & 0 & 0 \\ -12 & -6 & 12 & -6 & 0 & 0 \\ 6 & 2 & -6 & 4 & 0 & 0 \\ 0 & 0 & 0 & 0 & 0 & 0 \\ 0 & 0 & 0 & 0 & 0 & 0 \end{pmatrix} \quad k2 := \begin{pmatrix} 0 & 0 & 0 & 0 & 0 & 0 \\ 0 & 0 & 0 & 0 & 0 & 0 \\ 0 & 0 & 12 & 6 & -12 & 6 \\ 0 & 0 & 6 & 4 & -6 & 2 \\ 0 & 0 & -12 & -6 & 12 & -6 \\ 0 & 0 & 6 & 2 & -6 & 4 \end{pmatrix}$$

$$k := k1 + k2, \quad k = \begin{pmatrix} 12 & 6 & -12 & 6 & 0 & 0 \\ 6 & 4 & -6 & 2 & 0 & 0 \\ -12 & -6 & 24 & 0 & -12 & 6 \\ 6 & 2 & 0 & 8 & -6 & 2 \\ 0 & 0 & -12 & -6 & 12 & -6 \\ 0 & 0 & 6 & 2 & -6 & 4 \end{pmatrix}$$

$$K := \frac{E \cdot I}{L^3} \cdot k$$

Now we apply the boundary conditions ($v_1 = \theta_1 = 0$) and form the sub-stiffness matrix K_{sub}. Then we solve for the displacement vector $\{v\}$ by inversion of the matrix equation $\{F\} = [K_{\text{sub}}]\{v\}$:

$$Ksub := \frac{E \cdot I}{L^3} \cdot \begin{pmatrix} 24 & 0 & -12 & 6 \\ 0 & 8 & -6 & 2 \\ -12 & -6 & 12 & -6 \\ 6 & 2 & -6 & 4 \end{pmatrix} \quad F := \begin{pmatrix} -3000\,\text{N} \\ -1000\,\text{N} \\ 0 \\ 0 \end{pmatrix}$$

$$v := Ksub^{-1} \cdot F, \quad v = \begin{pmatrix} -2.165 \times 10^{-4} \\ -3.608 \times 10^{-4} \\ -5.772 \times 10^{-4} \\ -3.608 \times 10^{-4} \end{pmatrix} \text{m}$$

Thus the transverse displacements at nodes 2 and 3 are 0.217 and 0.577 mm, respectively. (Note that the units are not quite right in the preceding solution, as Mathcad® will not accept a vector with different units in this way. The rotational displacements should have units of radians.) Now let's calculate the maximum stress, which for a cantilever beam occurs at the root or node 1 here:

$$M1 := \left[-6\,\text{m} \cdot \left(-2.16510^{-4}\text{m} \right) + 2\,\text{m}^2 \cdot (-3.608) \cdot 10^{-4} \right] \cdot \frac{E \cdot I}{L^3}$$

$$\text{sigmamax} := \frac{M1 \cdot 10 \cdot 10^{-2}\,\text{m}}{I}$$

$$\text{sigmamax} = 1.213 \times 10^7\,\text{Pa}$$

We see that the maximum stress is 12.1 MPa, well below the yield stress for many steels.

5.7 Materials Selection in Flexural Structures

5.7.1 Material Index for Beams

We imagine a beam design where the loading and boundary conditions are specified, and thus the internal moment is known. We let the free geometric DDOF be the cross-sectional shape, and material DDOFs be the density and strength, in order to complete the design of a light, strong beam. For a beam of solid square cross-section of side b and length L, the stress is given by

$$\sigma = \frac{M^b/2}{b^4/12} = \frac{6M}{b^3} = \frac{S_f}{FS} \Rightarrow b = \sqrt[3]{\frac{6M \cdot FS}{S_f}}. \tag{5.33}$$

The beam weight w is given by

$$w = \rho A L g = \rho b^2 L g = w_0 \Rightarrow b = \sqrt{\frac{w_0}{\rho L g}}. \tag{5.34}$$

Eliminate the DDOF b by setting Eqs. (5.33) and (5.34) equal to one another. Then square both sides and solve for the material index:

$$\underline{\underline{\frac{S_f^{2/3}}{\rho}}} = \frac{(6M \cdot FS)^{2/3} L g}{w_0}. \tag{5.35}$$

Similarly for a light, stiff beam, we have

$$\underline{\underline{\frac{E^{1/2}}{\rho}}} = L\sqrt{12EI}. \tag{5.36}$$

5.7.2 Bending Shape Factor

EXAMPLE 5-11:

As we did previously, we take the square cross-section of side b as the reference shape. Then $A_0 = b^2$, $I_0 = b^4/12 = A_0^2/12$, and $c = b/2$. The maximum stress for this cross-section is then

$$\sigma_{max} = \sigma_o = \frac{M_{max} c_o}{I_o} = \frac{M_{max} b/2}{A_0^2/12} = \frac{6M_{max} b}{A_0^2} = \frac{6M_{max}}{b^4}.$$

Let's compare the reference square section to a circular section of radius R but equivalent area A_0 as seen in Figure E5-11.

EXAMPLE 5-11: *Cont'd*

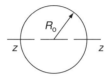

FIGURE E5-11 *Circular beam cross-section.*

Then

$$A = \pi R^2 = A_0 = b^2 \Rightarrow R = \frac{b}{\sqrt{\pi}}$$

$$I_{zz} = I = \frac{\pi}{4} R^4 = \frac{b^4}{4\pi}, \quad c = R$$

Finally,

$$\beta_o = \frac{\sigma}{\sigma_o} = \frac{M_{max} c/I}{6 M_{max}/b^3} = \frac{\dfrac{b}{\sqrt{\pi}} \dfrac{4\pi}{b^4}}{6/b^3} = \frac{2}{3}\frac{\pi}{\sqrt{\pi}} = 1.18, \quad A = A_o.$$

This result says that the circular cross-section is not as efficient in bending as the square cross-section of the same area, by about 18%. This makes sense if you think in terms of the amount of load-carrying fibers available where the stress is highest.

The cross-sectional shape of a beam is incorporated into material selection the same as it was for the axial and torsion structures—as an effective material property. Returning to Eq. (5.33) and using now the definition of β:

$$\sigma = \beta \frac{M^{b/2}}{b^4/12} = \beta \frac{6M}{b^3} = \frac{S_f}{FS} \Rightarrow b = \sqrt[3]{\beta \frac{6M \cdot FS}{S_f}}. \tag{5.38}$$

Now setting the squares of Eqs. (5.38) and (5.34) (which remains unchanged since we keep the areas equal) equal to one another, we find the material index with shape factor to be

$$\frac{S_f^{2/3}}{\rho} = \beta^{2/3} \frac{(6M \cdot FS)^{2/3} Lg}{w_0}. \tag{5.39}$$

Thus an efficient shape ($\beta < 1$) knocks down the material index, allowing a greater selection of materials to fulfill the design requirements.

5.8 Design of Flexural Structures

5.8.1 Effect of Boundary Conditions

Beam boundary conditions are often a design degree of freedom. Let us consider the difference in beam response between pinned and clamped boundary conditions. There is an interesting trade between stiffness and stress here. In Example 5-5 we can see that the maximum moment for the simply supported beam with concentrated load at the center is $M_{max} = PL/4$ (at $x = L/2$). For the same loading but with clamped boundaries, the maximum moment is $M_{max} = PL/8$ (at $x = L/2$), although the moment at the clamped boundaries is also $PL/8$. The maximum deflections for the two beam configurations are:

- Simply supported: $y_{max} = -PL^3/48EI$
- Clamped: $y_{max} = -PL^3/192EI$

Thus the clamped boundary condition acts to stiffen the beam, at the same time lowering the maximum stress (since the maximum strain has been decreased), albeit trading for some stress now at the boundaries that was not present in the simply supported case.

EXAMPLE 5-12:

The highway crew needs to place a new digital "Caution" sign over a four-lane highway. The desired configuration is shown in Figure E5-12A.

The sign is 3.67 m wide by 1.83 m high (12 ft by 6 ft) and should be placed centered between the four nominal 3.67 m wide traffic lanes. The sign weighs (with electronics and display) 44.5 kN. So as to not add any more weight to the sign support column, it is desired to keep the beam weight to no more than 1/10 the sign weight.

Design a light and strong beam that meets the requirements.

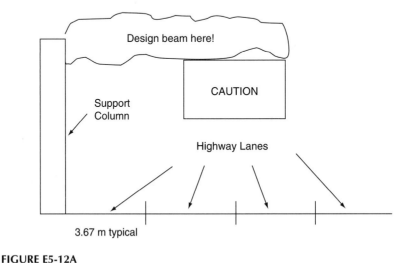

FIGURE E5-12A

EXAMPLE 5-12: *Cont'd*

I. *Problem Definition*

Performance requirements:
- Factor of safety = 2. The factor of safety is not specified. We choose this value even though the potential for loss of human life should the columns fail catastrophically is significant, the technology to implement the sign and the loads applied to the sign are well understood.
- No other performance requirements are given.

Service environment:
- Loads—are assumed to act statically downward. The load is 44.5 kN assumed to be distributed evenly along the beam at the sign location. (The assumption of downward static load is made to keep the problem simple enough for educational purposes. In the real situation, the loads could easily be dynamic and applied in other directions—e.g., from the wind, and thus require additional consideration.)
- The nominal beam length is required to be 9.15 m (30 ft).
- No other service environment specifications are given. (In a real situation, cold temperatures, corrosion from moisture and salt, and other considerations would need to be made.)

Project constraints:
- The beam weight must not exceed 4.45 kN so as to minimize the support loading.
- No other constraints are given. (Obviously in a real problem, cost would be a constraint.)

II. *Preliminary Design*

- Identify DDOFs:
  1. Geometry ⇒ minimize volume to keep weight low ⇒ area (A) and cross-section shape (I)
  2. Density ⇒ keep weight low (ρ)
  3. Strength ⇒ (S_f)
  4. Boundary conditions ⇒ in this design a cantilever with clamped boundary is indicated so boundary conditions are not a design degree of freedom
  5. Stiffness ⇒ not a primary design variable but good design suggests we check for excessive deflection just to be safe
- Trade study on cross-section configuration: The desire to keep costs low (and, as always, to keep things as simple as possible!) suggests to start with a simple solid square cross-section of side b. We have not been asked to consider any specific cross-sections.

  Material considerations suggest simple, readily available materials to keep costs low, but analysis needs to confirm the specific choice. Keeping material density low will also lead to less total weight and potentially less cost.
- FMEA: Failure expected to be from simple strength failure of the beam, but we will check for excessive deflection just to be sure.

Continued

EXAMPLE 5-12: *Cont'd*

III. *Detailed Design*

Before we can proceed, we need to find the maximum moment in the beam. We take as the origin of coordinates the fixed end of the beam. Then the reactions are

$$\sum F_y = 0: \; R_1 - 44.5 \, \text{kN} = 0 \Rightarrow R_1 = 44.5 \, \text{kN}$$

$$\sum M_0 = 0: \; M_1 - 44.5 \, \text{kN} \times 7.32 \, \text{m} = 0 \Rightarrow M_1 = 325.6 \, \text{kN} \cdot \text{m}$$

(Note that in the preceding we have neglected the self-weight of the beam itself, it being a small fraction of the supported load.) The distributed load function is

$$p(x) = -p_0 \langle x - 18 \rangle^0.$$

The shear and moment functions are

$$V(x) = -p_0 \langle x - 18 \rangle^1 + R_1$$
$$M(x) = \frac{p_0}{2} \langle x - 18 \rangle^2 - R_1 x + M_1$$

It is easy to see from $V(x) = 0$ that there are two locations where the shear goes to zero: $x = 0$ and $x = L$. Then $M_{max} = M(0) = M_1$ [$M(L)$ is, by the boundary conditions, zero].

From Eq. (5.37), the material index for a lightweight, solid square ($\beta = 1$) strong beam is

$$\frac{S_f^{2/3}}{\rho} = \beta^{2/3} \frac{(6M \cdot FS)^{2/3} Lg}{w_0} = (1)^{2/3} \left(6 \times 3.256 \times 10^5 \, \text{N} \cdot \text{m} \cdot 2 \right)^{2/3}$$

$$(9.146 \, \text{m}) \left(9.81 \, \text{m/s}^2 \right) \frac{1}{4.45 \times 10^3 \, \text{N}}$$

Now in order to plot this on the Ashby strength versus density chart, we need units of $(\text{MPa})^{2/3}/\text{Mg/m}^3$. A slight manipulation gets us there:

$$\frac{S_f^{2/3}}{\rho} = (3.92 \times 10^6 \, \text{N} \cdot \text{m})^{2/3} \left(\frac{\text{m}^3}{\text{m}^3} \right)^{2/3} \left(9.81 \, \text{m/s}^2 \right) \frac{1}{4.45 \, \text{MN}/9.146 \, \text{m}}$$

or

$$\frac{S_f^{2/3}}{\rho} = 49.97 \frac{\sqrt{\text{MPa}}}{\text{Mg/m}^3}.$$

We can plot this material index on the Ashby modulus versus density chart as follows:

$$50 \frac{(\text{MPa})^{2/3}}{\text{Mg/m}^3} = \frac{10000^{2/3}}{\rho} \Rightarrow \rho = 9.28 \approx 10$$

EXAMPLE 5-12: *Cont'd*

$$50 \frac{(\text{MPa})^{2/3}}{\text{Mg}/\text{m}^3} = \frac{S_\text{f}^{2/3}}{0.1} \Rightarrow S_\text{f} = 11.2 \approx 11.$$

Thus we plot a straight line on the chart between coordinates (0.1, 11) and (10, 10000) in Figure E5-12B. All materials above the line satisfy the light-strong criteria. Note for example that carbon fiber–reinforced polymers are an option.

Let's choose CFRP at a nominal failure strength of 1 GPa (from the chart). Then using Eq. (5.36) we can back solve for the side length b:

$$b = \sqrt[3]{\beta \frac{6M \cdot FS}{S_\text{f}}} = \ldots = \underline{0.158\,\text{m.}}$$

We can then provide the initial beam design specifications:

Carbon fiber reinforced polymer cantilever beam, with square cross-section of 0.2 m side

It will be left as an exercise for the student to improve on this design.

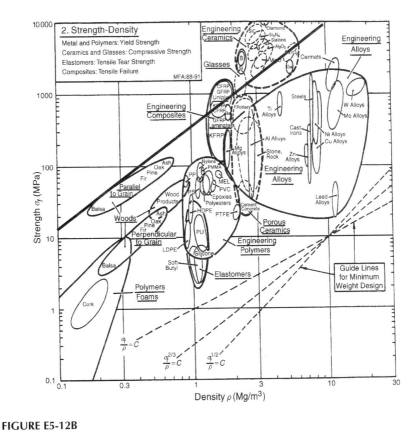

FIGURE E5-12B

Key Points to Remember

- Flexural structures are one of the most common structural types.
- We treat flexural structures at this level as one-dimensional beams.
- A beam has three coordinate degrees of freedom, but in practice using the linear theory these reduce to two: transverse displacement and rotation in the bending plane.
- Beams have internal shear and moment reactions to external transverse loading.
- The strain–curvature and moment–curvature relations are fundamental.
- Analysis (strength and stiffness) of beams proceeds from the method of sections for the simplest of beam loadings, to use of singularity functions for more complicated situations, and to the direct stiffness method (and FEM) for many practical applications.
- Material indices can be derived that include the beam shape.
- Boundary conditions have an effect on the overall beam response.

References

Crandall, S. H., Dahl, N. C., and Lardner, T. J. (1972). *An Introduction to the Mechanics of Solids*. McGraw-Hill, New York.

Shames, I. H., and Cozzarelli, F. A. (1992). *Elastic and Inelastic Stress Analysis*. Prentice-Hall, Upper Saddle River, NJ.

Problems

5-1. Look around your local environment and identify several flexural structures. Discuss the loads carried by each and the boundary conditions.

5-2. Set up a deck of playing cards as a simply supported laminated beam. Add some small load to the center of the cards and observe the flexural deformation. Comment on the shear deformation in the deck.

5-3. Recompute the summations in Eqs. (5.2), but this time from the right side of the free-body diagram.

5-4. A thin steel band with rectangular cross-section b by h passes over a pulley of radius $r = 635\,\text{mm}$. Determine the maximum extensional strain in the band if:
(a) the band thickness is $h = 1.27\,\text{mm}$ thick
(b) the band thickness is $h = 1.91\,\text{mm}$ thick
(Assume the neutral surface passes though the center of the band's cross-section.)

5-5. A steel strap of length L is bent to form a circle, and then its ends are butt welded together.
(a) Determine the shortest length of $h = 1\,\text{mm}$ thick strap that can be used if the maximum permissible strain in the strap is $2.0 \times 10^{-3}\,\text{mm/mm}$.
(b) If the length of the steel strap is $L = 3\,\text{m}$ and its thickness is $h = 1.5\,\text{mm}$, what is the maximum strain in the strap?

5-6. A simply supported beam is subjected to a centrally located concentrated load of P. Determine and plot the shear and moment expressions for the beam.

5-7. For the simply supported beam in Problem 5-6, determine by integration the maximum deflection of the beam if a concentrated downward load of P is acting at $L/2$. Also determine the maximum moment in the beam.

5-8. For the cantilever beam in Example 5-2, show that for small deflection, the elastic curve is essentially a circular arc. [Hint: first show that for small quantities, the equation of a circle centered on the origin is approximately the same as a parabola also centered on the origin—use the binomial theorem. Then, using Eq. (5.20) and (5.21), compute the elastic curve of a cantilever beam with a pure moment applied.]

5-9. For the beam in Example 5-6, compute the maximum deflection.

5-10. For the simply supported beam in Problem 5-7, determine by the direct stiffness method the maximum deflection of the beam.

5-11. For the beam in Example 5-6, compute the maximum moment, stress, and deflection for:
(a) $a = 0, \ b = L/2$
(b) $a = L/2, \ b = L$

5-12. For the beam in Figure P5-12, compute the expressions for shear and moment, and compute the maximum moment, maximum stress, and maximum deflection.

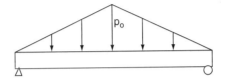

FIGURE P5-12

5-13. Explain the meaning of the shape factor in Example 5-11.

5-14. Derive the material index for a light, stiff beam. Include the shape factor.

5-15. Reconsider Example 5-12, but this time trade off the solid square section for a:
(a) thin-walled hollow square section
(b) "I" section

5-16. Consider the stadium light supports shown in Figure P5-16. Estimate the maximum bending stress in the support given the following assumptions:

- Transverse wind load of triangular distribution varying from zero at the bottom to $v = 60$ mph at the top ($F = 1/2 C_D^2 \rho v A$, where $C_D \approx 1$ for a smooth cylinder, $\rho =$ fluid density, $A =$ projected area)
- Ignore the self-weight of the support
- Support height is 50 ft
- Wall thickness is approximately 3/8 inch
- Taper is 1/8 in per foot

FIGURE P5-16

Design Project

5-17. An elevated trolley in a factory is used to carry loads from one location to another as shown in Figure P5-17. The slow moving trolley wheels ride on rails that are in turn supported by beams. Each 10 m long beam is supported by 2 columns. The bottoms of the beams are 6 m above the factory floor with the intervening space open to factory personnel. The design load for the trolley is 20 kN, but little safeguards are in place to insure the load carried. A small crane is available to lift the beams into place onto the columns that has a lifting capacity of 350 N.

Design a set of beams to support the trolley that can be placed with the existing crane. Check at least two different column cross-section configurations and at least two different material families.

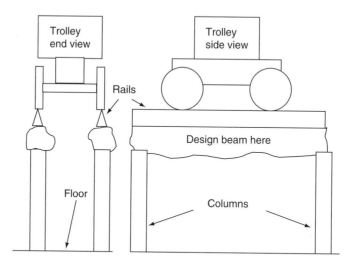

FIGURE P5-17

6   Combined Static Loading

> *You will notice that in three dimensions the situation can be much more complicated than in two.*
>
> —Richard Feynman, 1963

Objective: This chapter will introduce several design degrees of freedom in the context of the design of structures under combined loading.

In this chapter, the student will learn:

- How to analyze structures that are subjected to combined loading
- The general three-dimensional stress and strain response to combined loading
- The three-dimensional constitutive behavior of materials
- Material failure theories for combined loading
- Design of structures for combined loading

6.1   Introduction ℡🏛💻

Up to this point, we have considered structures and the loads they carry in a very constrained way. That is, we have constrained the loading to be one-dimensional: axial, torsion, and flexural loads applied in isolation from one another. In the case of flexural structures, where we applied both forces and moments, we made sure that the loads resulted in flexure in a single plane only. In each case, the resulting stress–strain state could reasonably be considered uniaxial. But you would be correct to ask: There must be situations where we would need a structure to carry various kinds of loads, or even a single type of load in various directions?

Yes, of course, that happens all of the time. Consider a drill bit: First and foremost the bit carries torsion loads as it twists into the workpiece. Yet at the same time, axial force is applied to advance the bit through the piece. Next time you're on the road, look at the large signs supported at the top of a pole (Figure 6-1), which carries the compressive axial loading of its own weight as well as the sign's weight, but also transverse wind loads, and possible seismic loads at the base.

FIGURE 6-1 *The tall column supporting the sign is subjected to three-dimensional loading.*

So what's the problem? Can't we just analyze each of these loads and their effects separately? Unfortunately, the answer is no. As we'll study further in this chapter, the stresses resulting from the loading are "tensor" quantities, and they can't simply be added or decoupled in an algebraic fashion. We have to look at the stress resulting from the combined loading in a special way.

Two other problems arise as well. First, the convenient coordinate system chosen for analysis, for example that we are naturally led to by the structure's geometry, will in general not lead us to the critical or maximum stress we need for a strength design. We need some way to hunt for the *principal stress*. Secondly, materials fail differently under multiaxial stress than in the uniaxial case, so again simply trying to analyze the structure separately for each kind of loading won't suffice.

6.1.1 General State of Stress at a Point

When forces are applied to a structural member, the forces are transmitted through the member as internal forces. Qualitatively, the intensity of the internal force at any point is called the stress at that point (see Chapter 3). Stresses in loaded members result from two basic types of forces, namely, surface forces and body forces (see Chapter 1). *Surface forces* are those that act on the surface of a body or member, for example when one body or member comes in contact with another body. *Body forces* act throughout the volume of the member; examples are gravitational, centrifugal, and magnetic forces. Gravitational body forces are the most common body forces in static or quasistatic structural members and are generally much smaller than surface forces. For this reason, body forces are often neglected in comparison to surface forces without introducing a significant error. (However, there are many applications where this is not true. Consider, for example, the mirror in a large space telescope. With diameters on the order of a few meters or a few tens of meters, the self-weight of the telescope can cause enough "gravity sag" to distort the mirror considerably from its intended shape.)

To determine the nature of internal forces, we divide the member into two parts by passing a cutting plane through the point of interest. Each of the two resulting parts may be considered a free body. The internal forces acting on the exposed cross-sectional area may have a distribution such that the internal force **F** (which is a vector and thus represented in boldface) varies in both magnitude and direction from point to point.

Consider a small area ΔA, around a point P of interest, on the surface generated by an arbitrary cutting plane. A system of internal forces acts on this small area, the resultant of which is $\Delta \mathbf{F}$, as is shown in Figure 6-2.

It should be noted that the resultant force vector $\Delta \mathbf{F}$ does not in general coincide with the outer normal **n** associated with the element of area ΔA. The resultant stress σ_F at point P is obtained by dividing $\Delta \mathbf{F}$ by ΔA, then taking the limit as ΔA approaches zero, and is given by

$$\sigma_F = \lim_{\Delta A \to 0} \frac{\Delta \mathbf{F}}{\Delta A} \tag{6.1}$$

The line of action of σ_F coincides with the line of action of resultant force $\Delta \mathbf{F}$, as shown in Figure 6-3.

It should be observed that the resultant stress σ_F is a function of the position of the point P in the member, the orientation of the cutting plane passing through the point P as identified by

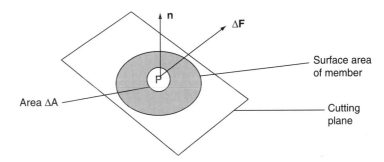

FIGURE 6-2 *Resultant of all forces acting over an area ΔA on the arbitrary surface of a member.*

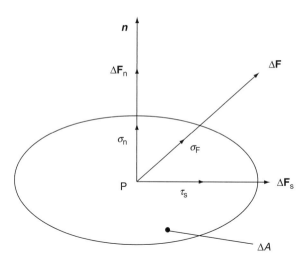

FIGURE 6-3 *Resolution of the forces and stresses transmitted through the area ΔA.*

the outer normal **n**, and the magnitude and direction of the resultant force on an infinitesimal area in the cutting plane around the point P. Thus to completely define the stress at a point we need not only the magnitude and direction of the force but also an additional direction of the outward normal **n** associated with the surface plane. Quantities such as stress (or strain) that require additional descriptors besides a magnitude and a direction (as with a vector quantity) for complete definition are called *tensors*. (This is not the exact definition of a tensor quantity, but it will suit our purposes here.) Thus a complete definition of stress or strain requires a magnitude and two directions. The magnitude of all vector or tensor quantities is influenced by the orientation of the coordinate system. However, the physical phenomena taking place at a point should not be dependent on the choice of a coordinate system. This implies that all operations performed with physical quantities should be independent of the orientation of the coordinate system, and the components in different coordinate systems must be obtainable from the components of the original coordinate system by appropriate transformation equations (described in Section 6.6.2).

A rigorous mathematical definition of a tensor is not being provided here for the sake of simplicity. However, it must be emphasized that any quantity (physical or mathematical) that transforms according to certain specific transformation laws when the original coordinate system changes its orientation is known as a *tensor*. Vector quantities (such as force and displacement) are first-order tensors, while stress and strain are second-order tensors. Higher-order tensors can also be defined mathematically, but they are generally more difficult to comprehend in physical terms.

The force $\Delta\mathbf{F}$ may be resolved into two components $\Delta\mathbf{F}_n$ and $\Delta\mathbf{F}_s$, along the normal to the small area and perpendicular to the normal **n**, respectively (see Figure 6.3). The force $\Delta\mathbf{F}_n$ is called the *normal force* on area ΔA and $\Delta\mathbf{F}_s$ is called the *shearing force* on ΔA. The normal and the shearing stress components at point P are obtained by letting ΔA approach zero (or become infinitesimal) and dividing the magnitude of the respective force by the magnitude of the area. Thus the magnitudes of the normal stress $\boldsymbol{\sigma}_n$ and shear stress $\boldsymbol{\tau}_s$ are given by

$$\boldsymbol{\sigma}_n = \mathrm{Lim}_{\Delta A \to 0} \frac{\Delta\mathbf{F}_n}{\Delta A} \qquad (6.2a)$$

$$\tau_s = \lim_{\Delta A \to 0} \frac{\Delta \mathbf{F}_s}{\Delta A} \tag{6.2b}$$

The unit vectors associated with σ_n and τ_s are perpendicular and tangent, respectively, to the cutting plane.

Cartesian components of stress for any orientation of a rectangular-Cartesian coordinate system (x, y, z) at P can also be obtained from the resultant stress. Consider the small or incremental area ΔA, whose outer normal now coincides with the positive z-direction as shown in Figure 6-4. If the resultant stress σ_F is resolved into components along the x-, y-, and z-axes, the Cartesian components σ_{zz}, τ_{zx}, and τ_{zy} are obtained. (From this point forward, the boldface will be dropped from the stress notation for convenience; however, we wish to emphasize that the student keep in mind that stress is a tensor quantity.)

The first subscript in the stress symbol refers to the outer normal and defines the plane upon which the stress component acts. The second subscript gives the direction in which the stress acts. In σ_{xx}, the first subscript x means that the outer normal of the infinitesimal area on which the force acts points along the x-axis. The second subscript x means that the force or the stress on the infinitesimal area acts in the x-direction. Thus the outward normal of the area and the force are in the same direction for normal stresses. In the term τ_{xy}, the outward normal of the area is along the x-axis, while the force is along the y-axis. Hence for shear stresses, the outward normal and force are perpendicular to each other.

Normal stresses will be positive (+ sign) when they produce tension and negative (− sign) when they produce compression. The sign of the shear stresses is a bit more complicated. Shear stresses are termed positive (+ sign) if they:

- Act in a positive coordinate direction on a surface whose normal also acts in a positive coordinate direction, or
- Act in a negative coordinate direction on a surface whose normal also acts in a negative coordinate direction.

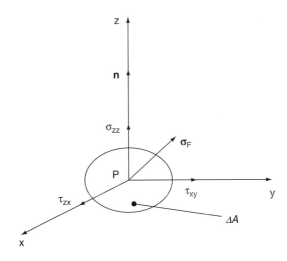

FIGURE 6-4 *Resolution of the resultant stress on an incremental area into three rectangular-Cartesian stress components.*

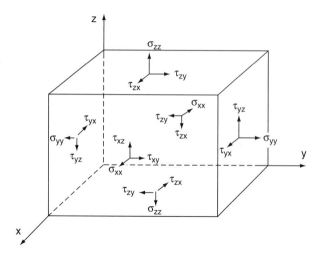

FIGURE 6-5 *Rectangular-Cartesian stress components acting along the faces of a small cubic element around a point in a loaded member. All stresses are positive as shown.*

Otherwise, the shear stress is negative ($-$ sign).

If the same procedure is followed using infinitesimal areas whose outer normals are in the positive x and y directions, two more sets of Cartesian components, $(\sigma_{xx}, \tau_{xy}, \tau_{xz})$ and $(\sigma_{yy}, \tau_{yx}, \tau_{yz})$, respectively, can be obtained. Hence at any point P in a member under an arbitrary load, nine Cartesian stress components can be identified. These nine components of stress constitute the *general state of stress* and are tabulated in the form of an array as follows:

$$\begin{bmatrix} \sigma_{xx} & \tau_{xy} & \tau_{xz} \\ \tau_{yx} & \sigma_{yy} & \tau_{yz} \\ \tau_{zx} & \tau_{zy} & \sigma_{zz} \end{bmatrix} \begin{matrix} \rightarrow \text{Outer normal parallel to x-axis} \\ \rightarrow \text{Outer normal parallel to y-axis} \\ \rightarrow \text{Outer normal parallel to z-axis} \end{matrix}$$

It is conventional to show these nine stress components on the faces of a small cubic element around a point in the loaded member, as shown in Figure 6-5.

Thus the three-dimensional state of stress in a body loaded by surface and body forces and couples is defined by three normal stress components and six shear stress components.

6.1.2 General State of Strain

In the previous section the state of stress at any point in a body was determined. The relationships obtained were based on conditions of equilibrium and no assumptions were imposed regarding the deformation in the object or the physical properties of the material that constituted the object. Hence those results are valid for any material and any amount of deformation in the object. In this section, the state of deformation and the associated strains will be analyzed. Because strain is a pure geometric quantity, no restrictions on the object material will be required.

When a body is subjected to a system of forces, if individual points in the body change their relative positions the body is said to be in a state of *deformation*. The movement of any point is a vector quantity known as *displacement*. Translation or rotation of a body as a whole, with no change in the relative positions of points in the body, is known as *rigid-body motion* and does not result in strain within the body.

Two types of strains are present: (i) extensional or *normal strain*, and (ii) *shear strain*. Figure 6-6 illustrates these two types of strains for a two-dimensional case.

In introductory mechanics, *normal strain* is defined as the change in length of a line segment between two points divided by the original length of the line segment. Thus in Figure 6-6a the normal strain along the *x*-direction can be written as

$$\varepsilon_{xx} = \frac{\Delta u}{\Delta x}.$$

In the limit as $\Delta x \to 0$ (that is, using an infinitesimal line segment), the normal strain equation can be written as follows. (Note: *u*, *v*, and *w* are a function of *x*, *y*, and *z*.)

$$\varepsilon_{xx} = \frac{\partial u}{\partial x}. \qquad (6.3a)$$

Similarly from Figure 6-6b, we can write

$$\varepsilon_{xx} = \frac{\partial v}{\partial y}. \qquad (6.3b)$$

Also, along the *z*-direction the normal strain can be expressed as

$$\varepsilon_{zz} = \frac{\partial w}{\partial z}. \qquad (6.3c)$$

The *shear strain* is defined as the angular change between the two line segments that were originally perpendicular and parallel, for example to the *x*- and *y*-axes. In Figure 6-6c the total change in the angle is $\theta_1 + \theta_2$. In the limiting case if θ_1 and θ_2 are very small, the total angle change can be expressed as

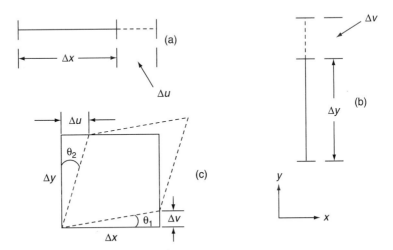

FIGURE 6-6 *(a) Normal strain in the x-direction. (b) Normal strain in the y-direction. (c) Shear strain in the xy-plane. The dashed-line segments indicate the deformed position.*

$$\theta_1(\text{radians}) + \theta_2(\text{radians}) = \tan\theta_1 + \tan\theta_2$$

where

$$\tan(\theta_1) = \frac{\partial v}{\partial x} \quad \text{and} \quad \tan(\theta_2) = \frac{\partial u}{\partial y}.$$

Thus the shearing strain is

$$\gamma_{xy} = \theta_1 + \theta_2 = \frac{\partial v}{\partial x} + \frac{\partial u}{\partial y}. \qquad (6.3\text{d})$$

When shear deformation involves a reduction of the right angle between two line segments oriented respectively along (say) the positive x- and y-axes, the shearing strain γ_{xy} is said to be positive; otherwise it is negative.

Similarly, if two line segments are considered parallel to the y- and z-axes, and another two parallel to the x- and z-axes, respectively, two more expressions for the corresponding shear strains can be written as

$$\gamma_{yz} = \frac{\partial v}{\partial z} + \frac{\partial w}{\partial y} \qquad (6.3\text{e})$$

and

$$\gamma_{xz} = \frac{\partial u}{\partial z} + \frac{\partial w}{\partial x}. \qquad (6.3\text{f})$$

Equations (6.3a–f) constitute the *strain-displacement equations* and are valid only for small strains of the order of 0.2% or less. For larger strains, higher-order terms have to be added to Eqs. (6.3). The more general equations are not provided here as we will only consider problems with small strains in this book. Analogous to the general state of stress the *general state of strain* at a point can be written in matrix notation as

$$\begin{bmatrix} \varepsilon_{xx} & \gamma_{xy} & \gamma_{xz} \\ \gamma_{yx} & \varepsilon_{yy} & \gamma_{yz} \\ \gamma_{zx} & \gamma_{zy} & \varepsilon_{zz} \end{bmatrix}. \qquad (6.4)$$

Equation (6.4) will be called the *strain tensor* and the components of the strain tensor are sometimes called *tensorial strains*. (See Section 6.6.7.)

EXAMPLE 6-1:

If the displacement field for a body is given by

$$\mathbf{u} = (x^2 + y^2)\hat{\mathbf{i}} + (3 + z)\hat{\mathbf{j}} + (x^2 + 2y^2)\hat{\mathbf{k}}$$

what is the deformed position of a point originally at $(3, 2, -1)$?

Answer: If the displacement field is known throughout the volume of the body, the displacement vector $\mathbf{u}$ of any point P is known once its coordinates are known. If $\mathbf{r}$ is the position vector of point P and $\mathbf{r}'$ that of point P′ (Figure E6-1), then

$$\mathbf{r}' = \mathbf{r} + \mathbf{u}$$

or

$$\mathbf{u} = \mathbf{r}' - \mathbf{r}.$$

The displacement vector $\mathbf{u}$ at point $(3, 2, -1)$ is

$$\mathbf{u} = (3^2 + 2^2)\hat{\mathbf{i}} + (3 - 1)\hat{\mathbf{j}} + (3^2 + 2 \cdot 2^2)\hat{\mathbf{k}}$$
$$= 13\hat{\mathbf{i}} + 2\hat{\mathbf{j}} + 17\hat{\mathbf{k}}$$

The initial position vector $\mathbf{r}$ of point P is

$$\mathbf{r} = 3\hat{\mathbf{i}} + 2\hat{\mathbf{j}} - 1\hat{\mathbf{k}}.$$

Thus the final position vector $\mathbf{r}'$ of point P′ is

$$\mathbf{r}' = \mathbf{r} + \mathbf{u} = \underline{16\hat{\mathbf{i}} + 4\hat{\mathbf{j}} + 16\hat{\mathbf{k}},}$$

which is the deformed position of the point P.

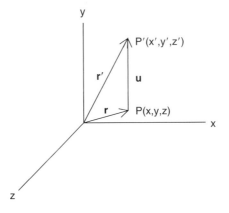

FIGURE E6-1

6.2   Equilibrium and Deformation Υ

In Section 6.1 we have seen that a general state of stress and strain are completely defined by nine respective components in any material. However, if body and surface couples are zero, it can be shown for the case of stresses that

$$\tau_{yz} = \tau_{zy}$$
$$\tau_{xy} = \tau_{yx} \tag{6.5}$$
$$\tau_{xz} = \tau_{zx}.$$

Equations (6.5) can be proven as follows. Consider the x–y plane as shown in Figure 6-7. Writing the moment equilibrium equation about the z-axis, one obtains the equation

$$\tau_{xy}(dydz)dx - \tau_{yx}(dxdz)dy = 0$$

or

$$\tau_{xy} = \tau_{yx}.$$

Note that the equal and opposite normal force (stress times area) components cancel out and hence have not been shown in the equation.

Hence the stress tensor can be written as

$$\begin{bmatrix} \sigma_{xx} & \tau_{xy} & \tau_{xz} \\ \tau_{xy} & \sigma_{yy} & \tau_{yz} \\ \tau_{xz} & \tau_{yz} & \sigma_{zz} \end{bmatrix}. \tag{6.6}$$

Similarly, we can write the strain components in the form

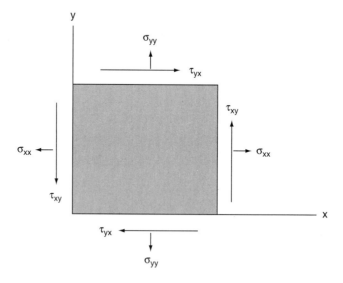

FIGURE 6-7   *Stresses acting on an element in the* x–y *plane to demonstrate the equality* $\tau_{xy} = \tau_{yx}$.

$$\begin{bmatrix} \varepsilon_{xx} & \gamma_{xy} & \gamma_{xz} \\ \gamma_{xy} & \varepsilon_{yy} & \gamma_{yz} \\ \gamma_{xz} & \gamma_{yz} & \varepsilon_{zz} \end{bmatrix}. \tag{6.7}$$

Note that in both the stress tensor equation (6.6) and strain tensor equation (6.7), the normal components lie along the matrix diagonal and the shear components are symmetric about the diagonal.

If the only forces that act on the cubic element in Figure 6-5 are surface forces and body forces, the summation of forces in the x, y, and z directions consecutively produce three equations of equilibrium as follows. Consider first a cubic element in which only the stress and body force components acting in the x-direction exist (Figure 6-8).

A summation of forces in the x-direction results in the expression:

$$\left(\sigma_{xx} + \frac{\partial \sigma_{xx}}{\partial x} dx - \sigma_{xx} \right) dy dz + \left(\tau_{yx} + \frac{\partial \tau_{yx}}{\partial y} dy - \tau_{yx} \right) dx dz \\ + \left(\tau_{zx} + \frac{\partial \tau_{zx}}{\partial z} dz - \tau_{zx} \right) dx dy + f_x dx dy dz = 0 \tag{6.8}$$

where f_x is the body force per unit volume in the x-direction. Dividing Eq. (6.8) throughout by the incremental volume $dx dy dz$ results in

$$\frac{\partial \sigma_{xx}}{\partial x} + \frac{\partial \tau_{yx}}{\partial y} + \frac{\partial \tau_{zx}}{\partial z} + f_x = 0. \tag{6.9a}$$

Similarly, summations in the y- and z-directions result in

$$\frac{\partial \tau_{xy}}{\partial x} + \frac{\partial \sigma_{yy}}{\partial y} + \frac{\partial \tau_{zy}}{\partial z} + f_y = 0$$

$$\frac{\partial \tau_{xz}}{\partial x} + \frac{\partial \tau_{yz}}{\partial y} + \frac{\partial \sigma_{zz}}{\partial z} + f_z = 0 \tag{6.9b, c}$$

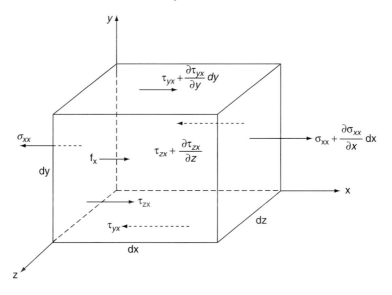

FIGURE 6-8   *Stresses and body forces acting in the x-direction only.*

where f_y and f_z are the body forces per unit volume in the y- and z-directions, respectively. The *stress equilibrium equations* (6.9) can be rewritten using Eqs. (6.5) as

$$\frac{\partial \sigma_{xx}}{\partial x} + \frac{\partial \tau_{xy}}{\partial y} + \frac{\partial \tau_{xz}}{\partial z} + f_x = 0$$

$$\frac{\partial \tau_{xy}}{\partial x} + \frac{\partial \sigma_{yy}}{\partial y} + \frac{\partial \tau_{yz}}{\partial z} + f_y = 0.$$ (6.10)

$$\frac{\partial \tau_{xz}}{\partial x} + \frac{\partial \tau_{yz}}{\partial y} + \frac{\partial \sigma_{zz}}{\partial z} + f_z = 0$$

6.3 Constitution: Two-Dimensional and Three-Dimensional Elastic Material ✿

In the elastic region, a linear relationship exists between stress and strain, called *Hooke's law* (see Section 3.3). For a three-dimensional stress–strain state in homogeneous isotropic material, the tensile stress–strain relations given in Eqs. (6.11) are expressions for obtaining the strains in terms of stress components and elastic constants.

$$\varepsilon_{xx} = \frac{1}{E}[\sigma_{xx} - v(\sigma_{yy} + \sigma_{zz})]$$

$$\varepsilon_{yy} = \frac{1}{E}[\sigma_{yy} - v(\sigma_{zz} + \sigma_{xx})]$$

$$\varepsilon_{zz} = \frac{1}{E}[\sigma_{zz} - v(\sigma_{xx} + \sigma_{yy})]$$

$$\gamma_{xy} = \frac{2(1+v)}{E}\tau_{xy}$$ (6.11)

$$\gamma_{yz} = \frac{2(1+v)}{E}\tau_{yz}$$

$$\gamma_{zx} = \frac{2(1+v)}{E}\tau_{zx}$$

Expressions for obtaining stresses in terms of strain components and elastic constants are given in Eqs. (6.12).

$$\sigma_{xx} = \frac{E}{(1+v)(1-2v)}[(1-v)\varepsilon_{xx} + v(\varepsilon_{yy} + \varepsilon_{zz})]$$

$$\sigma_{yy} = \frac{E}{(1+v)(1-2v)}[(1-v)\varepsilon_{yy} + v(\varepsilon_{zz} + \varepsilon_{xx})]$$

$$\sigma_{zz} = \frac{E}{(1+v)(1-2v)}[(1-v)\varepsilon_{zz} + v(\varepsilon_{xx} + \varepsilon_{yy})]$$

$$\tau_{xy} = \frac{E}{2(1+v)}\gamma_{xy}$$ (6.12)

$$\tau_{yz} = \frac{E}{2(1+v)}\gamma_{yz}$$

$$\tau_{zx} = \frac{E}{2(1+v)}\gamma_{zx}$$

If pure shear stresses are applied, the elastic response of the material is characterized by the last three equations in Eqs. (6.11) or (6.12). For any material, in general three normal stresses and six shear stresses exist. However, three relations between the shear stresses can be obtained,

$$\tau_{xy} = \tau_{yx}, \ \tau_{yz} = \tau_{zy}, \ \tau_{zx} = \tau_{xz}, \tag{6.13}$$

and thus only three shear stresses need to be obtained independently.

A third material property, the shear modulus G, can also be introduced. For an isotropic material, only two of the material properties G, E, and v are independent; the shear modulus, for example, can be calculated from the relation

$$G = \frac{E}{2(1+v)}. \tag{6.14}$$

EXAMPLE 6-2:

The strain components are measured at a point in a steel component in a machine and are listed below. Determine the stress at this point. (For steel, $E = 207$ GPa and $v = 0.3$.)

$$\varepsilon_{xx} = 300\,\mu\varepsilon, \varepsilon_{yy} = 300\,\mu\varepsilon, \varepsilon_{zz} = 300\,\mu\varepsilon$$
$$\gamma_{xy} = 200\,\mu\varepsilon, \ \gamma_{yz} = 200\,\mu\varepsilon, \gamma_{zx} = 200\,\mu\varepsilon$$

Answer: From Eq. (6.12)

$$\sigma_{xx} = \frac{E}{(1+v)(1-2v)}[(1-v)\varepsilon_{xx} + v(\varepsilon_{yy} + \varepsilon_{zz})]$$

$$= \frac{207 \times 10^9}{(1+0.3)(1-2\times 0.3)}[(1-0.3)(300\times 10^{-6}) + 0.3(200+100)10^{-6}]$$

$$= 119.4 \times 10^6\,\text{N/m}^2$$

$$= \underline{119.4\,\text{MPa}}.$$

Similarly, using the second and third expressions from Eqs. (6.12), we obtain the two remaining normal stress components as

$$\sigma_{yy} = \frac{E}{(1+v)(1-2v)}[(1-v)\varepsilon_{yy} + v(\varepsilon_{zz} + \varepsilon_{xx})]$$

$$= \frac{207 \times 10^9}{(1+0.3)(1-2\times 0.3)}[(1-0.3)(200\times 10^{-6}) + 0.3(100+300)10^{-6}]$$

$$= 103.5 \times 10^6\,\text{N/m}^2$$

$$= \underline{103.5\,\text{MPa}}$$

Continued

EXAMPLE 6-2: *Cont'd*

$$\sigma_{zz} = \frac{E}{(1+v)(1-2v)}[(1-v)\varepsilon_{zz} + v(\varepsilon_{xx} + \varepsilon_{yy})]$$

$$= \frac{207 \times 10^9}{(1+0.3)(1-2\times0.3)}[(1-0.3)(100\times10^{-6}) + 0.3(300+200)10^{-6}]$$

$$= 87.6 \times 10^6 \, \text{N/m}^2$$

$$= \underline{87.6 \, \text{MPa}}.$$

The shear stress components are obtained by using the last three expressions in Eq. (6.12) as

$$\tau_{xy} = \frac{E}{2(1+v)}\gamma_{xy}$$

$$= \frac{207 \times 10^9}{2(1+0.3)}200\times10^{-6}$$

$$= 15.9 \times 10^6 \, \text{N/m}^2$$

$$= \underline{15.9 \, \text{MPa}}$$

$$\tau_{yz} = \frac{E}{2(1+v)}\gamma_{yz}$$

$$= \frac{207 \times 10^9}{2(1+0.3)}100\times10^{-6}$$

$$= 7.9 \times 10^6 \, \text{N/m}^2 = \underline{7.9 \, \text{MPa}}$$

$$\tau_{zx} = \frac{E}{2(1+v)}\gamma_{zx}$$

$$= \frac{207 \times 10^9}{2(1+0.3)}150\times10^{-6}$$

$$= 11.9 \times 10^6 \, \text{N/m}^2$$

$$= \underline{11.9 \, \text{MPa}}.$$

EXAMPLE 6-3:

A rubber cube is inserted into a cavity of the same size and shape in a thick steel block, as shown in Figure E6-3. The rubber cube is pressed by a steel block with a pressure of p. Considering the thick steel cavity to be rigid and there is no friction between the cube and the cavity walls, find the pressure exerted by the rubber against the cavity walls.

Answer: Since the cube is constrained in the x- and y-directions, the strain components along these directions are zero:

$$\varepsilon_{xx} = 0 \text{ and } \varepsilon_{yy} = 0. \tag{I}$$

EXAMPLE 6-3: *Cont'd*

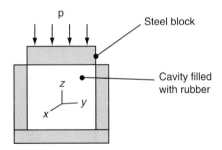

FIGURE E6-3

In the z-direction the stress in the rubber cube must balance the pressure applied to maintain equilibrium; thus

$$\sigma_{zz} = -p. \qquad (II)$$

Using Eqs. (I) and (II) and the first two expressions of Eq. (3.18) we can write

$$\varepsilon_{xx} = \frac{1}{E}[\sigma_{xx} - v(\sigma_{yy} - p)] = 0$$

$$\varepsilon_{yy} = \frac{1}{E}[\sigma_{yy} - v(\sigma_{xx} - p)] = 0 \qquad (III)$$

Simultaneously solving Eqs. (III), we obtain the pressure exerted by the cube against the cavity walls as:

$$\sigma_{xx} = \sigma_{yy} = -\frac{v}{1 - v}p$$

If the Poisson's ratio of rubber is known (often taken as 0.5), then the pressure exerted by the cube against the cavity walls can be determined.

6.4 Mechanics ⇔ Materials Link: The Combined Loading Tests 𝚼🕸

Aircraft structural components, bridge support sections, car and truck wheels, and boat hulls are all examples of structures that, in service, are normally loaded in more than one direction at once, that is, they are *multiaxially loaded*. It can be intuitively recognized that the evaluation of material characteristics by uniaxial tests only can lead to a misrepresentation of the behavior of a material in an actual engineering structure where multiaxial loading is the rule rather than an exception. Thus, using more realistic loading conditions during laboratory testing, for example under *biaxial* or *triaxial* conditions, will lead to more accurate representation of the expected behavior of the structure in service, which could in turn lead to safer and wider use of structural materials.

Metals have been well investigated for their mechanical behavior under biaxial loading, whereas more recent materials such as composites and layered materials have yet to be fully investigated under multiaxial loading. Because of the inherently anisotropic structure of composites, their strength under biaxial conditions critically depends on how well the loads match the directions of the reinforcements in the material under test. If they are well matched, the biaxial strength may exceed the value that might be expected from simple uniaxial tensile and compressive tests. Conversely, if they are poorly matched, the strength can be quite low.

Metals and polymer composites in the form of tubes have been tested under biaxial loading conditions leading to the creation of a large body of biaxial test data and a good understanding of the failure criteria. However, for many applications the materials are used in the form of flat or gently curved panels, so tubular test specimens are unsuitable. For this reason planar specimens have been developed (Welsh *et al.*, 2002) called cruciform test specimens (Figure 6-9), which can be used to cover the complete biaxial failure envelope—that is, tension–tension, tension–compression, compression–compression, and all the intermediate states. The cruciform specimens have been able to overcome the limitation of other types of flat specimens that would fail prematurely at the re-entrant corners and also not give a large region of uniform strain at the specimen center. This is not a limitation of tube testing, which is why tubes have had an essential part to play in the development of biaxial failure criteria.

Several types of biaxial and triaxial machines have been developed by researchers in universities and by the industry. One such triaxial testing machine is shown in Figure 6-10, which can be used to perform biaxial tests or traixial tests. This electromechanical test facility was developed specifically to evaluate the biaxial (in-plane) and triaxial (three-dimensional) response of composite materials, though this experimental test facility is capable of testing any

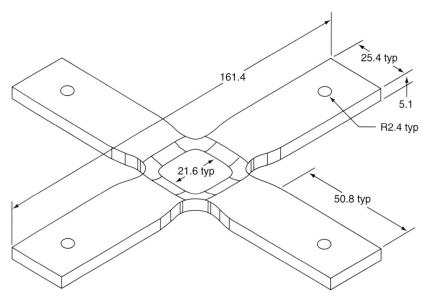

FIGURE 6-9 *Thickness-tapered cruciform specimen geometry schematic. All dimensions in milli-meters.*

FIGURE 6-10 *Triaxial Testing Facility (courtesy of Dr. Jeffry S. Welsh, Air Force Research Labs, Kirtland Air Force Base, New Mexico, USA).*

material and can generate any combination of tensile or compressive stresses in σ_1, σ_2, and σ_3 stress space.

6.5 Energetics Ŷ

Strain energy is the potential energy stored in a body by virtue of an elastic deformation, equal to the work done by the applied forces, to provide both normal and shear strains. In a body loaded within the elastic limit, the work done during loading is stored as recoverable strain energy in the material of the body. If the body is unloaded, the body does work (on the loading frame or the human or other interfacing objects) and releases all its energy. The work done to deform the body elastically depends only on the state of strain at the end of the test; it is independent of the history of loading. Strain energy stored in a material is a very useful metric for evaluating the suitability of a material for structural applications. The higher the strain energy stored, or the higher the strain energy absorbed under a given loading condition, the better. In this regard a quantity called *strain energy density, U*, is defined as the work done per unit volume to deform the material from a stress free state to a loaded state. Defining this quantity for a unit volume of material eliminates the effect of size of the body. If the strain energy is divided by the density, we obtain a quantity called *specific strain energy*. Strain energy density has the dimensions of J/m^3 in the SI metric units, or lb/in^2 in the US system of units. The strain energy density is equal to the area under the stress–strain curve measured from zero strain to a given strain value. The mathematical form is given in Eq. (6.15), which is graphically represented by the shaded area in Figure 6-11.

$$U = \int_{\text{volume}} \sigma d\varepsilon. \qquad (6.15)$$

If the specimen is loaded beyond its elastic limit and plastic deformation occurs and then the specimen is unloaded, only the energy represented by the shaded region in Figure 6-12 is recovered; the remainder of the energy is spent in deforming the material and is dissipated as heat energy. Note that line $P\varepsilon_p$ is the unloading path and is parallel to the linear or elastic

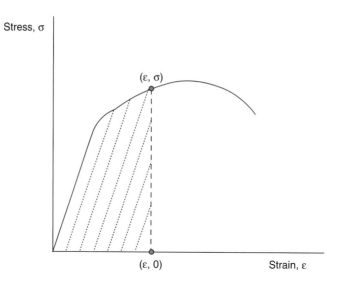

FIGURE 6-11 *The shaded region represents the strain energy per unit volume stored in the material.*

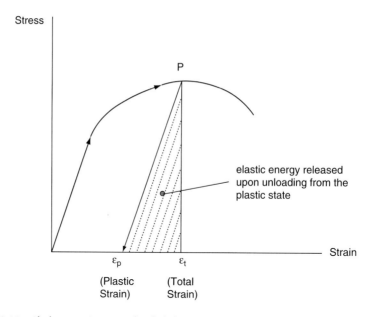

FIGURE 6-12 *If the specimen is loaded beyond its elastic limit and plastic deformation occurs and then the specimen is unloaded, only the energy represented by the shaded region is recovered. Loading and unloading path shown by arrows.*

region of the stress–strain curve. The area under the stress–strain curve up to fracture is called the *modulus of toughness* of the material.

The strain energy, U, for a member under a uniaxial normal stress state is given by Eq. (3.37), and for a member under shear stress only by Eq. (4.22). If a body is subjected to a general state of stress, that is, when at any point at least two or more nonzero stress components from among the six stress components, σ_{xx}, σ_{yy}, σ_{zz}, τ_{xy}, τ_{yz}, or τ_{xz} exist, the strain energy density can be obtained by adding the expressions given in Eqs. (3.37) and (4.22), as well as the four other expressions obtained through a permutation of the subscripts of the stress and strain components. The six expressions can be added to obtain the total strain energy density because strain energy is a scalar quantity. Thus, assuming elastic deformations in the body, we can write the total strain energy density as

$$U = \frac{1}{2}\left(\sigma_{xx}\varepsilon_{xx} + \sigma_{yy}\varepsilon_{yy} + \sigma_{zz}\varepsilon_{zz} + \tau_{xy}\gamma_{xy} + \tau_{yz}\gamma_{yz} + \tau_{xz}\gamma_{xz}\right). \tag{6.16}$$

Substituting for the strain components from Eq. (6.11) into Eq. (6.16), we obtain an expression for the strain energy density in terms of stress components:

$$\begin{aligned} U &= \frac{1}{2E}\left[\sigma_{xx}^2 + \sigma_{yy}^2 + \sigma_{zz}^2 - 2\upsilon(\sigma_{xx}\sigma_{yy} + \sigma_{yy}\sigma_{zz} + \sigma_{zz}\sigma_{xx})\right]. \\ &\quad + \frac{1}{2G}\left(\tau_{xy}^2 + \tau_{yz}^2 + \tau_{xz}^2\right) \end{aligned} \tag{6.17}$$

6.6   Analysis for Combined Loading Ύ

6.6.1   Stresses Acting on an Arbitrarily Oriented Plane

At any point within a body, the magnitude and direction of the resultant stress σ_F depends on the orientation of the plane passing through the point. Thus an infinite number of resultant-stress vectors can be used to represent the resultant stress at each point, since an infinite number of planes can be passed through the point of interest. If the nine Cartesian stresses (in the general case) at the point of interest P are known, the resultant stress acting on an arbitrary plane through point P can be determined.

Consider an oblique plane Q passing through point P (now at the origin of coordinates), shown in Figure 6-13 (the plane is shown slightly removed from point P for clarity).

The unit normal to plane Q is

$$\hat{\mathbf{n}} = l\hat{\mathbf{i}} + m\hat{\mathbf{j}} + n\hat{\mathbf{k}} \tag{6.18}$$

where l, m, n, are direction cosines of unit vector $\hat{\mathbf{n}}$. l is equal to the cosine of the angle between $\hat{\mathbf{n}}$ and the x-direction and may also be written as $\cos(n, x) = l$. Similarly, m and n can written as $\cos(n,y)$ and $\cos(n,z)$, respectively. $\hat{\mathbf{i}}$, $\hat{\mathbf{j}}$, $\hat{\mathbf{k}}$ are unit normal vectors relative to x, y, and z coordinate axes, respectively.

The stress components acting on plane PCB perpendicular to the x-axis are σ_{xx}, τ_{xy}, and τ_{xz}. Similarly, stress components on plane PAC perpendicular to the y-axis are σ_{yy}, τ_{yz}, and τ_{yx}, and stresses on plane PAB perpendicular to the z-axis are σ_{zz}, τ_{zy}, and τ_{zx}. Among these nine stress components, the stresses acting in the x-direction are σ_{xx}, τ_{yx}, and τ_{zx}, which act on the planes PBC, PAC, and PAB, respectively. Let the area of plane ABC be A. Then area of plane PCB is equal to the projection of area A on the yz-plane and is Al and so on. Then the equilibrium equation in the x-direction (neglecting any body forces) is

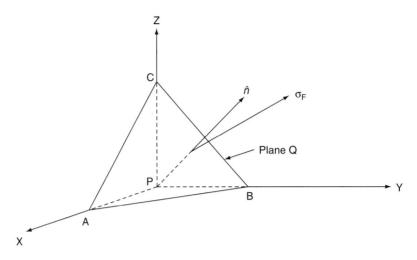

FIGURE 6-13 *Resultant stress σ_F acting on an arbitrarily oriented plane ABC.*

$$\sigma_{Fx}A - \sigma_{xx}Al - \tau_{yx}Am - \tau_{zx}An = 0$$
$$\Rightarrow \sigma_{Fx} = \sigma_{xx}l + \tau_{yx}m + \tau_{zx}n \tag{6.19a}$$

where σ_{Fx} is the component of the resultant stress in the x-direction.

Similarly, by considering equilibrium in the y and z directions we obtain

$$\sigma_{Fy} = \tau_{xy}l + \sigma_{yy}m + \tau_{zy}n \tag{6.19b}$$

$$\sigma_{Fz} = \tau_{xz}l + \tau_{yz}m + \sigma_{zz}n. \tag{6.19c}$$

The resultant stress vector $\boldsymbol{\sigma_F}$ can be written as

$$\boldsymbol{\sigma_F} = \sigma_{Fx}\hat{\mathbf{i}} + \sigma_{Fy}\hat{\mathbf{j}} + \sigma_{Fz}\hat{\mathbf{k}}. \tag{6.20}$$

The magnitude of the stress $\boldsymbol{\sigma_F}$ is

$$|\sigma_F| = \sqrt{\sigma_{Fx}^2 + \sigma_{Fy}^2 + \sigma_{Fz}^2}. \tag{6.21}$$

The projection of σ_F along the normal is the normal stress σ_{Fn} and is also equal to the sum of the projections of its components σ_{Fx}, σ_{Fy}, σ_{Fz}, along outer normal $\hat{\mathbf{n}}$. Hence,

$$\sigma_{Fn} = \sigma_{Fx}l + \sigma_{Fy}m + \sigma_{Fz}n. \tag{6.22}$$

Substituting for σ_{Fx}, σ_{Fy}, and σ_{Fz} from Eqs. (6.19) into Eq. (6.22), we obtain

$$\sigma_{Fn} = \sigma_{xx}l^2 + \sigma_{yy}m^2 + \sigma_{Fz}n^2 + 2\tau_{yz}mn + 2\tau_{xz}nl + 2\tau_{xy}lm. \tag{6.23}$$

The component of $\boldsymbol{\sigma_F}$ that is in the plane Q is the shear stress τ_{FS} and is given by

$$\tau_{FS} = \sqrt{\sigma_F^2 - \sigma_{Fn}^2}. \tag{6.24}$$

EXAMPLE 6-4:

A rectangular steel bar has a cross-section of $2\,\text{cm} \times 1\,\text{cm}$ and is subjected to a tensile load of $F = 5000\,\text{N}$, as shown in Figure E6-4. Determine the normal and shear stress on a plane whose outer normal $\hat{n}$ is inclined to the coordinate axes at 45° to the x-axis, 45° to the y-axis, and 90° to the z-axis:

Answer:

The direction cosines of the outer-normal to the plane are:

$$l = \cos 45° = \frac{1}{\sqrt{2}}, \; m = \cos 45° = \frac{1}{\sqrt{2}},$$

and $n = \cos 90° = 0$. The uniaxial stress $\sigma_{xy} = 5000\,\text{N}/(2 \times 1 \times 10^{-4}\text{m}^2) = 25\,\text{MPa}$.
From Eq. (6.19):

$$\sigma_{Fx} = \frac{1}{\sqrt{2}} \times 0 + \frac{1}{\sqrt{2}} \times 0 + 0 \times 0 = 0$$

$$\sigma_{Fy} = \frac{1}{\sqrt{2}} \times 0 + \frac{1}{\sqrt{2}} \times 25 + 0 \times 0 = \frac{25}{\sqrt{2}} = 17.7\,\text{MPa}.$$

$$\sigma_{Fz} = \frac{1}{\sqrt{2}} \times 0 + \frac{1}{\sqrt{2}} \times 0 + 0 \times 0 = 0$$

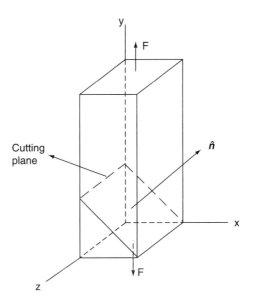

FIGURE E6-4

Continued

EXAMPLE 6-4: *Cont'd*

From Eq. (6.22) the normal stress on the specified plane is

$$\sigma_{Fn} = 0 + 17.7 \times \frac{1}{\sqrt{2}} + 0 = \frac{25}{2} = \underline{12.5\,\text{MPa}}$$

From Eq. (6.24) the shear stress on the specified plane is:

$$\tau_{Fn} = \sqrt{\left\{ 0 + \left(\frac{25}{\sqrt{2}}\right)^2 + 0 \right\} - (12.5)^2} = \underline{12.5\,\text{MPa}}.$$

6.6.2 Transformation of Stresses

For a strength design, at some point we will wish to compare a maximum stress component (or combination of maximum stress components) to failure strength. However, the coordinate system used initially for the stress analysis is often chosen for convenience, and we have no guarantee that it will lead to the 'maximum' stress components. In this section we develop the capability to determine stress components in any coordinate system given stress components in any other coordinate system. This is called the *transformation of stress*. Keep in mind that the stress at a point is invariant, but the components of stress depend on the coordinate system chosen at that point.

The six stress components used in the previous sections were obtained with reference to an *xyz* Cartesian coordinate system. If the coordinate system *xyz* is rotated through a certain angle about the same origin O, to a new transformed coordinate axis system *x'y'z'*, as shown in Figure 6-14, the six stress components in the *x'y'z'* coordinate system can be designated as $\sigma_{x'x'}$, $\sigma_{y'y'}$, $\sigma_{z'z'}$, $\tau_{x'y'}$, $\tau_{x'z'}$, and $\tau_{y'z'}$.

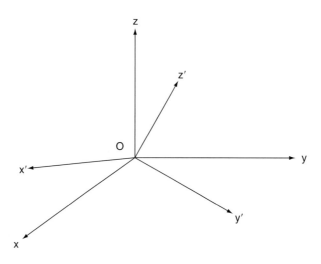

FIGURE 6-14 *Two colocated rectangular Cartesian coordinate systems.*

TABLE 6-1 Direction cosines of angles between various axis of coordinate systems xyz and $x'y'z'$ shown in Figure 6-14

	x	y	z
x'	l_1	l_2	l_3
y'	m_1	m_2	m_3
z'	n_1	n_2	n_3

The *direction cosines* of the angles between the coordinate axes xyz and the coordinate axes $x'y'z'$ are listed in Table 6-1. Recall that l_1 is the cosine of the angle between axis x and axis x', and so on.

Consider an arbitrary oblique plane abc (Figure 6-15) in the coordinate system with the outer normal $\hat{n}$. When the coordinate axes are rotated about the origin to a new orientation $x'y'z'$, consider the oblique plane a'b'c' (Figure 6-15). These planes are similar to the plane ABC in Figure 6-13, and that face oac in the plane xz is perpendicular to axis y, and plane oa'c' in the plane $x'z'$ is perpendicular to axis y', and so on. Also that stresses σ_{yy}, $\tau_{yz}(=\tau_{zy})$ and $\tau_{xy}(=\tau_{yx})$ act on the plane oac, while stresses $\sigma_{y'y'}$, $\tau_{x'y'}$ and $\tau_{y'z'}$ act on the plane oa'c', and so on. The resultant stress on plane oa'c' with outer normal coinciding with the y'-axis when projected on the y'-axis results in the normal stress $\sigma_{y'y'}$ and then projected on the n'-axes to obtain $\tau_{y'x'}$ and projected on the z'-axes to obtain $\tau_{y'z'}$. Hence by using Eq. (6.23), the transformed normal stresses can be written as

$$\sigma_{x'x'} = l_1^2\sigma_{xx} + l_2^2\sigma_{yy} + l_3^2\sigma_{zz} + 2l_1l_2\tau_{xy} + 2l_2l_3\tau_{yz} + 2l_3l_1\tau_{zx}$$
$$\sigma_{y'y'} = m_1^2\sigma_{xx} + m_2^2\sigma_{yy} + m_3^2\sigma_{zz} + 2m_1m_2\tau_{xy} + 2m_2m_3\tau_{yz} + 2m_3m_1\tau_{zx} \qquad (6.25)$$
$$\sigma_{z'z'} = n_1^2\sigma_{xx} + n_2^2\sigma_{yy} + n_3^2\sigma_{zz} + 2n_1n_2\tau_{xy} + 2n_2n_3\tau_{yz} + 2n_3n_1\tau_{zx}.$$

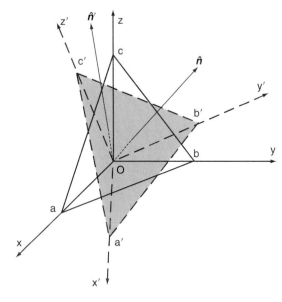

FIGURE 6-15 *The oblique plane abc shown in both the original and rotated coordinate systems.*

By using the results from Problem P6-3, the transformed shear stresses can be written as

$$\tau_{x'y'} = l_1 m_1 \sigma_{xx} + l_2 m_2 \sigma_{yy} + l_3 m_3 \sigma_{zz} + (l_1 m_2 + l_2 m_1)\tau_{xy} + (l_2 m_3 + l_3 m_2)\tau_{yz} + (l_3 m_1 + l_1 m_3)\tau_{zx}$$

$$\tau_{y'z'} = m_1 n_1 \sigma_{xx} + m_2 n_2 \sigma_{yy} + m_3 n_3 \sigma_{zz} + (m_1 n_2 + m_2 n_1)\tau_{xy} + (m_2 n_3 + m_3 n_2)\tau_{yz} +$$
$$(m_3 n_1 + m_1 n_3)\tau_{zx} \tag{6.26}$$

$$\tau_{z'x'} = n_1 l_1 \sigma_{xx} + n_2 l_2 \sigma_{yy} + n_3 l_3 \sigma_{zz} + (n_1 l_2 + n_2 l_1)\tau_{xy} + (n_2 l_3 + n_3 l_2)\tau_{yz} + (n_3 l_1 + n_1 l_3)\tau_{zx}.$$

Equations (6.25) and (6.26) determine the stress components relative to axes $x'y'z'$ in terms of the stress components relative to the axes xyz.

6.6.3 Principal Stresses

It was mentioned earlier that the resultant stress σ_F at a point P depends on the orientation of the cutting plane Q on which the stress acts. If the plane Q is such that the outer normal to the plane coincides with the resultant stress σ_F, the shear stress vanishes on the plane Q. In such a case the resultant stress σ_F is the normal stress σ_P on the plane (Figure 6-16). It can be shown that for any point P in a member there exist three mutually perpendicular planes at the point P on which the shear stress vanishes. (For two-dimensional stress states, it will be evident from Mohr's circle in what follows that there are two mutually perpendicular planes.) The normal stress acting on these particular planes is called a *principal stress*, the plane is known as the *principal plane*, and the outer normal to the principal plane is called the *principal direction* or *principal axis*.

The components of σ_F along the coordinate axes can be written as:

$$\sigma_{Fx} = \sigma_p \cos(n, x) = \sigma_p l$$
$$\sigma_{Fy} = \sigma_p \cos(n, y) = \sigma_p m \tag{6.27}$$
$$\sigma_{Fz} = \sigma_p \cos(n, z) = \sigma_p n.$$

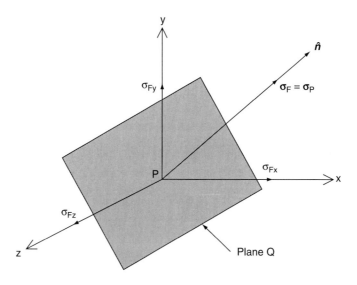

FIGURE 6-16 σ_F *shown coincident with the outer normal* $\hat{n}$*. Thus the shear stress vanishes and* σ_p *becomes equal in magnitude to* σ_F*.*

Substituting Eqs. (6.19) into (6.27) and upon rearranging the resulting expressions one obtains:

$$(\sigma_{xx} - \sigma_p)l + \tau_{yx}m + \tau_{zx}n = 0$$
$$\tau_{xy}l + (\sigma_{yy} - \sigma_p)m + \tau_{zy}n = 0 \qquad (6.28)$$
$$\tau_{xz}l + \tau_{yz}m + (\sigma_{zz} - \sigma_p)n = 0.$$

Since Eqs. (6.28) are linear homogeneous equations in (l, m, n) and since the trivial solution $l = m = n = 0$ is impossible since $l^2 + m^2 + n^2 = 1$ (see Example 6-5), nontrivial solutions for the direction cosines (l, m, n) of the principal plane will exist if and only if the determinant of the coefficients of (l, m, n) vanishes identically. Thus, we have

$$\begin{vmatrix} \sigma_{xx} - \sigma_p & \tau_{yx} & \tau_{zx} \\ \tau_{xy} & \sigma_{yy} - \sigma_p & \tau_{zy} \\ \tau_{xz} & \tau_{yz} & \sigma_{zz} - \sigma_p \end{vmatrix} = 0. \qquad (6.29)$$

On expanding the determinant in Eq. (6.29), one obtains a cubic characteristic equation as

$$\sigma^3 - I_1\sigma^2 + I_2\sigma - I_3 = 0 \qquad (6.30)$$

where the first, second, and third *stress invariants*, I_1, I_2, and I_3, respectively, are

$$I_1 = \sigma_{xx} + \sigma_{yy} + \sigma_{zz} = \text{constant} \qquad (6.31a)$$

$$I_2 = \begin{vmatrix} \sigma_{xx} & \tau_{xy} \\ \tau_{xy} & \sigma_{yy} \end{vmatrix} + \begin{vmatrix} \sigma_{xx} & \tau_{xz} \\ \tau_{xz} & \sigma_{zz} \end{vmatrix} + \begin{vmatrix} \sigma_{yy} & \tau_{yz} \\ \tau_{yz} & \sigma_{zz} \end{vmatrix}$$

$$= \sigma_{xx}\sigma_{yy} + \sigma_{yy}\sigma_{zz} + \sigma_{zz}\sigma_{xx} - \tau_{xy}^2 - \tau_{yz}^2 - \tau_{zx}^2 = \text{constant} \qquad (6.31b)$$

$$I_3 = \begin{vmatrix} \sigma_{xx} & \tau_{xy} & \tau_{xz} \\ \tau_{xy} & \sigma_{yy} & \tau_{yz} \\ \tau_{xz} & \tau_{yz} & \sigma_{zz} \end{vmatrix}$$

$$= \sigma_{xx}\sigma_{yy}\sigma_{zz} + 2\tau_{xy}\tau_{yz}\tau_{zx} - \sigma_{xx}\tau_{yz}^2 - \sigma_{yy}\tau_{zx}^2 - \sigma_{zz}\tau_{xy}^2$$

$$= \text{constant}. \qquad (6.31c)$$

The three roots of the cubic equation are the three *principal stresses* at point P in Figure 6-16. Generally, the three principal stresses are represented by σ_1, σ_2, and σ_3. Typically, these three principal stresses are algebraically ordered as $\sigma_1 > \sigma_2 > \sigma_3$ and implies that σ_1 has the largest algebraic value and σ_3 has the smallest algebraic value. Remember that in this ordering process, tensile stresses are considered as positive and compression stresses are considered negative.

The magnitude and directions of σ_1, σ_2, and σ_3 for any given equilibrium system of forces applied to a body, are uniquely determined and are independent of the orientation of the Cartesian coordinate axes. Thus the coefficients in Eq. (6.27) are called *stress invariants* (or constant) and must have the same magnitude for all orientations of the coordinate axes. Hence σ_1, the largest principal stress, is the maximum normal stress that can occur on any plane passing through the point.

After the principal stresses σ_1, σ_2, and σ_3 are determined, they can be substituted individually into Eq. (6.28) to give three sets of simultaneous equations, any two of which together with the expression $l^2 + m^2 + n^2 = 1$ can be solved to obtain three sets of direction cosines defining the three principal planes.

EXAMPLE 6-5:

Let **N** be a vector in the coordinate axes X, Y, Z as shown in Figure E6-5.

If l, m, n are the *direction cosines* of the vector $\hat{N}$, that is, l is the cosine of the angle between the x-axis and **N**, and so on, show that $l^2 + m^2 + n^2 = 1$.

Answer:

$$x^2 + y^2 + z^2 = R^2$$

$$\Rightarrow \frac{x^2}{R^2} + \frac{y^2}{R^2} + \frac{z^2}{R^2} = 1. \tag{a}$$

x/R is the cosine of the angle between $\hat{N}$ and the X-axis, which has been represented by the direction cosine l. Therefore

$$\frac{x^2}{R^2} = l^2. \tag{b}$$

Similarly

$$\frac{y^2}{R^2} = m^2, \quad \frac{z^2}{R^2} = n^2. \tag{c}$$

On substituting Eqs. (b) and (c) into (a) we obtain

$$l^2 + m^2 + n^2 = 1.$$

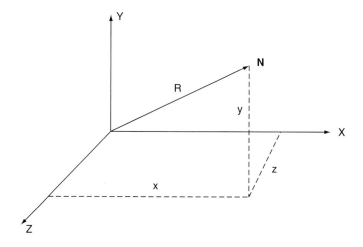

FIGURE E6-5

EXAMPLE 6-6:

The 1 in. diameter, L-shaped steel lever is loaded as shown in Figure E6-6. Determine the critical point where the stress is expected to be highest. Determine the Cartesian stress components and the principal stresses at that point.

   Answer: The point marked O is the critical point as the tensile bending stresses are highest here. This point is also subjected to torsion-induced shear stresses.

$$\sigma_{xx} = \frac{Mc}{I} = \frac{(10\,\text{lb} \times 14\,\text{in})\dfrac{1.0}{2}\,\text{in}}{\dfrac{\pi(1.0\,\text{in})^4}{64}}$$

$$= \underline{1426\,\text{lb/in}^2}$$

$$\tau_{xz} = \frac{Tr}{J} = \frac{(10\,\text{lb} \times 10\,\text{in})\dfrac{1.0}{2}\,\text{in}}{\pi\dfrac{1.0^4}{32}}$$

$$= \underline{509\,\text{lb/in}^2}.$$

All other stress components on a cubic element at point O are zero, that is, $\sigma_{yy} = \sigma_{zz} = \tau_{xy} = \tau_{yz} = 0$. (Note that $\tau_{xy} = \tau_{yx} = 0$ and $\tau_{yz} = \tau_{zy} = 0$.)

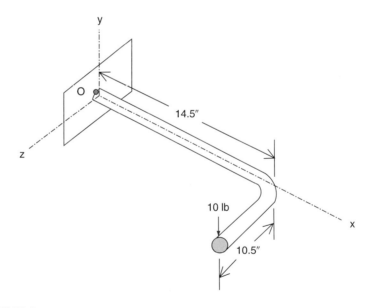

FIGURE E6-6

Continued

EXAMPLE 6-6: *Cont'd*

The principal stresses are determined from the cubic Eq. (6.30):

$$\sigma^3 - \sigma^2(1426) + \sigma(-509)^2 = 0$$
$$\Rightarrow \sigma(\sigma^2 - 1426\sigma - 259081) = 0$$
$$\Rightarrow \sigma_1 = \underline{1589\,\text{lb}/\text{in}^2}$$
$$\sigma_2 = \underline{0}$$
$$\sigma_3 = \underline{-163\,\text{lb}/\text{in}^2}.$$

Note: σ_1, σ_2, σ_3 are ordered such that $\sigma_1 > \sigma_2 > \sigma_3$. Note also that the original bending stress found was 'not' the maximum stress.

The maximum shear stress is given by

$$\tau_{\max} = \frac{1}{2}(\sigma_1 - \sigma_3) = \frac{1}{2}(1589 + 163) = 876\,\text{lb}/\text{in}^2.$$

EXAMPLE 6-7:

At a point in a machine component the stress state is

$$\begin{bmatrix} 1 & 2 & 1 \\ 2 & -2 & -3 \\ 1 & -3 & 4 \end{bmatrix}$$

All stresses are in MPa. Find the principal stresses.

Answer: From the given stress tensor, it is implied that $\sigma_{xx} = 1$, $\sigma_{yy} = -2$, $\sigma_{zz} = 4$, $\tau_{xy} = \tau_{yx} = 2$, $\tau_{zy} = \tau_{yz} = -3$, $\tau_{xz} = \tau_{zx} = 1$.

The three stress invariants are determined using Eq. (6.31), as

$$I_1 = 1 - 2 + 4 = \underline{3\,\text{MPa}}$$
$$I_2 = (1 \times -2) + (-2 \times 4) + (4 \times 1) - 2^2 - (-3)^2 - (1)^2 = \underline{-20\,\text{MPa}^2}$$
$$I_3 = (1 \times -2 \times 4) + 2(2 \times -3 \times 1) - 1(-3)^2 - (-2)(1)^2 - 4(2)^2 = \underline{-43\,\text{MPa}^3}$$

Using Eq. (6.30) we obtain the cubic equation

$$\sigma^3 - 3\sigma^2 - 20\sigma + 43 = 0.$$

The three roots of the cubic equation are the three principal stresses, which when algebraically ordered $(\sigma_1 > \sigma_2 > \sigma_3)$ are $\sigma_1 = \underline{5.25\,\text{MPa}}$, $\sigma_2 = \underline{1.95\,\text{MPa}}$, $\sigma_3 = \underline{-4.2\,\text{MPa}}$.

EXAMPLE 6-8:

The following matrix represents the state of stress at a point. Determine the principal stresses and their associated directions. All stresses are in MPa.

$$\begin{bmatrix} 1 & 2 & 1 \\ 2 & 1 & 1 \\ 1 & 1 & 1 \end{bmatrix}$$

Answer: Using Eq. (6.31) the stress invariants can be determined as

$$I_1 = \underline{3\,\text{MPa}}$$
$$I_2 = \underline{-3\,\text{MPa}^2}$$
$$I_3 = \underline{-1\,\text{MPa}^3}.$$

Therefore the cubic equation is

$$\sigma^3 - 3\sigma^2 - 3\sigma + 1 = 0.$$

The three roots of this equation are -1, $2 + \sqrt{3}$, and $2 - \sqrt{3}$. Thus the three principal stresses are

$$\sigma_1 = \underline{2 + \sqrt{3}\,\text{MPa}}, \quad \sigma_2 = \underline{2 - \sqrt{3}\,\text{MPa}}, \quad \sigma_3 = \underline{-1\,\text{MPa}}.$$

To obtain directions of principal stress the following procedure should be used:
For $\sigma_1 = 2 + \sqrt{3}\,\text{MPa}$, using Eq. (6.28) we obtain (where σ_p is σ_1)

$$(-1 - \sqrt{3})l + 2m + n = 0$$
$$2l + (-1 - \sqrt{3}) + n = 0\ . \qquad\qquad \text{(a)}$$
$$l + m + (-1 - \sqrt{3}) = 0$$

Using any two expressions from (a), along with the equation $l^2 + m^2 + n^2 = 1$, and solving simultaneously for l, m, and n, we obtain

$$l = \frac{1}{2}\sqrt{1 + \frac{1}{\sqrt{3}}} \Rightarrow \theta_x = \cos^{-1}(0.627) = 51.1°$$

$$m = \frac{1}{2}\sqrt{1 + \frac{1}{\sqrt{3}}} \Rightarrow \theta_y = \cos^{-1}(0.627) = 51.1°.$$

$$n = \sqrt{3 + \sqrt{3}} \Rightarrow \theta_z = \cos^{-1}(0.46) = 62.6°$$

Similarly, using $\sigma_2 = 2 - \sqrt{3}\,\text{MPa}$, it is left for the student to show that the angles are:

$$\theta_x = 71°, \quad \theta_y = 71°, \quad \theta_z = 27°.$$

Using $\sigma_3 = -1\,\text{MPa}$, show that the angles are:

$$\theta_x = 45°, \quad \theta_y = 45°, \quad \theta_z = 90°.$$

6.6.4 Principal Shear Stress

In Section 6.6.3, we saw that for a particular orientation of a plane through a point, the shear stresses vanish, while the normal stress achieves an extremum value. It could be intuitively imagined that on any other plane through the point both normal and shear stress components will exist. Also it could be imagined that on at least one plane, the shear stress will attain an extremum value. These extreme values of shear stress are called *principal shear stresses* and the planes on which they act are called *planes of principal shear*.

To determine the magnitudes of the principal shear stresses and the direction of the planes on which they act, consider an infinitesimal element oriented with its faces parallel to the principal planes (planes on which principal normal stresses act), as shown in Figure 6-17, cut by an arbitrary plane ABC. Notice that the reference axes are designated as 1, 2, 3 (instead of x, y, z) as they coincide with the principal directions. In this orientation all shear stress components on the coordinate planes are zero and only the principal normal stresses σ_1, σ_2, and σ_3 act. Based on the development in Section 6.6.1, the resultant stress on the oblique plane ABC can be written using Eq. (6.21) as

$$\sigma_F^2 = \sigma_{Fx}^2 + \sigma_{Fy}^2 + \sigma_{Fz}^2. \tag{6.32}$$

Using Eqs. (6.19), Eq. (6.32) can be written as (noting that all shear stresses are zero and $\sigma_{xx} = \sigma_1$, $\sigma_{yy} = \sigma_2$, and $\sigma_{zz} = \sigma_3$)

$$\sigma_F^2 = \sigma_1^2 l^2 + \sigma_2^2 m^2 + \sigma_3^2 n^2. \tag{6.33}$$

From Eq. (6.23),

$$\sigma_{Fn} = \sigma_1 l^2 + \sigma_2 m^2 + \sigma_3 n^2. \tag{6.34}$$

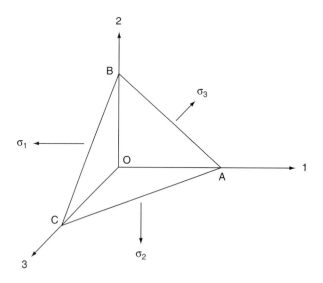

FIGURE 6-17 *Principal planes OAB, OBC, and OAC.*

Substituting Eqs. (6.33) and (6.34) into Eq. (6.24), the shear stress in the oblique plane ABC is obtained as

$$\tau_{Fs}^2 = \sigma_1^2 l^2 + \sigma_2^2 m^2 + \sigma_3^2 n^2 - (\sigma_1 l^2 + \sigma_2 m^2 + \sigma_3 n^2)^2. \tag{6.35}$$

This is the expression for the shear stress on any plane through the point in terms of the principal normal stresses and the direction cosines l, m, and n for the plane on which σ_F acts, if the coordinate axes are referenced to the principal axis.

The planes on which the *maximum and minimum shearing stresses* occur can be obtained from Eq. (6.35), by differentiating with respect to the direction cosines l, m, n. One of the direction cosines, l for example, in Eq. (6.35) can be eliminated by solving the expression

$$l^2 + m^2 + n^2 = 1 \tag{6.36}$$

for l and substituting into Eq. (6.35). Thus, on substituting $l^2 = 1 - (m^2 + n^2)$ in Eq. (6.35) we obtain

$$\tau_{Fs}^2 = \left[\sigma_1^2 - (\sigma_1^2 - \sigma_2^2)m^2 - (\sigma_1^2 - \sigma_3^2)n^2\right] - \left[\sigma_1 - (\sigma_1 - \sigma_2)m^2 - (\sigma_1 - \sigma_3)n^2\right]^2. \tag{6.37}$$

Taking the partial derivative of the Eq. (6.37) with respect to m and n yields the expressions

$$\frac{\partial \tau_{Fs}}{\partial m} = m\left[\frac{1}{2}(\sigma_1 - \sigma_2) - (\sigma_1 - \sigma_2)m^2 - (\sigma_1 - \sigma_3)n^2\right] = 0 \tag{6.38}$$

$$\frac{\partial \tau_{Fs}}{\partial n} = n\left[\frac{1}{2}(\sigma_1 - \sigma_3) - (\sigma_1 - \sigma_2)m^2 - (\sigma_1 - \sigma_3)n^2\right] = 0. \tag{6.39}$$

One solution of these two equations is $m = n = 0$, which together with Eq. (6.36) implies $l = \pm 1$—this represents the principal normal plane in which the shear stress is known to be zero. To obtain the plane on which the shear stress is maximum, or a minimum, first consider $m = 0$; then from Eq. (6.39) we obtain $n = \pm\frac{1}{\sqrt{2}}$, and from Eq. (6.36) $l = \pm\frac{1}{\sqrt{2}}$. Also if $n = 0$, then from Eq. (6.38) $m = \pm\frac{1}{\sqrt{2}}$. For these two cases from Eq. (6.36) $l = \pm\frac{1}{\sqrt{2}}$. Substituting the values of $m = \pm\frac{1}{\sqrt{2}}$ and $n = \pm\frac{1}{\sqrt{2}}$ into Eq. (6.37) yields

$$\tau_{Fs} = \frac{1}{2}(\sigma_1 - \sigma_3) = \tau_1. \tag{6.40a}$$

Repeating the procedure by eliminating m and n, in turn, from Eq. (6.35) yields other values for the direction cosines that make the shearing stresses maximum or minimum. This analysis will yield two more results:

$$\tau_{Fs} = \frac{1}{2}(\sigma_2 - \sigma_3) = \tau_2 \tag{6.40b}$$

$$\tau_{Fs} = \frac{1}{2}(\sigma_1 - \sigma_2) = \tau_3. \tag{6.40c}$$

Of these three possible results, the largest magnitude will be obtained from Eq. (6.40a), if the principal stresses are ordered such that $\sigma_1 > \sigma_2 > \sigma_3$. Thus the maximum shear stress is given by

$$\tau_{\max} = \frac{1}{2}(\sigma_{\max} - \sigma_{\min}) = \frac{1}{2}(\sigma_1 - \sigma_3). \tag{6.41}$$

From the foregoing derivation, it can be inferred that the maximum shearing stress acts in a plane that bisects the angle between the maximum and minimum normal stress vectors (note that the maximum and minimum normal stress vectors are mutually perpendicular). The three shearing stresses τ_1, τ_2, and τ_3 are called *principal shear stresses* and the planes on which they act are called *planes of principal shear*.

6.6.5 Mohr's Circle for Plane Stress Problems

Consider a transformation from the (x, y, z) coordinate axes to (x', y', z') coordinate axes with the requirement that the z-axis and the z'-axis remain coincident. Let the angle between the x and x' axes and the y and y' axes be θ after the transformation, as shown in Figure 6-18.

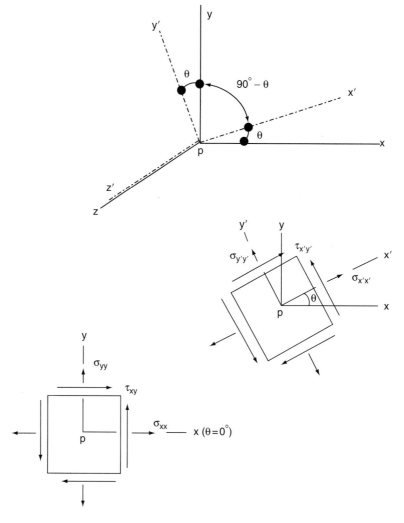

FIGURE 6-18 *Two-dimensional stress transformation.*

TABLE 6-2

	x	y	z
x'	$l_1 = \cos\theta$	$m_1 = \cos(90° - \theta)$	$n_1 = \cos 90° = 0$
y'	$l_2 = \cos(90° + \theta) = -\sin\theta$	$m_2 = \cos\theta$	$n_2 = \cos 90°$
z'	$l_3 = \cos 90° = 0$	$m_3 = \cos 90°$	$n_3 = \cos 0° = 1$

Thus, effectively Figure 6-18 represents a transformation in two dimensions or in a plane, namely, the x–y plane. The direction cosines between the axes under the above-mentioned transformation are listed in Table 6-2.

Using the direction cosines in Table 6-2 and Eqs. (6.25) and (6.26) the transformed stress components in the (x', y', z') coordinate axes can be written as

$$\sigma_{x'x} = \sigma_{xx}\cos^2\theta + \sigma_{yy}\sin^2\theta + 2\tau_{xy}\sin\theta\cos\theta$$

$$\sigma_{y'y'} = \sigma_{xx}\sin^2\theta + \sigma_{yy}\cos^2\theta - 2\tau_{xy}\sin\theta\cos\theta$$

$$\sigma_{z'z'} = \sigma_{zz} = 0, \quad \tau_{z'x'} = \tau_{zx} = 0, \quad \tau_{y'z'} = \tau_{yz} = 0 \tag{6.42}$$

$$\tau_{x'y'} = -(\sigma_{xx} - \sigma_{yy})\sin\theta\cos\theta + \tau_{xy}(\cos^2\theta - \sin^2\theta)$$

(The stress components depending on the z-direction are zero, and we are only considering a 2-D or plane stress condition.)

Using the trigonometric formulas

$$2\sin\theta\cos\theta = \sin 2\theta$$

$$\cos^2\theta - \sin^2\theta = \cos 2\theta$$

Eq. (6.42) can be written as

$$\sigma_{x'x'} = \frac{1}{2}(\sigma_{xx} + \sigma_{yy}) + \frac{1}{2}(\sigma_{xx} - \sigma_{yy})\cos 2\theta + \tau_{xy}\sin 2\theta$$

$$\sigma_{y'y'} = \frac{1}{2}(\sigma_{xx} + \sigma_{yy}) - \frac{1}{2}(\sigma_{xx} - \sigma_{yy})\cos 2\theta - \tau_{xy}\sin 2\theta \tag{6.43}$$

$$\tau_{x'y'} = -\frac{1}{2}(\sigma_{xx} - \sigma_{yy})\sin 2\theta + \tau_{xy}\cos 2\theta.$$

The first and last of the equations in (6.43) can be written as

$$\left\{\sigma_{x'x'} - \frac{1}{2}(\sigma_{xx} + \sigma_{yy})\right\}^2 = \left\{\frac{1}{2}(\sigma_{xx} - \sigma_{yy})\cos 2\theta + \tau_{xy}\sin 2\theta\right\}^2 \tag{a}$$

$$\tau_{x'y'}^2 = \left\{-\frac{1}{2}(\sigma_{xx} - \sigma_{yy})\sin 2\theta + \tau_{xy}\cos 2\theta\right\}^2. \tag{b}$$

Adding Eqs. (a) and (b) results in the expression

$$\left\{ \sigma_{x'x'} - \frac{1}{2}(\sigma_{xx} + \sigma_{yy}) \right\}^2 + \{\tau_{x'y'}\}^2$$

$$= \frac{1}{4}(\sigma_{xx} - \sigma_{yy})^2(\cos^2 2\theta + \sin^2 2\theta) + \frac{1}{2}(\sigma_{xx} - \sigma_{yy})\tau_{xy}\sin 2\theta \cos 2\theta$$

$$- \frac{1}{2}(\sigma_{xx} - \sigma_{yy})\tau_{xy}\sin 2\theta \cos 2\theta + \tau_{xy}^2(\sin^2 2\theta + \cos^2 2\theta)$$

$$\Rightarrow \left\{ \sigma_{x'x'} - \frac{1}{2}(\sigma_{xx} + \sigma_{yy}) \right\}^2 + \{\tau_{x'y'} - 0\}^2 = \frac{1}{4}(\sigma_{xx} - \sigma_{yy})^2 + \tau_{xy}^2.$$

(6.44)

Equation (6.44) represents the equation of a circle [i.e., of the form $(x - a)^2 + (y - b)^2 = r^2$] in the $(\sigma_{x'x'}, \tau_{x'y'})$ plane, whose center C has the coordinates $\left[\frac{1}{2}(\sigma_{xx} + \sigma_{yy}), 0\right]$ and has a radius of $r = \left\{ \frac{1}{4}(\sigma_{xx} + \sigma_{yy})^2 + \tau_{xy}^2 \right\}^{1/2}$. A graphical representation of this circle, popularly knows as *Mohr's circle* in honor of the German engineer Otto Mohr, who first employed it to study plane stress problems, is shown in Figure 6-19.

In the Mohr's circle diagram, the normal stress components σ are plotted on the horizontal axis while the shear stress components τ are plotted on the vertical axis. Tensile normal stresses are plotted along the positive σ-axis, that is, to the right of the τ-axis, while compressive normal stresses are plotted to the left of the τ-axis. Shear stress components that produce clockwise (cw) rotation of an infinitesimal element around the point P under consideration, as shown in Figure 6-20a, are plotted above the σ-axis. Shear stress components that produce a counterclockwise rotation (ccw) (Figure 6-20b) are plotted below the σ-axis.

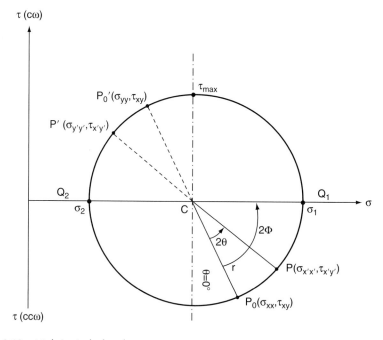

FIGURE 6-19 *Mohr's circle for plane stress.*

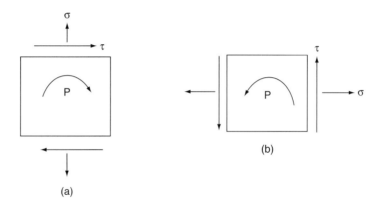

FIGURE 6-20 *Shear stress states: (a) clockwise shear, and (b) counterclockwise shear.*

From Figure 6-18, for the case $\theta = 0°$, the first and third of Eqs. (6.43) give

$$\sigma_{x'x'} = \sigma_{xx} \text{ and } \tau_{x'y'} = \tau_{xy}$$

which are the coordinates of point P_0 in Figure 6-19 (note from figure 6.18 that for $\theta = 0°$ the face with outer normal along the x-axis has a tensile stress σ_{xx} and a counterclockwise shear τ_{xy}). The coordinates of the center C are

$$\frac{\sigma_{xx} + \sigma_{yy}}{2} = \frac{\sigma_{x'x'} + \sigma_{y'y'}}{2}$$

which can be shown by adding the first and second of Eqs. (6.43). Thus using C as the center and length CP_0 as the radius, the Mohr's circle can be plotted. Another point $P'_0(\sigma_{yy}, \sigma_{xy})$ can be located, though it is not needed for drawing the Mohr's circle. This point represents the stress on the faces with outer normal along the y-axis or $\theta = \pi/2$. The points P and P' represent the stresses on the element with faces along the x'-axis and y'-axis as shown in Figure 6-18. In other words, stress components associated with each plane through a point are represented by a point on the Mohr's circle.

The principal stresses are located at points Q_1 and Q_2. By definition, in the principal planes the shear stress is zero. Thus from the last of Eq. (6.43) when $\tau_{x'y'} = 0$, we obtain

$$\tan 2\theta = \frac{2\tau_{xy}}{\sigma_{xx} - \sigma_{yy}}. \tag{6.45}$$

Solution of Eq. (6.45) will yield two values of θ, say $\theta = \Phi$ and $\Phi + \pi/2$, which are shown as 2Φ and $2\Phi + \pi$ on the Mohr's circle. The magnitude of the principal stresses can be obtained from the Mohr's circle as

$$\sigma_1 = \frac{\sigma_{xx} + \sigma_{yy}}{2} + \sqrt{\frac{1}{4}(\sigma_{xx} - \sigma_{yy})^2 + \tau_{xy}^2}$$
$$\sigma_2 = \frac{\sigma_{xx} + \sigma_{yy}}{2} - \sqrt{\frac{1}{4}(\sigma_{xx} - \sigma_{yy})^2 + \tau_{xy}^2} \tag{6.46}$$

Note that $\sigma_3 = 0$ as a plane stress case is being considered. Equation (6.46) can also be obtained from Eq. (6.30) for the plane stress case and is left as an exercise.

EXAMPLE 6-9:

For the state of plane stress shown in Figure E6-9A, determine (a) the principal planes, (b) the principal stresses, and (c) the maximum shearing stress and the corresponding normal stress using (i) the Mohr's circle method, and (ii) numerically without using the Mohr's circle diagram.

Answer:

(i) Mohr's Circle Method:

The stress state on the face with outer normal along the x-axis consists of a tensile normal stress and a shear stress producing counterclockwise rotation of the element. Thus this stress state is plotted to the right of the τ-axis (as tensile normal stress) and below the σ-axis (as ccw rotation of element due to shear) at point P_0. Similarly point P_0' is plotted to represent the stress state on the face with outer normal along the y-axis in Figure E6-9B. On drawing the line P_0P_0', it intersects the σ-axis at C, which is the center of the Mohr's circle. The abscissa of the center C is

$$\frac{\sigma_{xx} + \sigma_{yy}}{2} = \frac{80 + (-40)}{2} = 20 \text{ MPa}.$$

The radius of the circle is CP_0 and is given by

$$r = CP_0 = \sqrt{(CA)^2 + (P_0A)^2} = \sqrt{(80 - 20)^2 + (25)^2} = 65 \text{ MPa}.$$

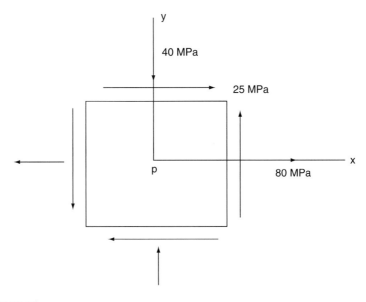

FIGURE E6-9A

EXAMPLE 6-9: *Cont'd*

(a,b) The principal stresses are represented by the points Q_1 and Q_2, where Q_1 represents the maximum principal stress σ_1 and Q_2 represents the minimum principal stress σ_2. Their magnitudes are given by

$$|\sigma_1| = OQ_1 = OC + CQ_1 = OC + CP_0$$
$$= 20 + 65 = 85 \text{ MPa}$$
$$|\sigma_2| = OQ_2 = OC - CQ_2 = OC - CQ_1 = OC + CP_0'$$
$$= 20 - 65 = \underline{-45 \text{ MPa}}$$

The angle Q_1CP_0 represents 2θ [see Figures E6-9B and E6-9C (b)], and is obtained as

$$\tan 2\theta = \frac{AP_0}{CA} = \frac{25}{60}$$
$$\Rightarrow 2\theta = 22.6° \Rightarrow \theta = 11.3°.$$

Thus in the Mohr's circle, line CP_0 must be rotated counterclockwise through an angle 22.6° to bring CP_0 into CQ_1. In the actual material the element should also be rotated counterclockwise through half the angle $\theta = 11.3°$ to obtain the principal stress state, as shown in Figure E6-9C(a).

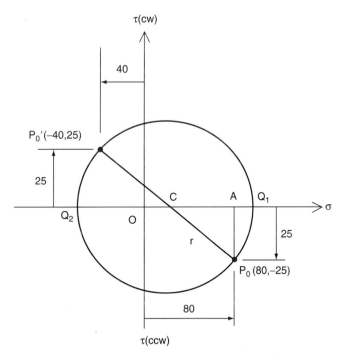

FIGURE E6-9B

Continued

EXAMPLE 6-9: *Cont'd*

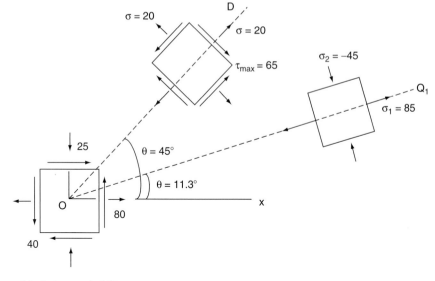

(a) all stresses in MPa

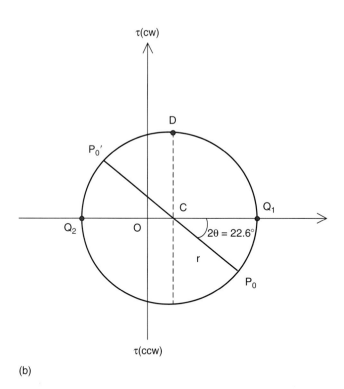

(b)

FIGURE E6-9C *(a) Orientation of principal normal stress and shear stress elements. (b) Magnitude of principal normal stresses and maximum shear stress.*

EXAMPLE 6-9: *Cont'd*

(c) Point D in Fig. E6-9C(b) represents the maximum shear stress state. CQ_1 can be rotated counterclockwise through $90°$ to bring CQ_1 into CD. The magnitude of the maximum shear stress is equal to the radius r of the Mohr's circle, that is, $\tau_{max} = 65$ MPa. In the real material the stress element is rotated counterclockwise through an angle of $\theta + 90°/2 = 11.3 + 45° = 56.3°$, to bring the axis Ox into the axis OD and the orientation of the element and the associated stress represent point D on the Mohr's circle. Since point D is located above the σ-axis (that is the $\tau(c\omega)$ axis), the shearing stress exerted on the faces of the element perpendicular to OD in Figure E6-9C(a) must be directed so that they will tend to rotate the element clockwise. The normal stress is the same as that at C, which is 20 MPa.

(ii) Numerical Solution

(a) Equation (6.45) can be used to obtain the orientation of the principal plane as

$$\tan 2\theta = \frac{2\tau_{xy}}{\sigma_{xx} - \sigma_{yy}} = \frac{2(+25)}{(+80) - (-40)} = \frac{50}{120}$$

$$\Rightarrow 2\theta = 22.6° \text{ and } 180° + 22.6° = 202.6°$$

$$\Rightarrow \theta = \underline{11.3°} \text{ and } 90° + 11.3° = \underline{101.3°}.$$

Thus the element is oriented at an angle of $11.3°$ (measured ccw) to the element containing the applied plane stress state. The second value of $\theta = 101.3°$ can also be used to define the element's σ orientation (Figure E6-9D). A plane that contains the face on which σ_1 acts perpendicularly is the principal plane, and similarly a plane that contains the face on which σ_2 acts is the other principal plane.

(b) Equation (6.46) reproduced here can be used to obtain the magnitude of the principal stresses as

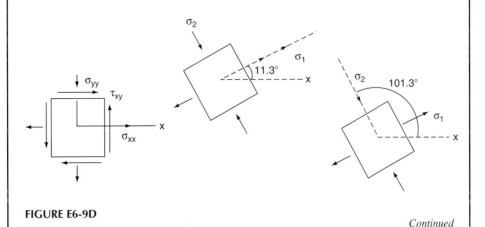

FIGURE E6-9D

Continued

EXAMPLE 6-9: *Cont'd*

$$\sigma_1 = \frac{\sigma_{xx} + \sigma_{yy}}{2} + \sqrt{\left(\frac{\sigma_{xx} - \sigma_{yy}}{2}\right)^2 + \tau_{xy}^2}$$

$$\sigma_2 = \frac{\sigma_{xx} + \sigma_{yy}}{2} - \sqrt{\left(\frac{\sigma_{xx} - \sigma_{yy}}{2}\right)^2 + \tau_{xy}^2}$$

$$\Rightarrow \sigma_1 = \frac{80 - 40}{2} + \sqrt{\left(\frac{80 - (-40)}{2}\right)^2 + 25^2} = \underline{85\,\text{MPa}}$$

$$\sigma_2 = \frac{80 - 40}{2} - \sqrt{\left(\frac{80 - (-40)}{2}\right)^2 + 25^2} = \underline{-45\,\text{MPa}}$$

(c) The maximum shear stress can be obtained analytically as follows:
Differentiating Eq. (6.45) with respect to $\theta = \theta_s$ and setting the results equal to zero, we obtain

$$\tan 2\theta_s = \frac{\left(\dfrac{\sigma_{xx} - \sigma_{yy}}{2}\right)}{\tau_{xy}}$$

This value of θ gives the orientation of the element corresponding to the maximum shear stress.

Solving the above equation for $\sin 2\theta_s$ and $\cos 2\theta_s$ and substituting into the last of Eq. (6.43) we obtain

$$\tau_{xy} = \tau_{max} = \left(\frac{\sigma_{yy} - \sigma_{xx}}{2}\right)\frac{\left(\dfrac{\sigma_{yy} - \sigma_{xx}}{2}\right)}{\sqrt{\left(\dfrac{\sigma_{yy} - \sigma_{xx}}{2}\right)^2 + \tau_{xy}^2}} + \tau_{xy}\frac{\tau_{xy}}{\sqrt{\left(\dfrac{\sigma_{yy} - \sigma_{xx}}{2}\right)^2 + \tau_{xy}^2}}$$

$$= \frac{\left(\dfrac{\sigma_{yy} - \sigma_{xx}}{2}\right)^2 + \tau_{xy}^2}{\sqrt{\left(\dfrac{\sigma_{yy} - \sigma_{xx}}{2}\right)^2 + \tau_{xy}^2}}$$

$$= \sqrt{\left(\dfrac{\sigma_{yy} - \sigma_{xx}}{2}\right)^2 + \tau_{xy}^2}$$

which is customarily written as

$$\tau_{max} = \sqrt{\left(\frac{\sigma_{xx} - \sigma_{yy}}{2}\right)^2 + \tau_{xy}^2}.$$

Note that the value of τ_{max} is also the radius of the Mohr's circle.

$$\therefore \tau_{max} = \sqrt{\left(\frac{80 - (-40)}{2}\right)^2 + 25^2} = \underline{65\,\text{MPa}}$$

EXAMPLE 6-9: *Cont'd*

To obtain the value of the normal stress acting on the element with the maximum shear stress, we substitute the value of $\sin 2\theta_s$ and $\cos 2\theta_s$ into first and second of Eq. (6.43) to obtain

$$\sigma_{xx}^s = \sigma_{yy}^s = \frac{\sigma_{xx} + \sigma_{yy}}{2}.$$

Thus the normal stress on each of the four faces of the maximum shear stress element is the same.

Therefore,

$$\sigma_{xx}^s = \sigma_{yy}^s = \frac{80 + (-40)}{2} = \underline{20\,\text{MPa}}.$$

6.6.6 Mohr's Circle for Three-Dimensional Stress State

The Mohr's circle diagram is a useful graphical representation of the stress state at any point. At a given point P (say) in a material, let the frame of reference be chosen such that the origin is at P and the coordinate axes (x, y, z) are coincident with the principal stress axes. Consider a plane with a fixed outer normal at point P. Let the normal and shear stresses be σ and τ, respectively, on this plane. Let Q be the point that represents the stress state (σ, τ) at P on a stress plane with axes σ and τ. For each differently oriented plane passing through a point P, a corresponding point Q can be identified on the stress plane. This plane with axes σ and τ is called the stress plane Π. It can be shown (without proof here) that all the points $Q(\sigma, \tau)$ for all possible orientations of the plane through point P, lie within the shaded area in Figure 6-21. The shaded region is called the Mohr's stress plane Π and is constructed as follows. The three principal stresses are algebraically arranged as $\sigma_1 \geq \sigma_2 \geq \sigma_3$. Three Mohr's circles are constructed with centers C_1, C_2, and C_3 having the coordinates $[1/2(\sigma_2 + \sigma_3), 0]$, $[1/2(\sigma_1 + \sigma_3), 0]$, and $[1/2(\sigma_1 + \sigma_2), 0]$ and radii $R_1 = 1/2(\sigma_2 - \sigma_3)$, $R_2 = 1/2(\sigma_1 - \sigma_3)$, and $R_3 = 1/2(\sigma_1 - \sigma_2)$, respectively.

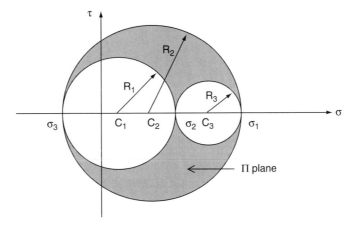

FIGURE 6-21 *Mohr's circles for 3-D stress states.*

It may be noted that the maximum shear stress is equal to $1/2(\sigma_1 - \sigma_3)$ and the associated normal stress is $1/2(\sigma_1 + \sigma_3)$. The other two extremum values for the shear stresses are $1/2(\sigma_2 - \sigma_3)$ and $1/2(\sigma_1 - \sigma_2)$ with the associated normal stresses of $1/2(\sigma_2 + \sigma_3)$ and $1/2(\sigma_1 + \sigma_2)$, respectively.

EXAMPLE 6-10:

If $\sigma_1 = \sigma_2 \neq \sigma_3$, draw the Mohr's circles and determine the maximum shear stress.

Answer: In this case, the point σ_1 coincides with σ_2 in Figure 6-21. Thus the Mohr's circle looks as shown in Figure E6-10, as the circle between points σ_1 and σ_2 collapses to a point, and the circle between σ_1 and σ_3 coincides with the circle between σ_2 and σ_3.

The maximum shear stress is $\tau_{max} = \frac{\sigma_1 - \sigma_3}{2}$ or $\frac{\sigma_2 - \sigma_3}{2}$. The associated normal stress is $1/2(\sigma_1 + \sigma_3)$ or $1/2(\sigma_2 + \sigma_3)$.

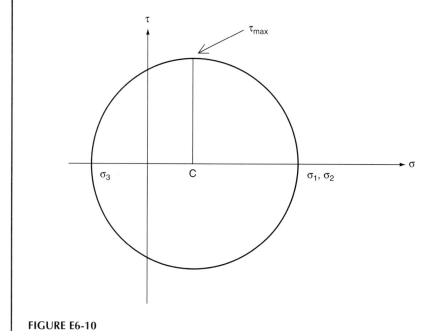

FIGURE E6-10

6.6.7 Strain Equations of Transformation

In this section, equations will be obtained to transform six Cartesian components of strain $\varepsilon_{xx}, \varepsilon_{yy}, \varepsilon_{zz}, \gamma_{xy}, \gamma_{yz}$, and γ_{zx} relative to an xyz coordinate system to six other Cartesian components of strain relative to an $x'y'z'$ coordinate system with the same origin. However, instead of deriving the transformation equations, it can be shown that the stress transformation Equations (6.25) and (6.26) can be converted to strain transformation equations using the following conversion rules (Dally and Riley, 1991):

$$
\begin{aligned}
\sigma_{xx} &\leftrightarrow \varepsilon_{xx} & 2\tau_{xy} &\leftrightarrow \gamma_{xy} \\
\sigma_{yy} &\leftrightarrow \varepsilon_{yy} & 2\tau_{yz} &\leftrightarrow \gamma_{yz} \ . \\
\sigma_{zz} &\leftrightarrow \varepsilon_{zz} & 2\tau_{zx} &\leftrightarrow \gamma_{zx}
\end{aligned}
\tag{6.47}
$$

Here the symbol $\leftrightarrow$ indicates the conversion from stress to strain or vice versa. Thus using Eq. (6.47) the *strain tensor* can be written as

$$
\begin{bmatrix}
\varepsilon_{xx} & \dfrac{1}{2}\gamma_{xy} & \dfrac{1}{2}\gamma_{xz} \\[2mm]
\dfrac{1}{2}\gamma_{xy} & \varepsilon_{yy} & \dfrac{1}{2}\gamma_{yz} \\[2mm]
\dfrac{1}{2}\gamma_{xz} & \dfrac{1}{2}\gamma_{yz} & \varepsilon_{zz}
\end{bmatrix}.
\tag{6.48}
$$

Equation (6.48) represents the second-order, symmetric strain tensor. The factor of 1/2 associated with the shearing strains allows the three-dimensional strain components to obey the transformation equations for a general second-order tensor.

Thus using Eq. (6.47), the Eqs. (6.25) and (6.26) can be converted to the strain transformation equations as

$$
\begin{aligned}
\varepsilon_{x'x'} &= l_1^2 \varepsilon_{xx} + l_2^2 \varepsilon_{yy} + l_3^2 \varepsilon_{zz} + l_1 l_2 \gamma_{xy} + l_2 l_3 \gamma_{yz} + l_3 l_1 \gamma_{zx} \\
\varepsilon_{y'y'} &= m_1^2 \varepsilon_{xx} + m_2^2 \varepsilon_{yy} + m_3^2 \varepsilon_{zz} + m_1 m_2 \gamma_{xy} + m_2 m_3 \gamma_{yz} + m_3 m_1 \gamma_{zx} \\
\varepsilon_{z'z'} &= n_1^2 \varepsilon_{xx} + n_2^2 \varepsilon_{yy} + n_3^2 \varepsilon_{zz} + n_1 n_2 \gamma_{xy} + n_2 n_3 \gamma_{yz} + n_3 n_1 \gamma_{zx} \\
\tfrac{1}{2}\gamma_{x'y'} &= l_1 m_1 \varepsilon_{xx} + l_2 m_2 \varepsilon_{yy} + l_3 m_3 \varepsilon_{zz} + \tfrac{1}{2}(l_1 m_2 + l_2 m_1)\gamma_{xy} + \tfrac{1}{2}(l_2 m_3 + l_3 m_2)\gamma_{yz} \\
&\quad + \tfrac{1}{2}(l_3 m_1 + l_1 m_3)\gamma_{zx} \\
\tfrac{1}{2}\gamma_{y'z'} &= m_1 n_1 \varepsilon_{xx} + m_2 n_2 \varepsilon_{yy} + m_3 n_3 \varepsilon_{zz} + \tfrac{1}{2}(m_1 n_2 + m_2 n_1)\gamma_{xy} + \tfrac{1}{2}(m_2 n_3 + m_3 n_2)\gamma_{yz} \\
&\quad + \tfrac{1}{2}(m_3 n_1 + m_1 n_3)\gamma_{zx} \\
\tfrac{1}{2}\gamma_{z'x'} &= n_1 l_1 \varepsilon_{xx} + n_2 l_2 \varepsilon_{yy} + n_3 l_3 \varepsilon_{zz} + \tfrac{1}{2}(n_1 l_2 + n_2 l_1)\gamma_{xy} + \tfrac{1}{2}(n_2 l_3 + n_3 l_2)\gamma_{yz} \\
&\quad + \tfrac{1}{2}(n_3 l_1 + n_1 l_3)\gamma_{zx}.
\end{aligned}
\tag{6.49}
$$

6.6.8 Principal Axes of Strain and Principal Strains

From the similarity between the laws of stress and strain transformation [Eq. (6.47)] it can be postulated that there exist three principal strains and three associated principal axes of strain or principal strain directions. Using Eq. (6.47) along with Eq. (6.30), the characteristic cubic equation whose roots give the principal strains can be written as

$$
\varepsilon_n^3 - J_1 \varepsilon_n^2 + J_2 \varepsilon_n - J_3 = 0
\tag{6.50}
$$

where J_1, J_2, J_3, are respectively the first, second, and third *invariants of strain*, which are given by the following expressions:

$$J_1 = \varepsilon_{xx} + \varepsilon_{yy} + \varepsilon_{zz}$$

$$J_2 = \varepsilon_{xx}\varepsilon_{yy} + \varepsilon_{yy}\varepsilon_{zz} + \varepsilon_{zz}\varepsilon_{xx} - \frac{\gamma_{xy}^2}{4} - \frac{\gamma_{yz}^2}{4} - \frac{\gamma_{zx}^2}{4}$$

$$= \begin{vmatrix} \varepsilon_{xx} & \frac{1}{2}\gamma_{xy} \\ \frac{1}{2}\gamma_{xy} & \varepsilon_{yy} \end{vmatrix} + \begin{vmatrix} \varepsilon_{yy} & \frac{1}{2}\gamma_{yz} \\ \frac{1}{2}\gamma_{yz} & \varepsilon_{zz} \end{vmatrix} + \begin{vmatrix} \varepsilon_{xx} & \frac{1}{2}\gamma_{xz} \\ \frac{1}{2}\gamma_{xz} & \varepsilon_{zz} \end{vmatrix}$$

$$J_3 = \varepsilon_{xx}\varepsilon_{yy}\varepsilon_{zz} - \frac{\varepsilon_{xx}\gamma_{yz}^2}{4} - \frac{\varepsilon_{yy}\gamma_{zx}^2}{4} - \frac{\varepsilon_{zz}\gamma_{xy}^2}{4} + \frac{\gamma_{xy}\gamma_{yz}\gamma_{zx}}{4}$$

$$= \begin{vmatrix} \varepsilon_{xx} & \frac{1}{2}\gamma_{xy} & \frac{1}{2}\gamma_{xz} \\ \frac{1}{2}\gamma_{xy} & \varepsilon_{yy} & \frac{1}{2}\gamma_{yz} \\ \frac{1}{2}\gamma_{xz} & \frac{1}{2}\gamma_{yz} & \varepsilon_{zz} \end{vmatrix}$$

(6.51)

The principal strain directions are also obtained by an analysis similar to that used for obtaining principal stress directions. The principal directions associated with the principal strains ε_1, ε_2, and ε_3 are obtained by substituting successively for $\varepsilon_i (i = 1, 2, 3)$ in the following equations and solving for the direction cosines l, m, and n:

$$(\varepsilon_{xx} - \varepsilon_i)l + \frac{1}{2}\gamma_{xy}m + \frac{1}{2}\gamma_{xz}n = 0$$

$$\frac{1}{2}\gamma_{xy}l + (\varepsilon_{yy} - \varepsilon_i)m + \frac{1}{2}\gamma_{yz}n = 0 \qquad (6.52)$$

$$l^2 + m^2 + n^2 = 1.$$

The method has been further illustrated in Example 6-11.

If ε_1, ε_2, and ε_3 are distinct ($\varepsilon_1 \neq \varepsilon_2 \neq \varepsilon_3$), then the axes represented by the directions of n_1, n_2, n_3 are unique and mutually perpendicular. If $\varepsilon_1 = \varepsilon_2 \neq \varepsilon_3$, then the axis of n_3 is unique and every direction perpendicular to n_3 is a principal direction (associated with the condition $\varepsilon_1 = \varepsilon_2$). If $\varepsilon_1 = \varepsilon_2 = \varepsilon_3$, then every direction is a principal direction.

EXAMPLE 6-11:

The displacement field for a body is given by the expression

$$\mathbf{u} = (x^2 + y^2)\hat{\mathbf{i}} + (3 + z)\hat{\mathbf{j}} + (x^2 + 2y^2)\hat{\mathbf{k}}.$$

Determine the principal strains at the point $(3, 1, -2)$ and the direction of the maximum principal strain.

Answer: The displacement components in the x-, y-, and z-directions are

$$u = x^2 + y^2, \quad v = 3 + z, \quad w = x^2 + 2y^2.$$

EXAMPLE 6-11: *Cont'd*

The strain components can be obtained from Eq. (6.3) as

$$\varepsilon_{xx} = 2x, \ \varepsilon_{yy} = 0, \ \varepsilon_{zz} = 0$$

$$\gamma_{xy} = 1, \ \gamma_{yz} = 3, \ \gamma_{xz} = 2x.$$

Therefore at point $(x, y, z) = (3, 1, -2)$, the strain components are

$$\varepsilon_{xx} = 6, \ \varepsilon_{yy} = 0, \ \varepsilon_{zz} = 0$$

$$\gamma_{xy} = 1, \ \gamma_{yz} = 3, \ \gamma_{xz} = 6.$$

The strain invariants are obtained from Eq. (6.51) as:

$$J_1 = 6, \ J_2 = -11.5, \ J_3 = -9.$$

Thus the characteristic cubic Eq. (6.50) is

$$\varepsilon_n^3 - 6\varepsilon_n^2 - 11.5\varepsilon_n + 9 = 0.$$

The roots of this cubic equation are the three principal strain magnitudes, which are

$$\underline{\varepsilon_1 = 7.4, \ \varepsilon_2 = -2.0, \ \varepsilon_3 = 0.6.}$$

(Note: $\varepsilon_1 + \varepsilon_2 + \varepsilon_3 = 6.0 = J_1$, the first invariant of strain. This provides a check for the correctness of the principal strain values calculated above.)

To obtain the direction of the maximum principal strain, $\varepsilon_1 = 7.4$, the following system of equations is solved for l, m, and n.

$$(6 - 7.4)l + 0.5m + 3n = 0$$

$$0.5l + (0 - 7.4)m + 0.5 \times 3 \times n = 0$$

$$l^2 + m^2 + n^2 = 1.$$

The solution gives:

$$l = 0.906 \Rightarrow \text{angle with } x\text{-axis is } \cos^{-1}(0.906) = \underline{25.0°}$$

$$m = 0.142 \Rightarrow \text{angle with } x\text{-axis is } \cos^{-1}(0.142) = \underline{81.8°}$$

$$n = 0.397 \Rightarrow \text{angle with } x\text{-axis is } \cos^{-1}(0.397) = \underline{66.5°}$$

EXAMPLE 6-12:

At a point in a body, the Cartesian components of strain are measured to be

$$\varepsilon_{xx} = 450\,\mu\varepsilon, \quad \varepsilon_{yy} = 300\,\mu\varepsilon, \quad \varepsilon_{zz} = 150\,\mu\varepsilon$$

$$\gamma_{xy} = 150\,\mu\varepsilon, \quad \gamma_{yz} = 150\,\mu\varepsilon, \quad \gamma_{xz} = 300\,\mu\varepsilon.$$

Transform these strain components into a new set of strain components relative to $x'y'z'$ coordinate axes, where the $x'y'z'$ axes are defined by the following direction cosines:

	x	y	z
x'	0.059	0.706	0.706
y'	0.706	−0.529	0.470
z'	0.706	0.470	−0.529

Answer: From Eq. (6.49):

$$\varepsilon_{x'x'} = (0.059)^2 450 + (0.706)^2 300 + (0.706)^2 150 + (0.059)(0.706)150$$
$$+ (0.706)(0.706)150 + (0.706)(0.059)300$$
$$= 319.2\,\mu\varepsilon$$

Similarly, $\varepsilon_{x'x'} = 348\,\mu\varepsilon$, $\varepsilon_{x'x'} = 233\,\mu\varepsilon$.
  Again from Eq. (6.49)

$$0.5\gamma_{x'y'} = (0.059)(0.706)450 + (0.706)(-0.529)150 + (0.706)(0.706)450$$
$$+ 0.5[(0.059)(-0.529) + (0.706)(0.706)]150 + 0.5[(0.706)(0.470)$$
$$+ (0.706)(-0.5290]150 + 0.5[(0.706)(0.706) + (0.059)(0.470)]300$$
$$= 67.2\,\mu\varepsilon$$
$$\Rightarrow \gamma_{x'y'} = 134.4\,\mu\varepsilon$$

Similarly, $\gamma_{y'z'} = 281\,\mu\varepsilon$, $\gamma_{z'x'} = 337.4\,\mu\varepsilon$.

6.6.9 Mohr's Circle for Plane Strain State

Since the equations for strain transformation [Eq. (6.49)] are of the same form as that of stress transformation equations (6.25) and (6.26), the use of Mohr's circle for determining stress on different planes can be extended to the analysis of strain on different planes through a point in the body. In a *Mohr's circle for strain* we plot the normal strain on the abscissa and half the shear strain on the ordinate, as shown in Figure 6-22.

If a shear deformation causes a given side in an element to rotate clockwise (cw), the corresponding point on the Mohr's circle is plotted above the horizontal axis, and if the deformation causes the side to rotate counterclockwise (ccw), the corresponding point is plotted below the horizontal axis. If the square element of side Δ, shown in Figure 6-23, is

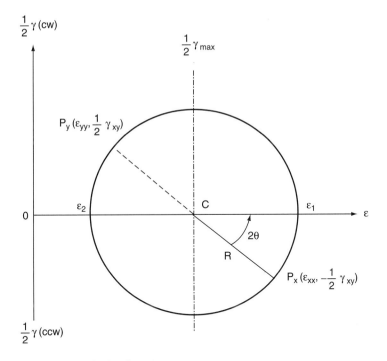

FIGURE 6-22 *Mohr's circle for plane strain.*

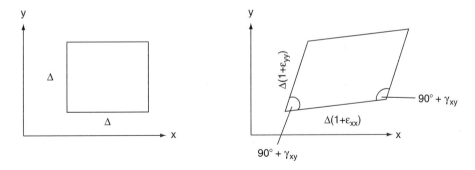

FIGURE 6-23 *Square element under combined normal and shear strain.*

deformed under the action of strain components ε_{xx}, ε_{yy}, γ_{xy}, then the sides associated with the strain ε_{yy} rotate cw while those sides associated with ε_{xx} rotate ccw.

The former strain is represented by point P_y on the Mohr's circle, while the latter is represented by point P_x. The center C has the coordinates $(\frac{\varepsilon_{xx} + \varepsilon_{yy}}{2}, 0)$ and the radius R of the circle is

$$ R = \sqrt{\left(\frac{\varepsilon_{xx} + \varepsilon_{yy}}{2}\right)^2 + \left(\frac{\gamma_{xy}}{2}\right)^2}. $$

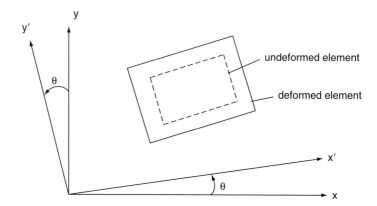

y

y'

θ

undeformed element

deformed element

θ

x'

x

FIGURE 6-24 *Undeformed and deformed elements in rotated coordinate system.*

If the line P_xP_y is rotated through an angle 2θ to coincide with the horizontal axis, then the shear strain is zero on the element oriented physically along the principal axes of strain $x'y'$, as shown in Figure 6-24. We thus obtain the extremum of the strain, namely, maximum strain ε_1 that is the maximum principal strain (since shear strain is zero), and ε_2 is the minimum principal strain. The maximum shear strain in the xy-plane is

$$\frac{1}{2}\gamma_{max} = \frac{\varepsilon_{max} - \varepsilon_{min}}{2} = \frac{\varepsilon_1 - \varepsilon_2}{2}. \tag{6.53}$$

The angle θ can also be obtained by changing the last of Eq. (6.42) from stress to strain using Eq. (6.47) to obtain

$$\frac{1}{2}\gamma_{x'y'} = -(\varepsilon_{xx} - \varepsilon_{yy})\sin\theta\cos\theta + \frac{1}{2}\gamma_{xy}(\cos^2\theta - \sin^2\theta) \tag{6.54}$$

and then setting $\gamma_{x'y'} = 0$ in the preceding equation, which results in the expression

$$\tan 2\theta = \frac{\gamma_{xy}}{\varepsilon_{xx} - \varepsilon_{yy}}. \tag{6.55}$$

6.7   Materials Selection for Combined Loading 🕸

6.7.1   Yielding under Multiaxial Stresses

Most engineering structures function in the elastic region for the majority of their life. However, the presence of stress concentrations and development of defects during service that can act as stress risers, or general degradation in the material for example due to corrosion, can cause the localized stress state to exceed the elastic limit of the material. This state represents the beginning of *inelastic deformation* or *plastic deformation* or the initiation of yielding at that point(s). In a very severe case, yielding may occur over the whole cross-section of a component. Generally, engineering designs are such that initiation of yielding will cause minimal loss of function or change in the deformation of a statically loaded structure.

However, initiation of yield can cause development of residual stresses in the structure (generally residual stresses are harmful if tensile), provide locations for crack initiation, provide locations for increased chemical attack, and so forth. Hence, if possible the structure should function in the elastic region over its whole life span.

In this chapter, we are primarily concerned with the phenomenon of the transition from the elastic state to the inelastic state or initiation of yield. Further straining may or may not result in more plastic flow. Brittle materials such as ceramics fracture with minimal yielding, whereas metals such as steel undergo large plastic deformation before fracture. Material behavior under increasing load after yielding, called *strain hardening* or *work hardening*, has been only briefly discussed. A few important hypotheses have been presented for the prediction of yielding under a given three-dimensional and two-dimensional (plane) state of stress. These hypotheses are called *yield criteria*.

In Chapter 4, Section 4.4.2 (also see *text web site*), it has been discussed that plastic or permanent deformation occurs by the process of slip on certain close-packed planes in a single crystal or polycrystalline material. Slip occurs on a certain plane only when the shear stress reaches a certain critical value τ_{CRSS} in a uniaxially loaded single crystal. In polycrystal materials the value of the critical shear stress is higher than for single crystals. It has been observed that application of uniform and equal tensile or compressive stress in all directions (called hydrostatic stress) does not contribute to the slip process. Only shearing stresses cause slip, which manifests itself macroscopically as plastic deformation. Thus it may be postulated that for any given loading condition it is only the shear stresses that will cause yielding. Hence, we first present a special state of stress called the *pure shear stress state* before discussing yield criteria.

6.7.1.1 Pure Shear

The state of stress at any point P can be characterized by six stress components referred to a coordinate frame of reference. The magnitudes of these stress components depend on the choice of the coordinate system. If we can find one or more particular frames of reference where $\sigma_x = \sigma_y = \sigma_z = 0$, then a state of pure shear exists at the point P. For this particular reference frame the stress matrix will be

$$\begin{bmatrix} 0 & \tau_{xy} & \tau_{xz} \\ \tau_{xy} & 0 & \tau_{yz} \\ \tau_{xz} & \tau_{yz} & 0 \end{bmatrix}. \tag{6.56}$$

For this coordinate system, the sum $\sigma_{xx} + \sigma_{yy} + \sigma_{zz}$ is equal to zero. The sum of the three normal stresses is also represented by I_1, which is the first stress invariant, and it does not change with the choice of the coordinate system. Hence if $I_1 = 0$ for any state of stress then that stress state is a pure shear stress state.

Let us now consider any arbitrary state of stress

$$\begin{bmatrix} \sigma_{xx} & \tau_{xy} & \tau_{xz} \\ \tau_{xy} & \sigma_{yy} & \tau_{yz} \\ \tau_{xz} & \tau_{yz} & \sigma_{zz} \end{bmatrix}$$

and let

$$s = \frac{1}{3}[\sigma_{xx} + \sigma_{yy} + \sigma_{zz}] = \frac{1}{3}I_1. \tag{6.57}$$

Now resolve the arbitrary stress state into two states:

$$
\begin{bmatrix} \sigma_{xx} & \tau_{xy} & \tau_{xz} \\ \tau_{xy} & \sigma_{yy} & \tau_{yz} \\ \tau_{xz} & \tau_{yz} & \sigma_{zz} \end{bmatrix} = \begin{bmatrix} s & 0 & 0 \\ 0 & s & 0 \\ 0 & 0 & s \end{bmatrix} + \begin{bmatrix} \sigma_{xx} - s & \tau_{xy} & \tau_{xz} \\ \tau_{xy} & \sigma_{yy} - s & \tau_{yz} \\ \tau_{xz} & \tau_{yz} & \sigma_{zz} - s \end{bmatrix}.
\tag{6.58}
$$

The first state on the right-hand side is a *hydrostatic stress state*, while the second is a *pure shear stress state* since the invariant quantity I_1 for this state is zero ($I_1 = 0$ is a sufficient condition). This is because, on using Eq. (6.57), we obtain for I_1

$$(\sigma_{xx} - s) + (\sigma_{yy} - s) + (\sigma_{zz} - s)$$

$$= (\sigma_{xx} + \sigma_{yy} + \sigma_{zz}) - 3s = (\sigma_{xx} + \sigma_{yy} + \sigma_{zz}) - 3\frac{1}{3}(\sigma_{xx} + \sigma_{yy} + \sigma_{zz}).$$

$$= 0$$

If the principal stress state is used, Eq. (6.58) can be written as

$$
\begin{bmatrix} \sigma_1 & 0 & 0 \\ 0 & \sigma_2 & 0 \\ 0 & 0 & \sigma_3 \end{bmatrix} = \begin{bmatrix} s & 0 & 0 \\ 0 & s & 0 \\ 0 & 0 & s \end{bmatrix} + \begin{bmatrix} \sigma_1 - s & 0 & 0 \\ 0 & \sigma_2 - s & 0 \\ 0 & 0 & \sigma_3 - s \end{bmatrix}
\tag{6.59}
$$

where $s = \frac{1}{3}[\sigma_1 + \sigma_2 + \sigma_3]$.

The quantity s is also called the *mean hydrostatic stress*, and the pure shear stress state is also called the *deviatoric stress state*. The first invariant of the deviatoric stress tensor is always zero. It has been experimentally shown that the influence of hydrostatic stress on yielding is insignificant and only the deviatoric stress state affects the yield condition. Thus, in the study of yielding of materials by plastic flow the effect of hydrostatic stress is neglected and only the pure shear stress state is considered. Hence the importance of the deviatoric stress state in the study of plasticity. The yield criteria developed next, however, have not been expressed explicitly in terms of the components of the deviatoric stress tensor for simplicity.

Note: A pure hydrostatic stress state produces volume changes only in a cubic element and no shape change. The pure shear state causes angular distortion in the cubic element but no volume change. There is a difference between pure shear and a simple shear state. A pure shear state will distort a circular element into an ellipse, while a simple shear will distort and rotate the circular element.

Similar to the analysis of stress, we can resolve any strain field into a hydrostatic or isotropic part and a deviatoric part. The strain tensor can be resolved into two parts as

$$
\begin{bmatrix} \varepsilon_{xx} & \frac{1}{2}\gamma_{xy} & \frac{1}{2}\gamma_{xz} \\ \frac{1}{2}\gamma_{xy} & \varepsilon_{yy} & \frac{1}{2}\gamma_{yz} \\ \frac{1}{2}\gamma_{xz} & \frac{1}{2}\gamma_{yz} & \varepsilon_{zz} \end{bmatrix} = \begin{bmatrix} e & 0 & 0 \\ 0 & e & 0 \\ 0 & 0 & e \end{bmatrix} + \begin{bmatrix} \varepsilon_{xx} - e & \frac{1}{2}\gamma_{xy} & \frac{1}{2}\gamma_{xz} \\ \frac{1}{2}\gamma_{xy} & \varepsilon_{yy} - e & \frac{1}{2}\gamma_{yz} \\ \frac{1}{2}\gamma_{xz} & \frac{1}{2}\gamma_{yz} & \varepsilon_{zz} - e \end{bmatrix}
\tag{6.60}
$$

where

$$
e = \frac{1}{3}(\varepsilon_{xx} + \varepsilon_{yy} + \varepsilon_{zz})
\tag{6.61}
$$

represents the mean elongation or hydrostatic strain at a given point. The first matrix on the right-hand side of Eq. (6.60) is the hydrostatic part of the strain tensor. The second matrix represents the deviatoric strain part of the strain tensor. If an element of the body is subjected to the deviatoric strain only, then the volumetric strain (see Problem 6-32) is equal to zero:

$$\frac{\Delta V}{V} = (\varepsilon_{xx} - e) + (\varepsilon_{yy} - e) + (\varepsilon_{zz} - e)$$

$$= \varepsilon_{xx} + \varepsilon_{yy} + \varepsilon_{zz} - 3e$$

$$= 0$$

This implies that if an element is subjected to deviatoric strain it undergoes deformation without a change in volume. Hence, this part is also known as the pure shear part of strain matrix.

If a principal strain state is used we can write

$$\begin{bmatrix} \varepsilon_1 & 0 & 0 \\ 0 & \varepsilon_2 & 0 \\ 0 & 0 & \varepsilon_3 \end{bmatrix} = \begin{bmatrix} e & 0 & 0 \\ 0 & e & 0 \\ 0 & 0 & e \end{bmatrix} + \begin{bmatrix} \varepsilon_1 - e & 0 & 0 \\ 0 & \varepsilon_2 - e & 0 \\ 0 & 0 & \varepsilon_3 - e \end{bmatrix} \tag{6.62}$$

where

$$e = \frac{1}{3}(\varepsilon_1 + \varepsilon_2 + \varepsilon_3). \tag{6.63}$$

6.7.2 Yield Criteria

The simplest state of stress is the uniaxial stress state. Such a stress condition is produced in a simple tension test experiment to obtain the uniaxial stress–strain curve of a material. This stress–strain curve provides information about the yield point of the material. Thus if an actual component, such as a tie-rod or a linkage in a four-bar mechanism, is in pure tension, then its failure is predictable at the yield point determined from a simple tension test.

If a structural component is subjected to a biaxial or triaxial state of stress, the prediction of failure (or yielding) is no longer as easy as the uniaxial loading case. In the multiaxial stress condition, we can no longer say that the material will yield when the largest normal stress reaches the yield point obtained from a uniaxial tension test, as the other normal stress components also influence yielding. Furthermore it is practically impossible to conduct experiments to obtain the yield condition for a whole range of stress combinations in all three possible orthogonal directions, and also consider factors such as stress concentration, temperature, and environmental effects. To overcome these difficulties, designers have relied on developing theories that relate failure behavior in the multiaxial stress situation to the failure behavior in a simple tension test in the same mode through a selected quantity such as stress, strain, or energy. Thus the failure theories will predict failure to occur when the maximum value of the selected mechanical quantity in the multiaxial stress state becomes equal to or exceeds the value of the same quantity that produces failure in a uniaxial tension test using the same material.

In general the elastic limit or the *yield stress* is a function of the state of stress represented by six stress components for an isotropic material. The yield condition can generally be written as

$$f(\sigma_{xx}, \ \sigma_{yy}, \ \sigma_{zz}, \ \tau_{xy}, \ \tau_{yz}, \ \tau_{zx}, \ M_1, \ M_2, \ \ldots.) = 0 \qquad (6.64)$$

where M_1 and M_2 are materials constants. For isotropic materials, the orientation of the principal stresses is immaterial, and the values of the three principal stresses are sufficient to describe the state of stress uniquely. The yield criteria therefore can be written as

$$f(\sigma_1, \ \sigma_2, \ \sigma_3, \ M_1, \ M_2, \ \ldots.) = 0. \qquad (6.65)$$

Based on this philosophy, several theories have been developed to predict the yield point when a component is subjected to multiaxial stresses. We will further discuss three of these theories in this chapter, namely the maximum normal stress theory, maximum shear stress theory, and distortion energy theory. The three theories are also known as Rankine's theory, Tresca–Guest theory, and Huber–Von Mises–Hencky theory, respectively. As described earlier a general multiaxial state of stress at any point can be fully described by three principal normal stresses and their directions. Hence all failure theories express the yielding criteria in terms of the principal normal stresses. It may be noted that there is only one nonzero principal normal stress in a uniaxial tension test situation.

6.7.2.1 Maximum Normal Stress Theory

The *maximum normal stress theory* based *yield criteria* states that yielding will occur under a multiaxial stress state when the maximum principal normal stress state becomes equal to or exceeds the normal stress at the yield point in a uniaxial tension test using a specimen of the same material.

Yielding is predicted by this theory to occur if any one of the following conditions is satisfied:

$$\begin{aligned} \sigma_1 &\geq S_Y \\ \sigma_2 &\geq S_Y \\ \sigma_3 &\geq S_Y \end{aligned} \qquad (6.66)$$

where S_Y is the uniaxial yield strength in tension, or if

$$\begin{aligned} \sigma_1 &\leq S_C \\ \sigma_2 &\leq S_C \\ \sigma_3 &\leq S_C \end{aligned} \qquad (6.67)$$

where S_C is the yield strength in compression.

These failure conditions can be represented graphically, as shown in Figure 6-25.

Experimental results show that this theory does not accurately predict yielding in ductile materials such as aluminum, copper, and steel. For example, consider a material subjected to a hydrostatic tension stress state $\sigma_1 = \sigma_2 = \sigma_3 = \sigma_{applied}$. This theory predicts that yielding will begin when $\sigma_{applied} = S_Y$, the tensile yield strength of the material. However, experiments have shown that a hydrostatic tension stress state does not produce any yielding. Thus the maximum normal stress theory fares poorly in predicting yield in ductile materials. However, it is more suited for accurate prediction of failure in brittle material such as cast iron.

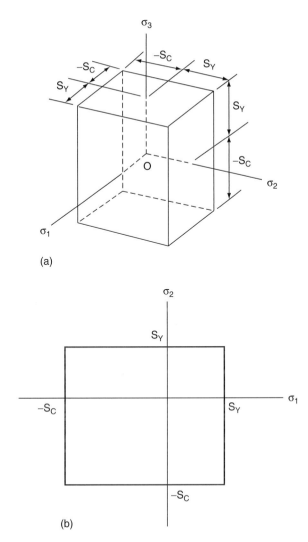

(a)

(b)

FIGURE 6-25 *Graphical representation of the maximum stress theory of yielding for (a) multiaxial stress state, (b) biaxial stress state.*

6.7.2.2 Maximum Shear Stress Theory

The *yield criteria* according to *maximum shear stress theory* states that yielding will occur under a multiaxial stress state when the maximum shear stress becomes equal to or exceeds the maximum shear stress at the yield point in a uniaxial tension test using a specimen of the same material. Recall that the principal shearing stresses are

$$\tau_1 = \pm \frac{1}{2}(\sigma_1 - \sigma_2)$$

$$\tau_2 = \pm \frac{1}{2}(\sigma_1 - \sigma_3). \qquad (6.40)$$

$$\tau_3 = \pm \frac{1}{2}(\sigma_1 - \sigma_2)$$

Also, for a uniaxial tension test, the only nonzero principal stress at the yield point is $\sigma_1 = S_Y =$ the yield strength, and hence the principal shearing stress at the yield point is

$$\tau_{tension} = \frac{S_Y}{2}.\qquad(6.68)$$

Thus according to the maximum shear stress yielding theory the yield conditions can be written as

$$|\sigma_2 - \sigma_3| \geq S_Y$$
$$|\sigma_2 - \sigma_3| \geq S_Y\,.\qquad(6.69)$$
$$|\sigma_1 - \sigma_2| \geq S_Y$$

Failure by yielding occurs if any one of the preceding expressions is satisfied. These yield conditions are represented graphically for a three-dimensional stress field in Figure 6-26. The yield surface is a hexagonal cylinder whose axis makes equal angles with the three principal stress axes. As with all yield theories, stress states that lie within the hexagonal cylinder (or the yield surface) do not result in yielding, while stress states lying outside the cylinder result in yielding. A stress state lying exactly on the yield surface signifies that the material is ready to yield.

For a *biaxial stress field*, that is, any one principal stress equals zero, say $\sigma_1 \neq 0$, $\sigma_2 \neq 0$, $\sigma_3 = 0$, the graphical representation is shown in Figure 6-27 assuming that the yield strength in

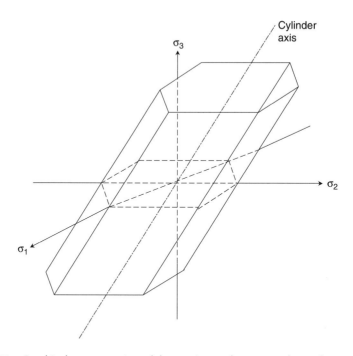

FIGURE 6-26 *Graphical representation of the maximum shear stress theory for a general three-dimensional stress state.*

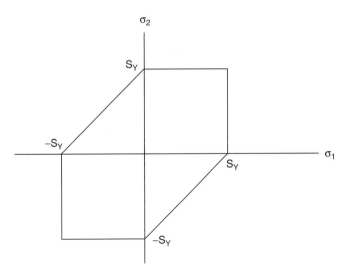

FIGURE 6-27 *Graphical representation of the maximum shear stress theory for a biaxial stress field.*

tension is equal to the yield strength in compression, which is approximately true for most structural metals.

Experimental results have shown that the maximum shear stress theory predicts yielding in ductile materials with reasonable accuracy. This theory also predicts the experimentally observed behavior of ductile materials under a hydrostatic stress state. If all principal stresses are equal, the shear stresses τ_1, τ_2, and τ_3 are all equal to zero and hence yielding will never begin regardless of the magnitude of the hydrostatic stress state. In the graphical representation, the hydrostatic stress state always lies on the axis of the hexagonal cylinder and hence within the yield surface, which implies yielding will never occur.

6.7.2.3 Distortion Energy Theory

According to the *distortion energy theory* the energy absorbed during distortion of an element due to pure shear stresses is responsible for failure by yielding, and not the total energy. The energy of distortion can be obtained by subtracting the energy of volumetric expansion due to hydrostatic stress from the total strain energy.

Using Hooke's law, it can be shown that the hydrostatic state of stress is related to the hydrostatic state of strain as

$$\begin{aligned}
\varepsilon_1 &= \frac{1}{E}[\sigma_1 - v(\sigma_2 + \sigma_3)] \\
\varepsilon_2 &= \frac{1}{E}[\sigma_2 - v(\sigma_3 + \sigma_1)] . \\
\varepsilon_3 &= \frac{1}{E}[\sigma_3 - v(\sigma_1 + \sigma_2)]
\end{aligned} \tag{6.70}$$

Adding and taking the mean, we obtain

$$e = \frac{1}{E}[1 - 2v]s \tag{6.71}$$

where s and e are defined in Eqs. (6.57) and (6.63), respectively. The work done or energy stored during volumetric change is given by

$$U_v = \frac{1}{2}es + \frac{1}{2}es + \frac{1}{2}es = \frac{3}{2}es.$$

On substituting for e from Eq. (6.71) we obtain

$$
\begin{aligned}
U_v &= \frac{3}{2E}(1 - 2v)s^2 \\
&= \frac{1 - 2v}{6E}(\sigma_1 + \sigma_2 + \sigma_3)^2
\end{aligned}
\tag{6.72}
$$

The total elastic strain energy is given by (using Eq. (6.70)):

$$
\begin{aligned}
U_T &= \frac{1}{2}\varepsilon_1\sigma_1 + \frac{1}{2}\varepsilon_2\sigma_2 + \frac{1}{2}\varepsilon_3\sigma_3 \\
&= \frac{1}{2E}[\sigma_1^2 + \sigma_2^2 + \sigma_3^2 - 2v(\sigma_1\sigma_2 + \sigma_2\sigma_3 + \sigma_3\sigma_1)]
\end{aligned}
\tag{6.73}
$$

Hence the energy of distortion U_d for a multiaxial stress state is obtained by subtracting energy due to volumetric change from the total energy, as

$$
\begin{aligned}
U_d &= U_T - U_V \\
&= \frac{1 + v}{3E}\left[\frac{(\sigma_1 - \sigma_2)^2 + (\sigma_2 - \sigma_3)^2 + (\sigma_3 - \sigma_1)^2}{2}\right].
\end{aligned}
\tag{6.74}
$$

(Note that distortion energy U_d is zero if $\sigma_1 = \sigma_2 = \sigma_3$ as in the case of hydrostatic stress states.)

The distortion-energy based yield criteria is that yielding will occur when the distortion energy in a unit volume under a general stress state equals the distortion energy in the same volume when uniaxially stressed to the yield strength.

For a uniaxial tension test, $\sigma_1 = S_Y$, $\sigma_2 = 0$, $\sigma_3 = 0$, where S_Y = yield strength, and the distortion energy is given by

$$U_d = \frac{1 + v}{3E}S_Y^2.
\tag{6.75}
$$

Equating the right-hand sides of Eqs. (6.74) and (6.75) results in the yield criteria

$$S_Y^2 \leq \frac{(\sigma_1 - \sigma_2)^2 + (\sigma_2 - \sigma_3)^2 + (\sigma_3 - \sigma_1)^2}{2}.
\tag{6.76}
$$

For a biaxial stress state $\sigma_3 = 0$ and Eq. (6.76) becomes

$$S_Y^2 \leq \sigma_1^2 - \sigma_1\sigma_2 + \sigma_2^2.
\tag{6.77}
$$

Figure 6-28 shows the elliptic yield locus graphically for the two-dimensional or biaxial stress state, using Eq. (6.77).

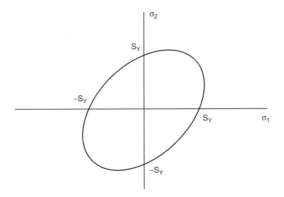

FIGURE 6-28 *Elliptic yield locus for biaxial stress state.*

EXAMPLE 6-13:

A point in a structural component has a stress state given by $\sigma_{xx} = 50$, $\sigma_{yy} = 70$, and $\tau_{xy} = 200$ MPa. The material of the component is ductile and has yield strength of 300 MPa. Will the material yield at the point under consideration according to the maximum shear stress theory?

Answer: The given stress state is a biaxial state of stress. Hence the two in-plane principal normal stresses can be determined using Eqs. (6.47) as

$$\sigma_1, \ \sigma_2 = \frac{\sigma_{xx} + \sigma_{yy}}{2} \pm \sqrt{\left(\frac{\sigma_{xx} - \sigma_{yy}}{2}\right)^2 + \tau_{xy}^2}$$

$\Rightarrow \sigma_1 = 260$ MPa, $\sigma_2 = -140$ MPa

The third principal stress $\sigma_3 = 0$. (Note that according to convention, the principal stress σ_1, σ_2, σ_3 should be ordered as $\sigma_1 > \sigma_2 > \sigma_3$. However, we have *not* followed the convention *only* for this example. You may follow the convention and renumber the principal stresses, and draw the yield plot axes accordingly.)

The principal stress state $(\sigma_1, \ \sigma_2) \equiv (260, -140)$ can be plotted on the biaxial yield plot to determine the yield condition. The stress state at the point lies outside the yield locus as shown in Figure E6-13. Hence yielding occurs.

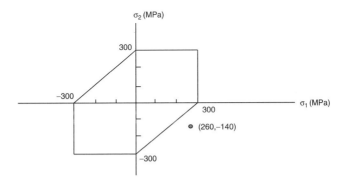

FIGURE E6-13

EXAMPLE 6-14:

Determine the yield condition at the point mentioned in Example 6-13, using the maximum normal stress theory.

Answer: The maximum normal stress theory states that yielding occurs only when any principal normal stress equals or exceeds the yield strength in a simple tension test. Hence this theory predicts that no yielding occurs. This result is also shown graphically in Figure E6-14, where the stress state lies within the yield locus.

This example further demonstrates that the maximum normal stress theory can give erroneous results for ductile materials.

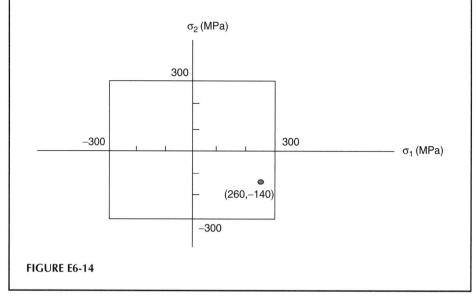

FIGURE E6-14

6.7.3 Materials Selection for Combined Loading: Strain Hardening or Work Hardening

When the yield limit of a metal is exceeded, plastic deformation under a constant load cannot proceed as it is blocked by significant changes within the crystal structure of the metal. Because of these changes an increasing load has to be applied to further the plastic deformation that fist occurred at the yield point. During further plastic deformation the indentation hardness and fracture strength also increase. Other properties such as density, electrical conductivity, magnetic properties, and resistance to wear may also change. This complex process of changing mechanical properties of a metal by plastic deformation is known as *strain hardening* or *work hardening*.

Several different mechanisms are responsible for the work hardening effect. The most prominent and effective mechanism is the resistance to dislocation motion. Increased resistance to dislocation motion can occur in different ways as explained later. Other mechanisms are diffusion, bending and distortion of crystal domains causing an increase in the resolved shear stress needed to initiate further flow due to the changed orientation of the crystal planes. We will first discuss work hardening in the context of resistance to dislocation motion in a single crystal and then in polycrystalline metals.

Plastic flow in single crystals: Initiation of plastic flow in single crystals has been discussed in Chapter 4, Section 4.4.2. Plastic flow occurs by the mechanism of "slip" along specific crystallographic directions and on closed packed atomic planes. The concept of *critically resolved shear stress* τ_{CRSS} was also introduced in Chapter 4 to predict the initiation of yielding in a single crystal. The τ_{CRSS} depends on the test conditions (e.g., temperature, strain rate) and the initial dislocation density and amount of impurities in the material.

When a material is subject to a shear stress, dislocations are generated. The number of dislocations generated and the rate of generation depends on the applied macroscopic strain (or stress) and on the rate at which the strain is applied. (For a review of dislocation theory, see Askeland and Phule, 2003, Chapter 4.) The higher the applied strain, the greater the number of dislocations generated or the dislocation density. (It may be noted that the mechanism for dislocation generation is the Frank–Read source.) The higher the strain rate, the higher the dislocation velocity. However, for increasing plastic strain there may be a sufficient increase in dislocation density and a reduction in the number of mobile dislocations. If the initial dislocation density is high and dislocations become immobile quickly, then work hardening is concurrent with the initiation of plastic deformation (and a conventional yield point is not observed). If the initial dislocation density is low, and an increase in dislocations is not accompanied by a significant reduction in mobile dislocations, strain softening manifested by a yield point is observed.

The additional dislocations generated during plastic deformation are responsible for work hardening, because the new dislocations themselves become obstacles to dislocation motion. When dislocations intersect they produce hard obstacles, which is a relative term implying a larger resistance to the motion of a dislocation. Thus if a material contains hard obstacles the critical resolved shear stress needed to deform the material continues to increase with the dislocation density. Hard obstacles can be created by dislocation intersections or nondeforming particles.

The post-yielding shear stress–shear strain response of a single-crystal material is schematically shown in Figure 6-29.

This post-yielding response consists of three stages with each stage representing a different behavior of the work-hardening rate. Work-hardening rate is the increase in stress necessary to produce an incremental increase in strain, or the slope of the shear stress–shear strain curve

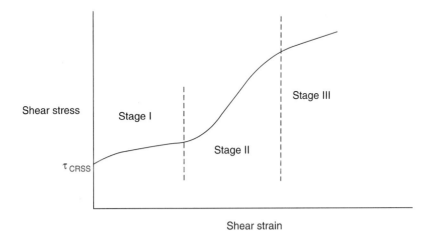

FIGURE 6-29 *Schematic diagram of the stress–strain curve for a work-hardening material.*

$d\tau/d\gamma$. In stage I, just after yielding, the work-hardening rate is low and practically constant. This stage is associated with slip beginning on a slip plane having the maximum value of $\cos\varphi\cos\lambda$. The stress required for increased strain rises very slowly due to low amount of dislocation interaction and jog formation. The work-hardening in this stage is mainly due to the overlap of dislocation stress fields among dislocations gliding on parallel planes.

At the end of stage I, the applied stress could be high enough so that τ_{CRSS} is exceeded on slip system(s) other than the one that is most favorably oriented. In stage II multiple slip systems can be activated and work-hardening results from interactions among dislocations on nonparallel planes. Work hardening is higher than that of stage I, indicating dislocation mobility is dropping with increasing strain.

Stage III is characterized by a lower work-hardening rate in comparison to stage II. The high stresses that are present are sufficient to cause cross-slip of screw dislocations and recombination of partial dislocations, thus removing obstacles to dislocation slip or bypassing obstacles.

The work-hardening behavior described above applies more closely to FCC type materials, such as copper. BCC materials such as iron have a comparatively larger number of slip systems, with some having easier slip than others. Thus dislocations in BCC crystals have a greater choice of glide planes. If several of the slip systems become active, dislocation interaction and jog formation will become more pronounced than in FCC crystals. Hence, in BCC crystals stage I is small (or stage II appears sooner) as compared to FCC crystals because of these dislocation interactions. The larger number of slip systems in BCC crystals also facilitates cross-slip of screw dislocations, causing stage III hardening to occur at lower strain than in FCC crystals.

In HCP type crystals, such as zinc, the crystal exhibits an extended stage I over most of its deformation. This is because the slip systems (occurring in the two basal planes or close packed planes) in HCP crystals do not intersect, and hence no dislocation interactions occur.

Plastic flow in polycrystalline materials: In addition to the slip mechanisms mentioned for single crystals another major obstacle to dislocation motion is internal boundaries in a polycrystalline material, also called grain boundaries. These boundaries impede dislocation motion along their entire slip plane length, in contrast to a dislocation or nondeforming particle, which impedes dislocation motion at a point. Hence grain boundaries create a greater resistance to slip. Also crystallographic factors do not allow a dislocation to pass from one grain to another through their boundary; displacements across grain boundaries must be matched, to permit grains to deform in unison, or else voids and cracks can appear at the boundaries. Hence, neighboring grains restrain plastic flow of each other resulting in a greater resistance to plastic flow, in other words increasing the strength of the material. (Grain size also dramatically affects strain hardening characteristics. Smaller the grain size higher the yield strength.) Each grain in a polycrystal has three shear and three tensile components of strain. The tensile strains are related through the constant volume condition as $\varepsilon_{xx} + \varepsilon_{yy} + \varepsilon_{zz} = 0$, during plastic flow. Thus it can be shown that five independent slip systems are required to ensure displacement compatibility across the grain boundaries.

It is this requirement that causes polycrystalline materials to have a yield stress that is greater than that of single crystals. In addition, the higher yield stress of polycrystals is due to the different orientation of slip planes in each grain, with some having a favorable orientation while others have the least favorable orientation in the grains. Other methods that can increase the yield strength of materials are:

(i) Combining two materials that are mutually soluble, called *solution hardening*.

Consider, for example, a Cu–Ni system with 65% Ni. This material has the highest yield strength compared to either pure copper or pure nickel. The addition of substitutional or interstitial atoms causes imperfections in the symmetry of a crystal, which hinders dislocation motion, thereby strengthening the metal.

Another example is the special behavior at yield of low-carbon steels and heat-treated aluminum–magnesium solid solution alloys that display an upper and a lower yield point in a tensile stress–strain curve, as shown in Figure 6-30. This is due to the presence of solute atoms in a solid solution either as interstitial atoms or as substitutional atoms. The interstitial atoms anchor the dislocations, whereas the substitutional atoms retard the motion of dislocations. The anchoring of dislocations results in the sharp display of yield in a tensile test. Pure iron does not display a sharp yield point, as it does not have interstitial carbon atoms, as does steel. In the absence of interstitials the material would yield at a stress of σ_{yl}. The presence of interstitials anchors the dislocation forming small clusters around them, which retard the slip process and hence raise the yield point to σ_{yu}. When the stress level rises further the dislocations begin to slip and move away from the cluster, allowing slip to occur at the lower stress level σ_{yl}.

   (ii) By the presence of impurities that act as pinning sites for dislocations, reducing their mobility, called *dispersion hardening*.

This, in turn, will increase the stress required to initiate slip, and the yield point of the material will increase. The smaller the average distance between pinning sites, the greater the shear stress required to cause slip or flow of metal. Typical examples are that of two-phase materials in which one phase provides the pinning sites while the second phase forms the matrix that surrounds the sites. This can be achieved by artificially adding pinning materials or by heat treatment. Examples are spheroidized steel consisting of cementite spheres acting as pinning sites embedded in a ferrite matrix; hard aluminum oxide (Al_2O_3) particles in a silver matrix used in electrical contacts; and dispersed thorium oxide (ThO_2) particles in a nickel matrix for excellent creep properties.

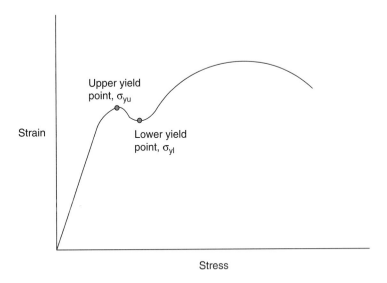

FIGURE 6-30   *Upper and lower yield points in a stress–strain plot.*

To summarize, any phenomenon that inhibits dislocation slip causes materials to become stronger. Dislocation pinning also generally increases the ultimate strength and the work-hardening rate.

6.8 Design of Structures under Combined Loads Ⓨ🏠💻

6.8.1 Stresses in Thin-Walled Pressure Vessels

When the wall thickness of the cylindrical or spherical pressure vessel (the two most commonly used) is 1/20th, or less, of its radius, the radial stress is very small compared to the tangential stress in the wall. The reason is that thin walls offer low resistance to bending and thus the internal forces exerted on a given portion of the wall are tangent to the surface of the pressure vessel. The resulting stresses on an element of the wall will be contained in a plane tangent to the surface of the vessel, as shown in Figure 6-31.

In the *cylindrical pressure vessel*, because of the axisymmetric geometry and loading, no shearing stresses are exerted and hence the 2-D stress state consists of principal stresses along the tangential direction called hoop stress σ_1 (Figure 6-31a) and along the axial direction (if the pressure vessel is closed at the ends) called longitudinal stress σ_2.

To determine the hoop stress σ_1, consider an element of length Δx as shown in Figure 6-32, removed from any location along the axis of the cylinder. This element is bounded by the x–y

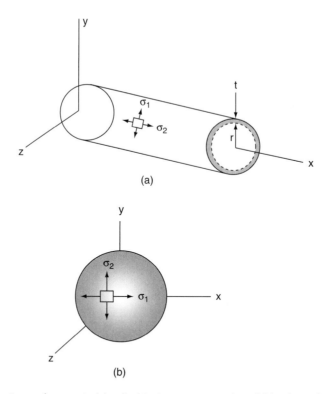

(a)

(b)

FIGURE 6-31 *Stress elements in (a) cylindrical pressure vessel, and (b) spherical pressure vessel.*

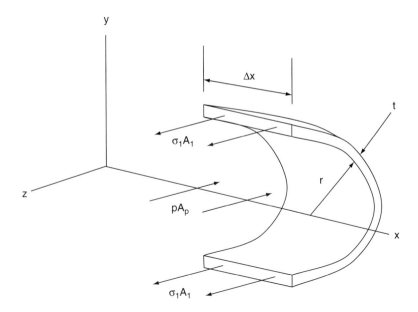

FIGURE 6-32 *First free-body diagram for the cylindrical pressure vessel.*

plane and two planes perpendicular to the x-axis separated by a distance Δx. Only half the cylinder is considered due to symmetry about the x–y plane.

Writing the equilibrium equation $\Sigma F_z = 0$, we have

$$\sigma_1 A_1 + \sigma_1 A_1 - pA_p = 0$$
$$\Rightarrow \sigma_1(t\Delta x) + \sigma_1(t\Delta x) - p(2r\Delta x) = 0$$
$$\Rightarrow \sigma_1 = \frac{pr}{t} \tag{6.78}$$

where p is the gage pressure inside the vessel, that is, the absolute pressure minus the atmospheric pressure, and Ap is the projected area in xy-plane.

To determine the longitudinal stress σ_2, cut a section perpendicular to the x-axis as shown in Figure 6-33. The forces acting (on the left side section) are due to the internal stress σ_2 acting on the area $2\pi rt$ and the internal pressure p on the area πr^2. Writing the equilibrium equation, $\Sigma F_x = 0$, we have

$$\sigma_2 A_2 - pA_p = 0$$
$$\Rightarrow \sigma_2(2\pi rt) - p(\pi r^2) = 0$$
$$\Rightarrow \sigma_2 = \frac{pr}{2t}. \tag{6.79}$$

Note from Eqs. (6.78) and (6.79) that

$$\sigma_1 = 2\sigma_2.$$

Design of Structures under Combined Loads   255

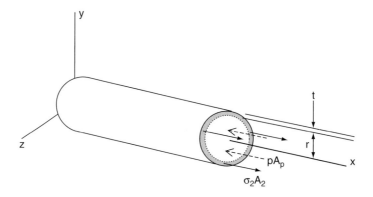

FIGURE 6-33 *Second free-body diagram for the cylindrical pressure vessel.*

EXAMPLE 6-15:

A closed cylindrical pressure vessel has an internal diameter of 1.8 m and is made of steel plate 0.018 m thick. The internal pressure is 1.4 MPa. Determine the shearing stress in a region close to the internal surface of the vessel wall.

Answer:

The hoop stress is given by Eq. (6.78) as

$$\sigma_1 = \frac{pd}{2t} = 70 \, \text{MPa}.$$

The longitudinal stress is given by Eq. (6.79) as

$$\sigma_2 = \frac{pd}{4t} = 35 \, \text{MPa}.$$

The radial stress on the internal surface is

$$\sigma_3 = -p = -1.4 \, \text{MPa}.$$

Thus the maximum shear stress is

$$\tau_{\text{max}} = \frac{1}{2}(\sigma_1 - \sigma_3) = \frac{1}{2}(70 + 1.4) = \underline{35.7 \, \text{MPa}}.$$

(Note: This expression for τ_{max} is valid if the principal stresses are algebraically arranged as $\sigma_1 > \sigma_2 > \sigma_3$.)

6.8.2 Bearing and Contact Stress

One unique aspect of compressive loading is the problems of *bearing stress* or *contact stress*. This arises from the fact that the compressive member may carry considerable load down to a

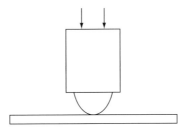

FIGURE 6-34   *Compressive loads carried to a support through reduced area.*

support through a small area, thus increasing the stress at the contact or bearing point over the nominal stress in the compressive element (Figure 6-34).

Bearing loads and the resulting bearing stresses are assumed to be distributed over the whole surface rather than being concentrated over a very small area. Bearing stress is due to a direct compression between members over a "projected area." For example, in pin-connected members the bearing load P is distributed over the projected area (Ld), as shown in Figure 6-35.

Contact stresses occur in a similar fashion, but over "contact areas" of much smaller dimensions than for bearing stresses. The contact area is more difficult to compute than the bearing area and will be discussed further later. As in all strength designs, the concern is to keep the bearing stress or the contact stress in the support less than or equal to the appropriate allowable strength.

Examples of bearing or contact stress occur in bolted joints, meshing of gear teeth, ball and roller bearings, and beams resting on foundations. In the last example, the American Institute of Steel Construction states: "When a beam is supported by a masonry wall or pilaster, it is essential that the beam reaction be distributed over an area sufficient to keep the average

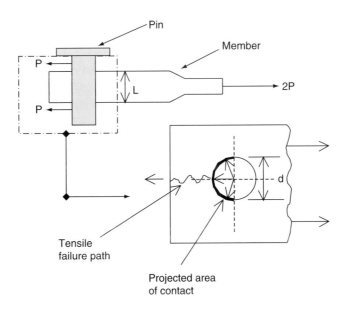

FIGURE 6-35   *Distributed bearing loading due to the contact between pin and the member.*

pressure on the masonry within allowable limits." Another example would be the C-channels contacting the floor in Figure 3-4.

The bearing stress, σ_b, is calculated as

$$\sigma_b = \frac{\text{bearing load}}{\text{bearing area}}.$$

Notice in Figure 6-35 that bearing loading gives rise to hoop tension, which can cause tensile failure as shown. Though the distribution of the bearing load on the bearing surfaces is assumed uniform, it may actually be quite different. A uniform distribution depends on the fit of the pin in the hole, rigidity of the members, and bending of the pin.

For complicated surfaces, exact determination of contact stress is difficult since it depends strongly on the geometry of the contacting surfaces and is three-dimensional. One approach to the problem is to assume the contact surfaces are either spherical or cylindrical. The resulting contact stresses are called *Hertzian stresses* (after the German engineer Heinrich Hertz, 1857–1894).

When two circular spheres of radius D_1 and D_2 are forced into contact (Figure 6-36), the resulting circular contact area of radius a can be shown to be (Shigley and Mischke, 2001)

$$a = \sqrt[3]{\frac{3F}{8} \frac{(1 - v_1^2)/E_1 + (1 - v_2^2)/E_2}{1/D_1 + 1/D_2}} \tag{6.80}$$

where F is the applied force, and E_1, v_1, E_2, v_2 refer to the elastic modulus and Poisson's ratio of sphere 1 and sphere 2, respectively.

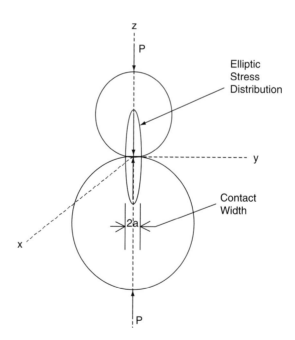

FIGURE 6-36 *Two circular spheres (or circular cylinders) placed in contact.*

The maximum pressure on each sphere occurs at the center of the contact area and is given by

$$p_{max} = \frac{3F}{2\pi a^2}. \tag{6.81}$$

The actual stress distribution is somewhat complicated, and in fact is three-dimensional (all six components of stress are present). But the maximum stress is in the z-direction and is sufficient to consider for our purposes here. (Keep in mind, however, that for a given material, the maximum normal stress may not be the appropriate stress to consider for failure.)

The z-direction normal stress (along the z-axis) is given by

$$\sigma_z(z) = \frac{-p_{max}}{1 + \frac{z^2}{a^2}}. \tag{6.82}$$

A plot of σ_z versus depth in the sphere looks like the one shown in Figure 6-37.

Note that at a distance of $z = 2a$ (one contact area diameter away from the contact surface), the maximum contact stress σ_z has diminished by about 80% from its maximum value.

It has been shown that the normal stresses are maximum at the contact surface, while the maximum shear stress $\tau_{max} = 0.3 P_{max}$ occurs slightly below the contacting surface at approximately $z = 0.4a$. It is believed that this maximum shear stress is responsible for surface fatigue failures of the contacting surface by the mechanisms of *spallation*. Spallation is the removal of a small, thin chip from the surface. A crack originates at the point of maximum shear stress below the surface and then propagates to the surface, resulting in the chip.

If instead of two spheres in contact, we consider the case of two circular cylinders of length L in contact (Figure 6-36 is now an edge view), then the contact area is a narrow rectangle of sides $2b$ by L, where b is given by

$$b = \sqrt{\frac{2F}{\pi L} \frac{(1 - v_1^2)/E_1 + (1 - v_2^2)/E_2}{1/D_1 + 1/D_2}} \tag{6.83}$$

and the maximum pressure is given by

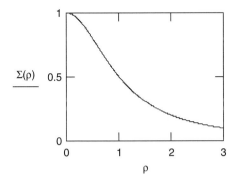

FIGURE 6-37   *Plot of* $\Sigma = \sigma_z/p_{max}$ *versus* $\rho = z/a$.

$$p_{max} = \frac{2F}{\pi bL}. \tag{6.84}$$

The maximum stress is given in this case by

$$\sigma_z = \frac{-p_{max}}{\sqrt{1 + \frac{z^2}{b^2}}}. \tag{6.85}$$

The magnitude of the maximum shear stress is $0.3P_{max}$ and it occurs at the point $z = 0.75b$ below the contacting surfaces for $v = 0.3$.

EXAMPLE 6-16:

Two steel balls of 25 mm diameter each are pressed together by a 1 kN force. Determine the maximum normal stress at the contact point. Given that $E = 210\,GPa$, $v = 0.3$.
   Answer: D = 0.025 m, F = 1000 N
   The radius of contact area is

$$a = \sqrt[3]{\frac{3F}{8}\frac{1 - v^2}{E}D}$$
$$\Rightarrow a = 3.438 \times 10^{-4}\,m$$

The maximum pressure over the contact area is

$$p_{max} = \frac{3F}{2\pi a^2}$$
$$\Rightarrow p_{max} = \underline{4.04 \times 10^9\,N/m^2}$$

Which is the maximum stress at the contact point.
   The resulting contact stress is seen to be very high, and a failure analysis would have to be considered carefully.

EXAMPLE 6-17:

A plate of steel 3 in. thick and 12 in. wide is placed over two rollers, as shown in Figure E6-17. The rollers are made of ASTM class 25 cast iron and have a diameter of 8 in. and length $L = 12\,in$. Find the maximum Hertz contact pressure and the maximum shear stress. If a uniform pressure was assumed over the contact area, determine the magnitude of the pressure. Is there a difference between the Hertzian and uniform pressure values? Spacing $l = 12\,in$, density of cast iron = $0.28\,lb/in^3$.
   The compressive load on each cylindrical roller due to the plate is equal to

$$2 \times 12\,in. \times 3\,in. \times 12\,in. \times 0.28\,lb/in^3 = 241.92\,lb$$

Weight of the roller $= \frac{\pi}{4} \times 8\,in.^2 \times 12\,in. \times 0.28\,lb/in^3 = 168.84\,lb.$

EXAMPLE 6-17: *Cont'd*

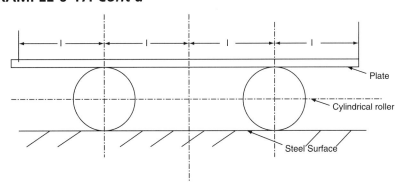

FIGURE E6-17

Hence the critical location of contact is at the roller–steel surface contact, where the compressive load is

$$F = 241.92 + 168.84 = 410.76 \, \text{lb}.$$

Hence we will consider the problem as a cast iron cylinder contacting a plane steel surface under a load of 410.76 lb uniformly distributed over the length of the cylinder.
Diameter of the roller $D_1 = 8$ in.
Diameter of surface $D_2 = \infty$

$$\text{Contact width } b = \sqrt{\frac{2 \times 410.76 \, \text{lb}}{\pi \times 12 \, \text{in}} \frac{\dfrac{1 - 0.3^2}{30 \times 10^6 \, \text{psi}} + \dfrac{1 - 0.26^2}{13 \times 10^6 \, \text{psi}}}{\dfrac{1}{8 \, \text{in}} + \dfrac{1}{\infty}}}$$

$$= 4.22 \times 10^{-3} \text{in}.$$

The maximum Hertz contact pressure is

$$p_{\max} = \frac{2F}{\pi b L} = \frac{2 \times 410.76 \, \text{lb}}{\pi \times 4.22 \times 10^{-3} \, \text{in.} \times 12 \, \text{in.}} = \underline{5165.7 \, \text{psi}}.$$

The max shear stress $= \tau_{\max} = 0.3 p_{\max} = 0.3 \times 5165.7 = 1549.7 \, \text{psi}$ at a location $z = 0.75b = 3.16 \times 10^{-3}$in. below the contacting steel surface or at $z = 3.61 \times 10^{-3}$ in above the contact surface inside the roller.

If a uniform pressure distribution is assumed, the force $F = 410.76 \, \text{lb}$ acts over the contact area of $(2bL \, \text{in.}^2) = 2 \times 4.22 \times 10^{-3} \times 12 \, \text{in}^2 = 101.28 \times 10^{-3} \, \text{in}^2$

Therefore

$$\text{Uniform pressure } p = \frac{F}{2bl} = \frac{410.76 \, \text{lb}}{101.28 \times 10^{-3} \, \text{in}^2} = 4055.7 \, \text{psi}.$$

$\varDelta$ pressure $=$ Hertz maximum contact pressure $P_{\max}$ $-$ Uniform pressure p $= 5165.7 - 4055.7 = \underline{1110 \, \text{psi}}.$
Hence, assuming a uniform pressure underestimates the contact pressure by 21%.

Key Points to Remember

- The general state of stress at any point in a stressed material is given by three normal stress components and six shear stress components. Three relationships exist between the six shear stress components and hence only three shear stress components need to be independently known.
- The general state of strain at any point in a deformed material due to applied loads is given by three normal strain components and six shear strain components. Three relationships exist between the six shear strain components and hence only three shear strain components need to be independently known.
- The Cartesian components of stress and strain depend up on the choice of the coordinate system, though this choice of the coordinate system does not affect the physical phenomenon occurring at the point.
- The three principal stresses and three principal strains at a point are independent of the choice of the coordinate system, or in other words do not vary with the orientation of the coordinate system.
- The Mohr's circle is an excellent graphical method for visualizing stresses and strains in two-dimensional and three-dimensional bodies.
- Yielding or plastic flow occurs due to the phenomenon of slip on certain closed packed planes in a crystal.
- Plastic flow is affected by the pure shear stress state, while the hydrostatic stress state has no effect.

References

Askeland, D. R., and Phule, P. P. (2003). *The Science and Engineering of Materials*, 4th ed. Thomson Brooks/Cole, San Francisco.

Dally, J. W., and Riley, W. F. (1991). *Experimental Stress Analysis*, 3rd ed. College House Enterprises, Knoxville, TN.

Shigley, J. E., and Mischke, C. R. (2001). *Mechanical Engineering Design*, 6th ed. McGraw-Hill, New York.

Welsh, J. S., Mayes, J. S., and Key, C. T. (2002). Damage initiation mechanics in glass fabric composites. *AIAA J*, 1742.

Problems

6-1. Consider a function $\Phi(x, y)$, which is called the stress function. If the stresses in a plane stress case are written in terms of the stress function as given below, show that these stresses satisfy the equations of equilibrium [Eq. (6.6)] if the body forces are neglected.

$$\sigma_{xx} = \frac{\partial^2 \Phi}{\partial y^2}, \quad \sigma_{yy} = \frac{\partial^2 \Phi}{\partial x^2}, \quad \tau_{xy} = -\frac{\partial^2 \Phi}{\partial x \partial y}.$$

6-2. Draw elements similar to that in Figure 6-7 in the $y-z$ and $x-z$ plane and prove the remaining equations in Eq. (6.5).

6-3. Write Eq. (6.24) in terms of the six stress components σ_{xx}, σ_{yy}, σ_{zz}, τ_{xy}, τ_{xz}, and τ_{yz}.

6-4. At a point in a machine component, the stress components are $\sigma_{xx} = 70\,\text{MPa}$, $\sigma_{yy} = -35\,\text{MPa}$, $\sigma_{zz} = 35\,\text{MPa}$, $\tau_{xy} = 40\,\text{MPa}$, $\tau_{yz} = \tau_{zx} = 0$. Determine the normal and shear stress on a plane passing through the point whose outer normal has the direction cosines

$$l = 0.428, \quad m = 0.514, \quad n = 0.742.$$

6-5. At a point in an m/c component the stress components are $\sigma_{xx} = 40\,\text{MPa}$, $\sigma_{yy} = 80\,\text{MPa}$, $\sigma_{zz} = 40\,\text{MPa}$, $\tau_{xy} = 90\,\text{MPa}$, $\tau_{yz} = 50\,\text{MPa}$, $\tau_{zx} = 30\,\text{MPa}$. Determine the normal and shear stress on a plane passing through the point whose outer normal has the direction cosines:

$$l = 0.444, \quad m = 0.555, \quad n = 0.777.$$

6-6. Consider the stress state given in problems 6-4 and 6-5. Transform these stress states into a new state relative to a transformed axes $x'y'z'$, where the direction cosines of the transformed axes relative to the original coordinate system xyz are given in the accompanying tables.

(a)

	x	y	z
x'	0.667	0.667	−0.333
y'	−0.667	0.333	−0.667
z'	−0.333	0.667	0.667

(b)

	x	y	z
x'	0.182	0.545	0.818
y'	0.545	0.636	−0.545
z'	0.818	−0.545	0.182

(c)

	x	y	z
x'	0.360	0.480	0.80
y'	0.48	0.64	−0.60
z'	0.80	−0.60	0

6-7. Solve Example 6-4 if the plane has an outer normal inclined to the coordinate axes at (a) $90°$ to the x-axis, $45°$ to the y-axis, and $45°$ to the z-axis; (b) equally inclined to all three coordinate axes.

6-8. At a point in a body $\sigma_{xx} = 10\,\text{MPa}$, $\sigma_{yy} = -5\,\text{MPa}$, $\sigma_{zz} = -5\,\text{MPa}$, $\tau_{xy} = \tau_{yz} = \tau_{zx} = 10\,\text{MPa}$. Determine the normal and shear stress on a plane that is equally inclined to all three coordinate axes.

6-9. Show that the stress invariants can be written in terms of the principal stresses as follows:

$$I_1 = \sigma_1 + \sigma_2 + \sigma_3$$
$$I_2 = \sigma_1\sigma_2 + \sigma_2\sigma_3 + \sigma_2\sigma_3$$
$$I_3 = \sigma_1\sigma_2\sigma_3.$$

6-10. At a point in a stressed member the Cartesian components are
$\sigma_{xx} = \sigma_{yy} = \sigma_{zz} = 50\,\text{MPa}$, $\tau_{xy} = 200\,\text{MPa}$, $\tau_{xz} = 0$, and $\tau_{yz} = 150\,\text{MPa}$. Determine the principal stresses and the maximum shear stresses at the point. Check the invariance of I_1, I_2, and I_3.

6-11. For the state of stress given below, determine the principal stresses and their associated directions. Explain the results.

$$\begin{bmatrix} 0 & 10 & 10 \\ 10 & 0 & 10 \\ 10 & 10 & 0 \end{bmatrix}$$

6-12. A stress state is given as $\sigma_{xx} = -5$, $\sigma_{yy} = 1$, $\sigma_{zz} = 1$, $\tau_{xy} = -1$, $\tau_{yz} = \tau_{zx} = 0$. All stresses in MPa. Determine the principal stresses, the direction cosines of the principal stress directions, and the maximum shear stress.

6-13. Shown in Figure P6-13 is a beam in which the shear stress τ_{xy} on the top surface is zero. If the beam is cut along the plane A, show that the shear stress on the new surface is zero.

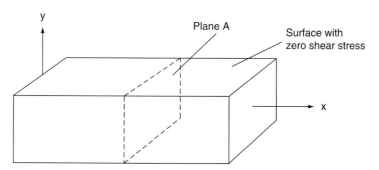

FIGURE P6-13

6-14. If the Cartesian coordinate axes are chosen to coincide with principal stress directions, the stress matrix can be written as

$$\begin{bmatrix} \sigma_1 & 0 & 0 \\ 0 & \sigma_2 & 0 \\ 0 & 0 & \sigma_3 \end{bmatrix}.$$

Determine the normal and shear stress on a plane having a normal **n**. Also write the stress invariants in terms of the principal stress σ_1, σ_2, and σ_3.

6-15. Consider a coordinate system *xyz* that coincides with the three principal directions (Figure P6-15) in which a plane is equally inclined to these three axes. Such a plane is called an *octahedral plane*. The normal and the shear stresses on these planes are called the octahedral normal stress σ_{oct}, and octahedral shear stress τ_{oct}, respectively.

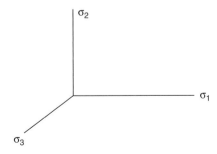

FIGURE P6-15

Show that there are eight octahedral planes, determine the direction cosines of their outward normals, and schematically draw them. Determine the expression for σ_{oct} and τ_{oct} in terms of the principal stresses.

6-16. Rewrite the expressions for σ_{oct} and τ_{oct} derived in Problem 6-15 in terms of the stress invariants, and in terms of Cartesian stress components.

6-17. A state of stress is given below:

$$\begin{bmatrix} 1 & 2 & 4 \\ 2 & 2 & 1 \\ 4 & 1 & 3 \end{bmatrix} \text{MPa.}$$

Determine (a) the octahedral normal and shear stresses, (b) the hydrostatic stress state, and (c) the pure shear stress state.

6-18. Show that subtracting a hydrostatic stress from a given arbitrary stress state does not change the principal stress directions.

6-19. If the stress state in a component is $\sigma_1 = \sigma_2 = \sigma_3$, draw the 3-D Mohr's circles. Comment on the shear stress state in an arbitrary plane.

6-20. In Example 6-6 use the Mohr's circle approach to determine the principal stresses and the maximum shear stress.

6-21. If $\sigma_{xx} = 45\,\text{MPa}$, $\sigma_{yy} = 90\,\text{MPa}$, and $\tau_{xy} = 45\,\text{MPa}$, use the Mohr's circle method to determine the in-plane principal stresses.

6-22. For each of the four states of stress shown in Figure P6-22, determine (a) the principal stresses, (b) the principal planes, and (c) the maximum shear stress.

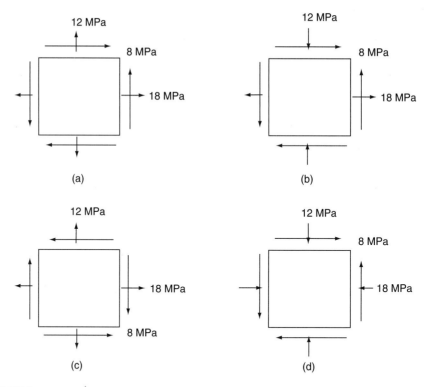

(a)   **(b)**

(c)   **(d)**

FIGURE P6-22

6-23. For the state of stress shown in Figure P6-23, determine the normal and shear stresses on the element if it is rotated through (a) 30° clockwise, and (b) 30° counterclockwise.

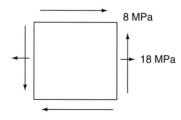

FIGURE P6-23

6-24. Two points P and P′ in an undeformed body have coordinates (1, 1, 1) and (2, 0, −1), respectively. If the displacement field given in Example 6-1 is imposed on the body, what is the distance between the points after deformation?

6-25. If the displacement field for a body is given by

$$\mathbf{u} = y^2\hat{\mathbf{i}} + 3yz\hat{\mathbf{j}} + (4 + 6x^2)\hat{\mathbf{k}},$$

write the strain matrix at the point (2, 3, −1).

6-26. The displacement field for a body is given by

$$\mathbf{u} = (x^2 + 2)\hat{\mathbf{i}} + (3x + 4y^2)\hat{\mathbf{j}} + (2x^3 + z)\hat{\mathbf{k}}.$$

What is the deformed position of a point originally at $(1, 2, 3)$?

6-27. For the displacement field given in Problem 6-26, what are the normal strain components at the point $(1, 2, 3)$?

6-28. If the normal strain is defined by the following equations, which are used for large strains, determine the normal strains at the point $(1, 2, 3)$. The displacement field is given as: $u = (x^2 + y^2 + 2)10^{-4}$, $v = (3x + 4y^2)10^{-4}$, and $w = (2x^3 + 4z)10^{-4}$. Compare these strains to those obtained using Eqs. (6.3) that use the small strain assumption and determine the error caused due to this assumption.

$$\varepsilon_{xx} = \frac{\partial u}{\partial x} + \frac{1}{2}\left[\left(\frac{\partial u}{\partial x}\right)^2 + \left(\frac{\partial v}{\partial x}\right)^2 + \left(\frac{\partial w}{\partial x}\right)^2\right]$$

$$\varepsilon_{yy} = \frac{\partial v}{\partial y} + \frac{1}{2}\left[\left(\frac{\partial u}{\partial y}\right)^2 + \left(\frac{\partial v}{\partial y}\right)^2 + \left(\frac{\partial w}{\partial y}\right)^2\right]$$

$$\varepsilon_{zz} = \frac{\partial w}{\partial z} + \frac{1}{2}\left[\left(\frac{\partial u}{\partial z}\right)^2 + \left(\frac{\partial v}{\partial z}\right)^2 + \left(\frac{\partial w}{\partial z}\right)^2\right]$$

6-29. Repeat Problems 6-27 and 6-28 using the displacement field given below:

$$\mathbf{u} = \left[(x^2 + 2)\hat{\mathbf{i}} + (3x + 4y^2)\hat{\mathbf{j}} + (2x^3 + z)\hat{\mathbf{k}}\right] \times 10^{-3}.$$

6-30. Transform the Cartesian components of strain listed in Example 6-12 to a new coordinate system $x'y'z'$, which is defined by the following direction cosines:

(a)

	x	y	z
x'	0.36	0.48	0.80
y'	0.48	0.64	−0.60
z'	0.80	−0.60	0.0

(b)

	x	y	z
x'	0.11	−0.89	0.44
y'	−0.89	0.44	0.78
z'	0.44	0.78	0.44

6-31. Determine the three principal strains, the maximum shear strain, and the direction of the maximum principal strain for the following cases:

(a) $\varepsilon_{xx} = 450\,\mu\varepsilon$, $\varepsilon_{yy} = 300\,\mu\varepsilon$, $\varepsilon_{zz} = 150\,\mu\varepsilon$
$\gamma_{xy} = 150\,\mu\varepsilon$, $\gamma_{yz} = 150\,\mu\varepsilon$, $\gamma_{xz} = 300\,\mu\varepsilon$.

(b) $\varepsilon_{xx} = 900\,\mu\varepsilon$, $\varepsilon_{yy} = 800\,\mu\varepsilon$, $\varepsilon_{zz} = 500\,\mu\varepsilon$
$\gamma_{xy} = 300\,\mu\varepsilon$, $\gamma_{yz} = 650\,\mu\varepsilon$, $\gamma_{xz} = 500\,\mu\varepsilon$.

(c) $\sigma_{xx} = 220\,\text{MPa}$, $\sigma_{yy} = 75\,\text{MPa}$, $\sigma_{zz} = 155\,\text{MPa}$
$\tau_{xy} = 110\,\text{MPa}$, $\tau_{yz} = 50\,\text{MPa}$, $\tau_{xz} = 60\,\text{MPa}$.

(d) $\sigma_{xx} = 220\,\text{MPa}$, $\sigma_{yy} = 75\,\text{MPa}$, $\sigma_{zz} = 0\,\text{MPa}$
$\tau_{xy} = 110\,\text{MPa}$, $\tau_{yz} = 0\,\text{MPa}$, $\tau_{xz} = 0\,\text{MPa}$.

6-32. Consider a small rectangular element in a deformed body that has its edges oriented along the principal axes. No shearing strains occur, only normal strains are applied. Hence the shape of the element does not change, however, its volume changes. Show that the quantity known as volumetric strain is given by

$$\frac{\Delta V}{V} = \frac{\text{Change in volume}}{\text{Original volume}}$$
$$= \varepsilon_1 + \varepsilon_2 + \varepsilon_3.$$

6-33. A cube of steel has an edge length of 25 mm when unstressed. A tensile stress of 50,000 Pa is applied normal to one pair of parallel faces, and a tensile stress of 60,000 Pa applied normal to the second pair of parallel faces. What is the distance between the third pair of parallel faces after the application of the load?

6-34. Rework Problem 6-33, except this time the stresses applied are compressive (pressures).

6-35. At a certain point in a steel machine component the strains are given below. Determine the stress matrix, given that for steel $E = 207\,\text{GPa}$, $v = 0.3$.

$$\varepsilon_{xx} = 450\,\mu\varepsilon, \quad \varepsilon_{yy} = 300\,\mu\varepsilon, \quad \varepsilon_{zz} = 150\,\mu\varepsilon$$
$$\gamma_{xy} = 150\,\mu\varepsilon, \quad \gamma_{yz} = 150\,\mu\varepsilon, \quad \gamma_{xz} = 300\,\mu\varepsilon.$$

6-36. A thin rectangular rubber sheet is enclosed between two thick steel plates and the rubber sheet is subjected to a compressive stress of σ_{xx} and σ_{yy} in the x- and y-directions, respectively. Determine the strains in the x- and y-directions and the stress along the z-direction (thickness direction) of the rubber sheet.

6-37. Determine the change in volume of a 10-mm cube of plastic ($E = 3\,\text{GPa}$ and $v = 0.35$) when dropped to the bottom of a lake 2000 m deep.

6-38. The flat surface of a machine component is found to have principal stresses of 200 MPa and 100 MPa. What will be the yield strength of the material of the component if (a) the maximum shear stress theory is used, (b) the distortion energy theory is used? What should be the tensile yield strength of the material if a factor of safety of 2 with respect to initial yielding is required?

6-39. A round shaft is subjected to a torque T as shown in Figure P6-39. Determine the diameter required to just avoid initial yielding using:

(a) maximum normal stress theory

(b) maximum shear stress theory

(c) distortion energy theory.

The material of the shaft is steel with a tensile yield strength of 60 ksi.

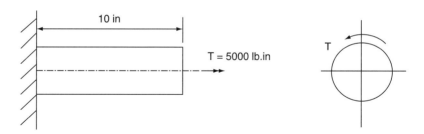

FIGURE P6-39

6-40. Solve Problem 6-39 again for the loading condition shown in Figure P6-40.

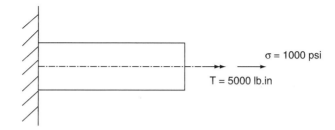

FIGURE P6-40

6-41. Solve Problem 6-39 again for the loading condition shown in Figure P6-41.

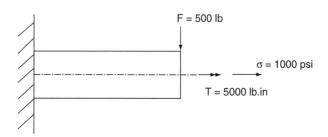

FIGURE P6-41

6-42. For each stress state shown in Figure P6-42, draw a Mohr's circle diagram, find all three principal normal stresses, the three principal shear stresses, and the angle between the x-axis and the maximum principal stress σ_1. All stresses are in MPa.

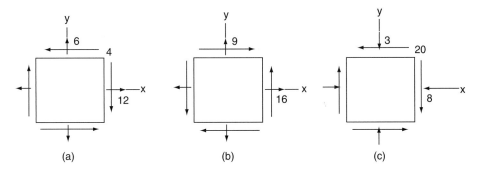

FIGURE P6-42

6-43. Three elements a, b, c, made of the same material and shown in Figure P6-43, are subjected to different states of stress. Which element will yield first according to the (i) maximum normal stress theory, (ii) maximum shear stress theory, and (iii) distortion energy theory? All stresses are in MPa.

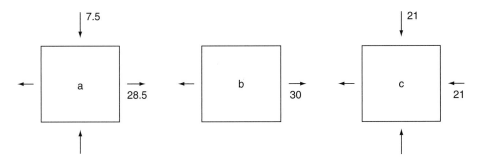

FIGURE P6-43

6-44. At a point in a machine component, the state of stress is as shown in Figure P6-44. The tensile yield strength of the materials is 420 MPa. If the shearing stress at the point is $\tau = 210$ MPa, when yielding initiates, what is the tensile stress σ at the point (a) according to the maximum shear stress theory, and (b) according to the distortion energy theory?

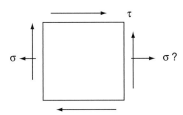

FIGURE P6-44

6-45. A cylindrical tube made of 2024-T4 aluminum has a diameter of 50 mm and wall thickness of 3 mm. An axial tensile load of 60 kN and a torque of 0.7 kN-m are applied.

Will the tube yield? If not, how much can the tensile load or the torque be increased before yielding occurs?

6-46. A thin-walled circular tube is made of AISI 1020 steel. It is subjected to a torque of 6 kN-m and a pure bending moment of 4.5 kN-m. If the diameter of the tube is 50 mm, what must be the thickness so that the factor of safety against yielding is 1.25?

6-47. A 1-D link is subjected to simultaneous axial compression and torsion such that $\sigma_{xx} = -30$ kpsi and $\tau_{xy} = 15$ kpsi, respectively. Use the principal stress and maximum shear stress equations to show that this structure will fail in tension, and that the maximum shear stress is greater than the applied shear stress. Verify using Mohr's circle.

6-48. Using the Mohr's circle diagram determine the maximum shear stress for a cylindrical pressure vessel.

6-49. Consider a thin-walled *spherical pressure vessel* of radius r and wall thickness t, with internal pressure p. Using the methodology of Section 6.8.1, determine the internal stress state at any point in the wall.

6-50. What is the third principal stress on the outer surface and the inner surface of a cylindrical closed pressure vessel with internal pressure p? Determine the maximum shear stress in each case.

6-51. A pressure vessel made of AISI 1020 steel has an inner diameter of 7.5 in. and a wall thickness of 0.25 in. Does this vessel qualify as a thin-walled pressure vessel? If the yield strength of the steel is 225 MPa, what is the maximum permissible pressure that can be used?

6-52. A cylindrical pressure vessel has an inner diameter of 240 mm and a wall thickness of 10 mm. The end caps are spherical and of thickness 10 mm. If the internal pressure is 2.4 MPa, find (a) the normal stress and the maximum shear stress in the cylindrical wall, (b) the normal stress and the maximum shear stress in the wall of the spherical end cap.

6-53. A cylindrical pressure vessel has an outer diameter of 1000 mm and is fabricated from AISI 1020 steel with a yield strength of 225 MPa. The maximum internal pressure during service is expected to be 3.5 MPa. If a factor of safety of 2 is desired, determine the smallest wall thickness that must be used.

6-54. A closed cylindrical pressure vessel is fabricated from steel sheets that are welded along a helix that forms an angle of 60° with the transverse plane (Figure P6-54). The outer diameter is 1 m and the wall thickness is 0.02 m. For an internal pressure of 1.25 MPa, determine the stress in directions perpendicular and parallel to the helical weld.

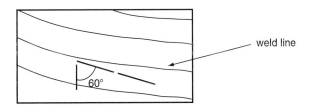

weld line

60°

FIGURE P6-54

6-55. Draw a schematic graph of true stress versus true strain for a material that shows no work hardening.

6-56. It was shown in Section 4.4.2 that when a specimen is loaded in tension shear stresses appear on different slip planes of the specimen. Find the angle between the force and a slip plane that maximizes the shear force acting on the plane. What is the angle if a compressive load is applied?

6-57. A tensile stress of 10,000 N/m^2 is applied to a cubic crystal in the [100] direction. What is the shear stress on the {110} plane in the [$\bar{1}$11] direction?

6-58. Rework Example 6-16, except one of the balls is steel and the other is aluminum.

6-59. Rework Problem 6-58, but now consider in contact one steel cylinder with one aluminum cylinder. Each cylinder has a diameter of 25 mm and length of 100 mm.

6-60. Plot the ratio of $\Sigma = \sigma_z / p_{max}$ versus z/b for two circular cylinders in contact. By how much is the stress diminished at a distance of $2b$ away from the contacting surfaces?

6-61. A railway wagon of mass 10,000 kg rests on two axles. Each axle has two cylindrical wheels of 600 mm diameter that rest on flat rails of 30 mm width. The wheels and the rails are made of AISI 1030 Q & T (205°C) grade steel with a yield strength of 648 MPa. Would you approve the choice of material? If not, suggest an appropriate material for the wheels and rails.

6-62. In a ball bearing, a spherical ball of 10 mm diameter presses against a semicircular groove of 11 mm diameter, as shown in Figure P6-62. Note that the groove diameter is considered an *internal diameter* (with negative value), while the ball diameter is considered an *external diameter*. Because of external loading, the force pressing the ball against the groove is 2500 N. What are the possible failure modes? Using Ashby's materials selection charts, create a list of possible materials for the ball and groove. Choose the best material and determine its suitability for this application. Give specification of the materials used.

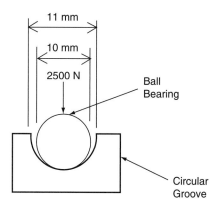

FIGURE P6-62

6-63. A square element is distorted under the action of a plane stress field, resulting in the strains shown in Figure P6-63.

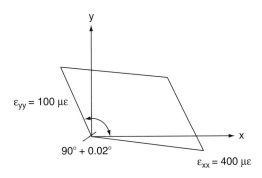

$\varepsilon_{yy} = 100 \ \mu\varepsilon$

$90° + 0.02°$

$\varepsilon_{xx} = 400 \ \mu\varepsilon$

FIGURE P6-63

Using the Mohr's circle method determine (a) the principal axes and strains, (b) the maximum shear strain and the corresponding normal strain.

6-64. Using the Ashby materials selection method make a list of materials that are suitable for fabricating a safe cylindrical pressure vessel with an internal pressure of 25,000 Pa. The diameter of the vessel is 500 mm, length is 2 m, and it should not weigh more than 45 kg. What are the suitable materials if the weight of the vessel is 15 kg or less?

7   Fracture

The worst sin in an engineering material is not lack of strength or lack of stiffness . . . , but lack of toughness, that is to say, lack of resistance to the propagation of cracks.

—J. E. Gordon, *The New Science of Strong Materials* (1976)

Objective: To understand cracking and fracture behavior of materials for designing safe and fracture resistant structures. Two DDOF are introduced: a, K_C.

What the student will learn in this chapter:

- Why we must consider fracture mechanics and its relationship with different materials in the design of structures
- Introduction to fracture mechanics
- The atomistic process of fracture
- Design of simple structural elements with crack-like defects
- An introduction to materials selection for fracture-resistant structures

7.1   Introduction to Fracture in Materials ⟁🏠💻

The aim of *fracture mechanics* is to provide engineers with a quantitative tool for designing against fracture in engineering structures. Fracture in a structure can lead to substantial financial loss and worse—loss of life. Some dramatic examples of fracture are the failure of the Liberty ship hulls in the early 1940s (see Figure 7-1), fracture of the Titanic passenger ship due to impact, the Comet aircraft in the 1950s, and more recently an Aloha Airlines aircraft whose outer skin fractured and peeled off in flight (see Figure 7-2). In all these examples, the structure underwent a catastrophic failure.

In a vast majority of the cases though, a structure develops cracks and continues to function normally until the crack is detected during periodic inspection. For example, Figure 7-3 shows a large crack in the aluminum 7075-T73511 strut of a large commercial airliner. This particular failure occurred as a result of a ductile overload and had no other prior defects that caused the cracking. Other components that are more near to us, such as pressure vessels (e.g., propane gas tanks), automobile components, and household appliances, can also develop defects and

FIGURE 7-1 *The ship* Schenectady *cracked open in 1941.*

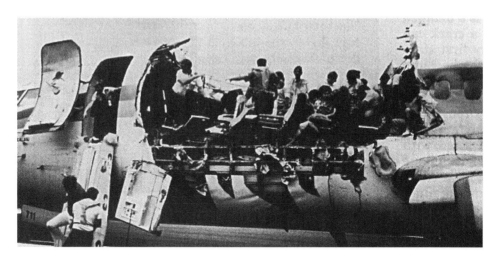

FIGURE 7-2 *Fractured and peeled skin of the fuselage of an Aloha airliner.*

cracks, sometimes causing severe injury and financial loss. Thus, the importance of design against fracture for an engineer can never be overemphasized. The designer should make the system as robust and damage resistant as possible within the constraints of cost, manufacturing

FIGURE 7-3 *A large crack in one of the struts of a commercial airliner (courtesy of Boeing Aerospace Corporation).*

capability, materials available, and so forth. The emphasis in this chapter is on the macroscopic aspects of fracture mechanics and its utilization in the design of fracture-resistant and durable structures. However, some insight has also been provided on the materials science and microscopic basis of failure mechanisms to help the reader better appreciate the relationships between macroscopic fracture and submicroscopic mechanisms that contribute to ultimate failure.

Separation in a material body may be the result of shearing along a slip plane or of separation by extension normal to the separation plane. The forces and stresses that produce the first type of separation are the same that produce slip; thus the *shear separation* process is but a continuation of the slip process, extending beyond the limits of the crystal. The extensions producing actual separation along planes normal to the lines of action of the forces, however, are unrelated to slip phenomena. The two separation processes are of a different nature.

Shearing separation can take place, however, only in metals, in which the interatomic bonds are continually reestablished in the course of the slip process. Failure in metals could be a competition between the two processes of shear flow (or plastic deformation) and fracture by separation along planes normal to line of action of force. In *ionic crystals* and *covalently bonded* crystals, slip cannot proceed without a partial disruption of bonds, unless the thermal activation of the atoms is particularly strong, making the reestablishment of the bonds possible. Such conditions are the exception rather than the rule. In *amorphous materials* (which do not have a regular, periodic arrangement of atoms, for example, glass) in which slip does not occur, separation is always associated with a force or an extension normal to the separation plane. That such separation may be preceded and accompanied by substantial inelastic deformation does not change the fact that separation and deformation are different, although not independent, processes.

When a metal is subjected to a tensile load, it can fail by *plastic deformation* or by fracture. If the stress state is not large enough to break the atomic bonds and cause permanent separation of atoms across a plane, but large enough to produce permanent deformation by shearing of atomic planes, then failure occurs by plastic flow and not by fracture. Generally failure by fracture is an undesirable event and it is relatively preferable that permanent deformation occur by plastic flow, though in real practice the designer would like to have materials that have a high resistance to both plastic flow and fracture. However, it has been found that factors contributing to an increase in a material's fracture resistance decrease its plastic flow resistance and vice versa. The stress state can also significantly affect a material's relative resistance to plastic flow and fracture. This is because fracture is affected by the dilatational stress, whereas onset of plastic flow is affected by the critical resolved shear stress. The presence of a biaxial or triaxial tensile stress state, as opposed to a uniaxial stress state, increases the level of applied stress necessary to initiate flow and simultaneously increases the dilatational stress promoting fracture. A quantitative measure of these two effects is provided by the *mean pressure*, $p = (\sigma_1 + \sigma_2 + \sigma_3)/3$, where σ_1, σ_2, and σ_3 are the principal (normal) stresses and can be either tensile or compressive. A high positive value of p promotes fracture while a negative p promotes plastic flow over fracture.

With the exception of pure shearing separation of a single metal crystal, the initiation of fracture represents a discontinuity in the deformation process. When a material is subjected to forces, the changes in its internal atomic arrangements are manifested by the deformation, which reaches a limit at the point of fracture, and leads to an instability of the internal structure. When this condition is reached, the continuous transformation of mechanical (potential) energy into heat energy is broken by the appearance of a new mechanism: transformation of the energy of the applied forces into the energy of the newly created surfaces along which fracture occurs. The manner in which this transition point is reached depends essentially on the process of change of the internal state of the atomic arrangement, that is, of the deformation preceding and leading up to the transition.

Fracture is a process of progressive separation of atomic bonds, which starts at points where the alternative mechanism of storing strain energy in the interatomic bonds is not available, and where the absolute value of the accumulated bond energy is highest. When some of the atomic bonds are broken, the energy accumulated in the remaining bonds increases, since the energy of the externally acting forces is balanced by an increase in the separation distance of the atomic particles from their position of equilibrium toward energy levels at which separation becomes increasingly possible. Atomic fracture is thus a "chain reaction" process. Fracture may be very rapid and lead to macroscopic fracture under a single force application of short duration, if the energy of the applied forces is relatively high or if its distribution over the interatomic bonds is highly nonuniform. On the other hand, macroscopic fracture may be slowly progressive and proceed under either repeated force application or long duration of forces of a relatively low energy potential, if the process is accompanied by a gradual change of the distribution of bond energies and by concentration of the response to the external forces within a decreasing number of bonds.

The level of energy of the applied forces at which the critical state is reached depends necessarily on the process of energy dissipation preceding it, that is, on the relative amounts of stored and dissipated energy. The energy that is not stored reversibly produces structural changes in the material, which are associated with its irrecoverable deformation. The close interrelation between fracture and inelastic deformation is the result of the fact that the greater the portion of the energy of the applied forces that is dissipated into heat energy in the course of inelastic deformation, the smaller the stored energy potential that remains available for fracture. Thus, under conditions that are not conducive to the dissipation of applied energy into heat energy (or other forms of energy dissipation) the main portion of the energy of the applied forces will be used to produce fracture. Such conditions are created by low temperatures and high rates of application of forces or of energy, as well as by a large share of the work of volumetric expansion in the total amount of the work of the applied forces. The difference in the effect on fracture of volumetric expansion and compression is due to the fact that, although volumetric energy cannot be effectively dissipated, volumetric compression clearly cannot cause separation. Elevated temperatures, low rates of application of force or of energy, and a large share of distortional work in the total work of the applied forces lead to a considerable intensification of the process of dissipation of applied energy into heat by place change of particles, and thus increase the permanent deformation that precedes and accompanies fracture.

If the instability of the position in which the particle finds itself as a result of changes in the external conditions cannot be resolved by a place change, the particle will tend to regain its stability by disrupting the bonds connecting it with some of the surrounding particles, initiating fracture on the atomic scale. Whether and how rapidly this atomic fracture (which is a local instability phenomenon) will multiply and spread into a visible crack depends on energy considerations, since local instability spreads only if by this process the potential of all the forces involved is diminished.

Although the probability of occurrence of local bond disruption within the submicro-scopic structure is highest at some intrinsic point of instability such as an imperfection of this structure—for instance, a single dislocation or a concentration of dislocations in a block boundary—there are factors of instability present in any structural arrangement, even in the absence of imperfections. These factors are related to the temperature oscillations within the atomic or molecular structure, as a result of which transient cracks of atomic size might be opened during particularly large low-temperature energy fluctuations. It is, however, equally probable that compensating fluctuations of thermal energy above a threshold would occur rapidly enough to heal these incipient *atomic cracks*. In fact, the heat energy set free at the same time at which atomic fracture occurs provides the heat just in the location where it is most needed to heal those fractures. Under such conditions, the levels of potential energy pertaining, respectively, to the cracked and uncracked state differ by such infinitely small amounts that no definite trend in the separation process can develop. A clear trend toward large-scale propagation of atomic fracture can develop only if the potential energies of the two states differ by finite amounts. Thus, if the interatomic distances are increased by the action of external forces, the gain in potential energy accompanying a spreading crack increases, as a result of the increased energy difference between the state of the material body before and after fracture.

Fracture cannot be produced by hydrostatic compression alone, as all particles are crowded together, not separated. If it is assumed that bond disruption is associated with a critical separation distance between the particles, a state of hydrostatic compression will increase the amount of the relative separation of particles necessary to cause bond disruption. A superimposed hydrostatic pressure, therefore, will increase the intensity of the forces required to produce fracture, whereas a hydrostatic tension will have the opposite effect. For moderate pressures, which do not cause changes in the interatomic distances of a magnitude comparable to the separation distance, these effects may be negligible. They will, however, become more pronounced the higher the volumetric compression, as well as the compressibility of the material.

Materials in which even a moderate change of hydrostatic pressure or tension appreciably affects fracture under general conditions of stress are the microscopically and macroscopically inhomogeneous materials. Microscopic inhomogeneities of the structure produce an inhomogeneous response within the material to the (homogeneous) state of hydrostatic stress. This fact explains the considerable influence on the condition of fracture of very moderate hydrostatic stresses in such materials as cast iron, concrete, and stone, in comparison to their negligible effect in the statistically homogeneous and relatively incompressible metals and amorphous substances.

Fracture is essentially the propagation from the submicroscopic into the macroscopic scale of instabilities within the potential field of the atomic or molecular forces. All effects that facilitate or inhibit the formation and propagation of such instabilities are of primary importance in influencing, if not determining, the occurrence and progress of fracture on a macroscopic scale.

7.2 Equilibrium and Deformation ϒ

7.2.1 Ideal Fracture Strength of a Solid

We have seen in Review Module R1, "Fundamental Materials Science Concepts," that the energy associated with the atoms separated by a distance equal to the equilibrium atomic spacing is a minimum. If the two atoms are separated by a large enough distance that the energy of the two-atom system and the force between the two atoms approaches zero, it may be said that the bond between the two atoms has been severed. This is the basis for initiation of fracture at the atomic scale. Based on this consideration it can be shown that fracture strength, which is the stress at which fracture occurs, is given by the equation

$$\sigma_{\text{fracture}} = \sqrt{\frac{E\Gamma_{\text{s}}}{r_{\text{o}}}} \tag{7.1}$$

where E is the Young's modulus, Γ_{s} is the surface energy per unit area of the fracture surface created, and r_{o} is the equilibrium spacing of the atoms in the lattice. Equation (7.1) can be derived as follows.

The interatomic force between two atoms varies with their separation distance as shown in Figure 7-4. The force reaches a maximum value at r_{m}, and at this point the theoretical fracture strength σ_{fracture} is attained. The force–displacement relationship due to increase of the interatomic distance from equilibrium spacing r_{o} to r_{m} can be approximated by a sinusoidal curve as shown by the dashed sine curve, and let this displacement be $\lambda/4$. In the elastic region, the stress (force per unit area, F/r_{o}^2) required to separate atoms across a plane of unit area is given by

$$\sigma = \sigma_{\text{fracture}} \sin\frac{2\pi r}{\lambda} \tag{7.2}$$

where σ_{fracture} is the theoretical *cohesive strength*.

The *fracture work* (per unit area) is given by the integral

$$\int_0^{\frac{\lambda}{2}} \sigma_{\text{fracture}} \sin\left(\frac{2\pi r}{\lambda}\right) dr = \frac{\lambda \sigma_{\text{fracture}}}{\pi}. \tag{7.3}$$

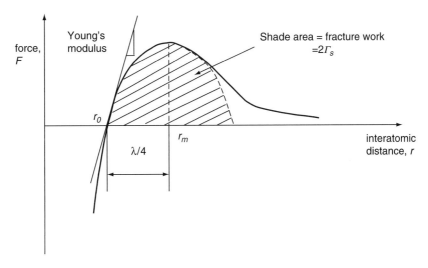

FIGURE 7-4 *Bond force versus interatomic spacing.*

The fracture work is equal to $2\Gamma_s$, where Γ_s is the surface energy per unit area created due to formation of the new fracture surface, and the factor of 2 is due to the formation of two surfaces; thus,

$$\frac{\lambda \sigma_{fracture}}{\pi} = 2\Gamma_s \tag{7.4}$$

An estimate of λ is obtained as follows:
For small displacements Eq. (7.2) can be approximated by the expression

$$\sigma = \sigma_{fracture} \frac{2\pi r}{\lambda} \tag{7.5}$$

For an elastic stretching of the bonds, stress σ is given by

$$\sigma = \varepsilon E$$

with the strain $\varepsilon = r/r_0$, where r is the change in interatomic spacing; thus

$$\sigma = \frac{r}{r_o} E. \tag{7.6}$$

Combining Eqs. (7.5) and (7.6) results in

$$\sigma_{fracture} = \frac{\lambda E}{2\pi r_o} \tag{7.7}$$

Eliminating λ between Eqs. (7.4) and (7.7), we obtain

$$\sigma_{fracture}^2 = \frac{E \Gamma_s}{r_o}$$

$$\Rightarrow \sigma_{\text{fracture}} = \sqrt{\frac{E\Gamma_s}{r_o}}. \tag{7.8}$$

(It may be noticed that the Young's modulus E can be obtained from the slope of the force–displacement curve at the equilibrium spacing r_o.)

Equation (7.8) assumes that all bonds are broken simultaneously across the fracture plane, and the stress at which this occurs is the fracture strength. The values of σ_{fracture} obtained experimentally are generally in the range shown in Eq. (7.9).

$$0.1\,E < \sigma_{\text{fracture}} < 0.2E \tag{7.9}$$

7.3 Constitution ✦

Experimental determination of fracture strength of brittle materials shows that this value is three to four orders of magnitude lower than the theoretical estimates given in Section 7.2. This discrepancy in the fracture strength is due to the inhomogeneity and imperfections of the real structure and flaws in the material, which lower the global fracture strength by magnifying the stress locally in the immediate vicinity of the crack tip. Also it is postulated that the interatomic forces across a potential fracture surface are not overcome simultaneously but in a certain sequence; a crack is thus initiated under an overall force that, by producing separation of a limited number of bonds only, is but a small fraction of the force that would be required to break all bonds simultaneously.

The first quantitative evidence for stress magnification or stress *concentration effect of flaws* was provided by Inglis (1913), who analyzed elliptical holes in a large flat plate subjected to tensile loads perpendicular to the major axis of the elliptic hole (see Figure 7-5). The stress at the tip of the major axis is given by

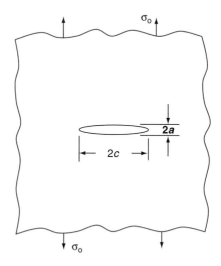

FIGURE 7-5 *Elliptic hole in a large flat plate subjected to tensile loading.*

$$\sigma = \sigma_0 \left(1 + \frac{2c}{a}\right). \tag{7.10}$$

When $c \gg a$, the elliptic hole begins to approach the configuration of a sharp crack. For this case, Inglis provided the following expression for the stress at the tip of the major axis of the ellipse:

$$\sigma = \sigma_0 \left(1 + 2\sqrt{\frac{c}{R}}\right) \tag{7.11}$$

where R is the radius of curvature at the tip of the major axis.

In the limiting case when the crack is infinitely sharp, $R = 0$, and the stress approaches infinity even for very small loads. However, this singular nature of the stress field is not feasible in real materials. The stress cannot become infinitely large, as the material should yield at a certain finite stress level. Hence fracture in materials cannot be based on the concept of the effect of stress concentration of flaws and yielding, based on yield theories such as the Von Mises or octahedral shear stress theories. Thus a different approach and formulation of fracture mechanics problems is used and is briefly described next.

A crack in any body can be subjected to one or a combination of three types of loading. The three *crack loading modes* are shown in Figure 7-6. The opening mode of loading, called *Mode I*, involves loads that produce displacements of the crack surfaces perpendicular to the plane of the crack. The stress intensity factor for opening mode (explained in more detail in Section 7.4) is represented by a subscript I as K_I. The shearing mode of loading, called *Mode II*, involves in-plane shear loads that produce displacements of the crack surfaces in the plane of the crack and perpendicular to the leading edge of the crack. The stress intensity factor for shearing mode (or sliding mode) is represented by a subscript II as K_{II}. The tearing mode of loading called *Mode III*, involves out-of-plane shear loads that produce displacements of the crack surfaces that are in the plane of the crack and parallel to the leading edge of the crack. The stress intensity factor for tearing mode is represented by a subscript III as K_{III}.

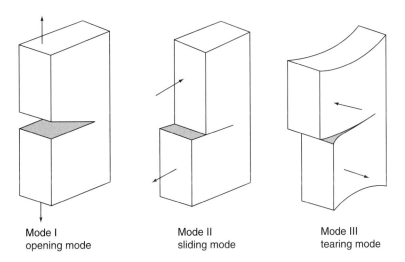

Mode I
opening mode

Mode II
sliding mode

Mode III
tearing mode

FIGURE 7-6 *Three modes of crack loading.*

Earlier we solved for the maximum stress in the vicinity of an elliptic hole in a large plate under remotely applied tensile stress. It was found that as the radius of curvature of the elliptic hole approaches zero, which implies a sharper crack, the maximum stress approaches infinity. In reality, however, the material at the crack tip will start yielding when the stress exceeds the yield strength under the prevalent stress conditions (for example, plane stress or triaxial stress state). Hence, if a stress concentration exists because of a limiting physical defect such as a sharp crack, this may result in very large local stresses under application of finite remote stress; a different approach is necessary to solve for the stress field around the crack tip.

7.4 Analysis for Fracture $\curlyvee$

The most suitable approach is to use a particular type of complex function, and ensure that the boundary conditions for the problem (for example, remote tensile stress applied to a large plate with a crack) are satisfied. However, in this text we will not provide the derivation of the stress field, but rather provide the final solution only. The derivation is available in textbooks devoted to fracture mechanics, such as those by Broek, Anderson, Dowling, and Knott, listed in the references.

We consider the problem of a center crack of length $2a$ in an infinite body under an applied uniform biaxial tensile stress, as shown in Figure 7-7.

Using the complex stress functions approach outlined in the references, the stresses at any point $P(r,\theta)$ in the vicinity of the Mode I or opening mode crack tip shown in Figure 7-7 are given by

$$\sigma_{xx} = \sigma_o\sqrt{\pi a}\left(\frac{1}{\sqrt{2\pi r}}\right)\cos\left(\frac{\theta}{2}\right)\left\{1 - \sin\left(\frac{\theta}{2}\right)\sin\left(\frac{3\theta}{2}\right)\right\}$$

$$\sigma_{yy} = \sigma_o\sqrt{\pi a}\left(\frac{1}{\sqrt{2\pi r}}\right)\cos\left(\frac{\theta}{2}\right)\left\{1 + \sin\left(\frac{\theta}{2}\right)\sin\left(\frac{3\theta}{2}\right)\right\}. \qquad (7.12)$$

$$\tau_{xy} = \sigma_o\sqrt{\pi a}\left(\frac{1}{\sqrt{2\pi r}}\right)\sin\left(\frac{\theta}{2}\right)\cos\left(\frac{\theta}{2}\right)\cos\left(\frac{3\theta}{2}\right)$$

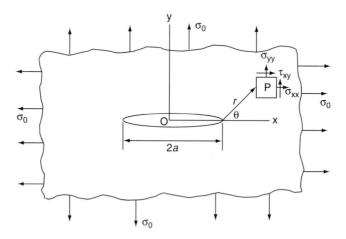

FIGURE 7-7 *Center crack of length 2a under biaxial tension in an infinite plate.*

These equations are valid only in the vicinity of the crack tip, that is, only when $r \ll a$. Along the x-axis, that is, $\theta = 0°$, σ_{xx} may be written as

$$\begin{aligned}
\sigma_{xx} &= \sigma_o \sqrt{\pi a}\left(\frac{1}{\sqrt{2\pi r}}\right) \\
&= K_I\left(\frac{1}{\sqrt{2\pi r}}\right)
\end{aligned} \tag{7.13}$$

where the quantity K_I is expressed as

$$K_I = \sigma_o \sqrt{\pi a} \tag{7.14}$$

and is called the *stress intensity factor* under Mode I loading, which has units such as $MN(m)^{1/2}$ or $ksi(in)^{1/2}$. The advantage of writing the stress close to the crack tip in terms of a single parameter K arises from the fact that the $1/\sqrt{r}$ dependency is followed whatever the applied stress system. This single-parameter characterization of the stress field in the vicinity of the crack tip is valid under a special condition, that the material around the tip undergoes *small-scale yielding*, that is, the size of the plastic zone around the crack tip is small compared to the dimensions of the crack and the specimen. In this case, the stresses around the crack tip can be developed on the assumption that the overall material behaves in a linear elastic fashion, except for the very small plastic zone around the crack tip. Such an approach is often termed *linear elastic fracture mechanics* (LEFM). Thus the stress field around a crack tip loaded in Mode I can be written as

$$\begin{aligned}
\sigma_{xx} &= K_I\left(\frac{1}{\sqrt{2\pi r}}\right)\cos\left(\frac{\theta}{2}\right)\left\{1 - \sin\left(\frac{\theta}{2}\right)\sin\left(\frac{3\theta}{2}\right)\right\} \\
\sigma_{yy} &= K_I\left(\frac{1}{\sqrt{2\pi r}}\right)\cos\left(\frac{\theta}{2}\right)\left\{1 + \sin\left(\frac{\theta}{2}\right)\sin\left(\frac{3\theta}{2}\right)\right\} \\
\tau_{xy} &= K_I\left(\frac{1}{\sqrt{2\pi r}}\right)\sin\left(\frac{\theta}{2}\right)\cos\left(\frac{\theta}{2}\right)\cos\left(\frac{3\theta}{2}\right) \\
\sigma_{zz} &= v(\sigma_{xx} + \sigma_{yy}) \text{ plane}-\text{strain} \\
\sigma_{zz} &= 0 \text{ plane} - \text{stress} \\
\tau_{xz} &= \tau_{yz} = 0
\end{aligned} \tag{7.15}$$

Similarly, the stress field around a crack tip loaded in Mode II (Figure 7-6) is given by

$$\begin{aligned}
\sigma_{xx} &= K_{II}\left(\frac{1}{\sqrt{2\pi r}}\right)\sin\left(\frac{\theta}{2}\right)\left\{2 + \cos\left(\frac{\theta}{2}\right)\cos\left(\frac{3\theta}{2}\right)\right\} \\
\sigma_{yy} &= K_{II}\left(\frac{1}{\sqrt{2\pi r}}\right)\sin\left(\frac{\theta}{2}\right)\cos\left(\frac{\theta}{2}\right)\cos\left(\frac{3\theta}{2}\right) \\
\tau_{xy} &= K_{II}\left(\frac{1}{\sqrt{2\pi r}}\right)\cos\left(\frac{\theta}{2}\right)\left\{1 - \sin\left(\frac{\theta}{2}\right)\sin\left(\frac{3\theta}{2}\right)\right\} \\
\sigma_{zz} &= v(\sigma_{xx} + \sigma_{yy}) \\
\tau_{xz} &= \tau_{yz} = 0
\end{aligned} \tag{7.16}$$

where the stress intensity factor in Mode II is given by

$$K_{\text{II}} = \tau_o \sqrt{\pi a}. \tag{7.17}$$

The stress field around a center crack in an infinite body under Mode III loading condition, as shown in Figure 7-6, is given by

$$\sigma_{xx} = \sigma_{yy} = \sigma_{zz} = 0$$

$$\tau_{xy} = 0$$

$$\tau_{xz} = \frac{-K_{\text{III}}}{\sqrt{2\pi r}} \sin \frac{\theta}{2} \tag{7.18}$$

$$\tau_{yz} = \frac{K_{\text{III}}}{\sqrt{2\pi r}} \cos \frac{\theta}{2}$$

where the stress intensity factor in Mode III is given by

$$K_{\text{III}} = \tau_0 \sqrt{\pi a}. \tag{7.19}$$

7.5 Energetics 'Y'

Griffith (1920) derived a thermodynamic criterion for fracture by considering the total change in energy of a cracked body as the crack length was increased. The analysis was applied to brittle materials that fracture with very little plastic deformation occurring around the crack tip. Energy changes in the body as a whole are considered to derive useful expressions for the fracture stress, without regard to fracture mechanisms in the immediate vicinity of the crack tip. To illustrate *Griffith's energy approach*, consider a semiinfinite plate of unit thickness that contains a through thickness center crack of finite length, $2a$, that is subjected to a uniform tensile stress, σ_∞, as shown in Figure 7-8.

Let us assume a fixed grip condition, that is, the uncracked plate is loaded to the desired load and then the two grips are locked or fixed, and then a crack is introduced. The total potential energy, U, can be written as

$$U = U_0 + U_a + U_\Gamma - W \tag{7.20}$$

where
   U = total potential energy of the system
   U_0 = elastic strain energy of the loaded uncracked plate (a constant value)
   U_a = change in elastic energy caused by introducing the crack in the plate
   U_Γ = change in elastic surface energy caused by formation of new crack surfaces
   W = work performed by external forces.
Griffith used a stress analysis approach developed by Inglis (1913) to show that for unit thickness

$$U_a = -\frac{\pi \sigma^2 a^2}{E}. \tag{7.21}$$

The negative sign is due to a decrease in the elastic strain energy of the plate owing to introduction of the crack because the plate loses stiffness and the load applied by the fixed grips drops.

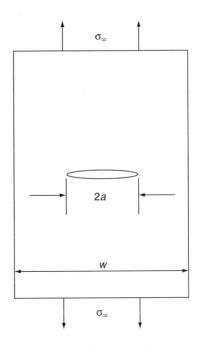

FIGURE 7-8 *Semiinfinite plate with elliptical center crack.*

Also,

$$U_\Gamma = 2(2a\Gamma_s) \tag{7.22}$$

where

Γ_s = elastic surface energy per unit area, and $W = 0$ as no work is done by external forces in the fixed grip condition.

Therefore, the total energy U of the cracked plate is

$$U = U_0 - \frac{\pi\sigma^2 a^2}{E} + 4a\Gamma_s \tag{7.23}$$

The equilibrium condition for crack extension is obtained by setting $\frac{dU}{da} = 0$:

$$\Rightarrow \frac{dU}{da} = \frac{dU_0}{da} + \frac{d}{da}\left(-\frac{\pi\sigma^2 a^2}{E}\right) + \frac{d}{da}(4a\Gamma_s) = 0$$

$$\Rightarrow \frac{2\pi\sigma^2 a}{E} = 4\Gamma_s, \tag{7.24}$$

which can be rearranged as

$$\sigma\sqrt{\pi a} = \sqrt{2E\Gamma_s} = \text{a constant.} \tag{7.25}$$

That is, crack extension in brittle materials occurs when the product $\sigma\sqrt{\pi a}$ attains a constant critical value.

The Griffith criteria for fracture can also be visualized, as in Figure 7-9.
We can rewrite Eq. (7.24) as

$$G = \frac{\pi\sigma^2 a}{E} = 2\Gamma_s.$$

(7.26)

In this form G is the material's *elastic energy release rate*, which represents the elastic energy per unit crack surface area that is available for infinitesimal crack growth.
From Eq. (7.25) the fracture stress can be written for the plane stress case as

$$\sigma = \sqrt{\frac{2\Gamma_s E}{\pi a}} = \sqrt{\frac{GE}{\pi a}},$$

(7.27)

which is the Griffith's fracture criterion.
Similarly it can be shown for the plane strain case that

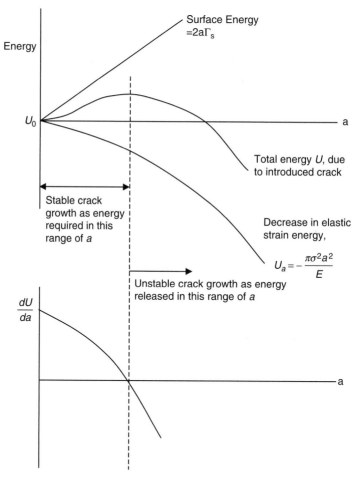

FIGURE 7-9   *Energy balance for crack growth in a brittle material.*

$$\sigma = \sqrt{\frac{2\Gamma_s E}{\pi a(1 - v^2)}} = \sqrt{\frac{GE}{\pi a(1 - v^2)}}. \tag{7.28}$$

Using Eq. (7.14) in Eq. (7.27), squaring, and rearranging, we obtain

$$K_I^2 = GE \quad \text{for plane stress} \tag{7.29}$$

and using Eq. (7.14) in Eq. (7.28), we obtain

$$K_I^2 = \frac{GE}{1 - v^2} \quad \text{for plane strain.} \tag{7.30}$$

Equations (7.29) and (7.30) are very useful, as they relate energy release rate directly to standard values of the stress intensity factor for Mode I crack loading, which are now available for a large number of specimen geometries and loading systems (Tada *et al.*, 1985).

Similar results can also be obtained for Mode II and Mode III loading of the crack.

7.6 Mechanics ⇔ Materials Link: The Fracture Test

7.6.1 Fracture Toughness, K_{IC}

Qualitatively, *fracture toughness* is a material's resistance to failure due to the presence of and growth of a crack. Fracture toughness controls failure by fracture in the same way as the yield strength controls failure by yielding.

Consider a Mode I edge crack in a plate, as shown in Figure 7-10. When the remotely applied stress σ_o increases, the stresses in the vicinity of the crack tip increase and eventually will approach the yield strength of the material. At this point, a certain volume of material around the crack tip schematically shown as a circle of diameter $2r_p$ in Figure 7-10 will begin to undergo plastic deformation.

The approximate size of the *crack tip plastic zone* r_p is given by Eq. (7.15) for $\theta = 0°$

$$\sigma_y = S_Y = \frac{K_I}{\sqrt{2\pi r_p}}$$
$$\Rightarrow r_p = \frac{1}{2\pi}\left(\frac{K_I^2}{S_Y^2}\right) \tag{7.31}$$

Depending upon the loading conditions, the dimensions of the specimen, the length of the crack, and the properties of the material, the size of the plastic zone will vary. If the plastic zone is small compared to the size of the crack and the dimensions of the specimen, then the region outside of the plastic zone can be characterized by elastic stress field equation [such as Eq. (7.12)] or by the stress intensity factor K_I. Generally, if the crack length, the length of uncracked ligament, and the thickness and length of the specimen are much greater than the size of the plastic zone, linear elastic fracture mechanics analysis is valid. Under such conditions, when the plastic zone size is small, if a specimen is loaded until the crack extends,

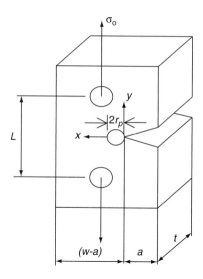

FIGURE 7-10 *An edge-cracked tensile fracture specimen with a small plastic zone of size $2r_p$ around the crack tip. The specimen width is w, crack length a, uncracked ligament (w − a), thickness t, and specimen length L.*

the K value obtained at this load level is a property of the material and is called the *fracture toughness*, represented by K_C.

Fracture takes place when any one of the stress intensity factors, K_I, K_{II}, or K_{III}, reaches its corresponding critical value, K_{IC}, K_{IIC}, or K_{IIIC}, respectively. Among the critical stress intensity factors, K_{IC} is typically most critical. Thus for most materials, Mode I loading is the governing mode of brittle fracture, and the small plastic zone absorbs very little energy. Most of the material behaves in linear elastic fashion before fracture, as in the case of brittle materials. More details on the methodology and experimental measurement of K_{IC} based on linear elastic fracture mechanics principles for metals can be obtained from ASTM Standard No. E399, and No. D5045 for polymers. For metals, compact tension specimens of the type shown schematically in Figure 7-10 are used.

Generally, if the thickness of the specimen is not large (in comparison to the plastic zone size), the stress in the z-direction is small because of lack of constraint in the z-direction, and hence larger deformation can occur in the z-direction. If the thickness is large, the plastic zone is surrounded by a large amount of low-stressed elastic material that effectively constrains the development of the plastic zone by causing a low strain ε_z and larger tensile stress σ_z as compared to the case of a specimen with a low thickness. Thus in the later case, also known as *plane strain*, the plastic zone size is small and should be used to obtain K_{IC}. In the plane strain condition, for example in thick plates, the value of K_I approaches the limiting and constant value of K_{IC} and does not depend on the geometry of the specimen. If a plane strain condition is absent, as in thin sheets, the toughness depends on both the material and the geometry. It is now generally accepted that plane strain conditions occur if the thickness of the specimen and other dimensions of the specimen satisfy the following relationship:

$$\left.\begin{array}{l}\text{thickness}\\\text{crack length}\\\text{width}\\\text{length}\\\text{uncracked ligament}\end{array}\right\} \geq 2.5\left(\dfrac{K_{\mathrm{I}}^2}{\sigma_0^2}\right). \tag{7.32}$$

Meeting all the requirements for plane strain conditions listed in Eq. (7.32) ensures that the plastic zone is small compared to all dimensions of the specimen.

If the plastic zone size becomes large and comparable to other characteristic dimensions of the specimen, the boundaries of the specimen have a greater influence on the crack tip stress field and hence the plastic zone size. If the plastic zone size $(2r_{\mathrm{p}})$ is not small compared to the crack length, specimen length, width, thickness, and the uncracked ligament, LEFM is no longer applicable and other methods such as J-integral and crack tip opening displacement (CTOD) should be used (see references).

Typical average values of K_{IC} for some materials are listed in Table 7-1. It may be noted that these values should only be used for an understanding of fracture mechanics and practice, and should not be used for actual design applications.

It should be noted that in general it is not necessary for the fracture stress to exceed the yield strength of the material for fracture to occur. Depending upon the size of the crack and critical stress intensity of the material, the fracture stress may or may not have to exceed the yield strength. For short cracks in a ductile material such as aluminum, the fracture stress may exceed the yield strength, but will be less than or equal to the ultimate strength.

7.6.2 Fracture Mechanisms in Materials

The fracture toughness of most engineering metals and alloys are in the range 10 to 200 MPa$\sqrt{\mathrm{m}}$. Polymers, which have also become popular as engineering materials, have K_{IC} values in the range 1 to 7 MPa$\sqrt{\mathrm{m}}$. Polymer matrix reinforced composites have found very significant use as structural and lightweight materials, and have fracture toughness values in the range 10 to 70 MPa$\sqrt{\mathrm{m}}$, which is comparable to some metals. Ceramics typically have low values of fracture toughness in the range 2 to 10 MPa$\sqrt{\mathrm{m}}$. This can be expected from their very low ductility. Though ceramics have very high yield strength, their application in tension is limited by their susceptibility to fracture even in the presence of small cracks, due to their low

TABLE 7-1   Yield strength and fracture toughness data for some common structural materials

Material	Yield strength, S_Y		K_{IC}	
	MPa	ksi	MPa$\sqrt{m}$	ksi$\sqrt{in}$
AISI 1045 steel	269	39	50	46
4340 (500°F temper) steel	1570	227	57	52
4340 (800°F temper) steel	1405	205	85	78
A538 steel	1722	250	111	100
2024–T351 aluminum	377	55	37	34
6061–T651 aluminum	296	43	28	26
7075–T651 aluminum	534	78	29	27
Ti-6Al-4V titanium	820	119	106	96

fracture toughness. Recent efforts at developing ceramic matrix composites have improved the fracture toughness, though they are still not comparable to metals or polymer composites or metal matrix composites.

The fracture behavior of metals, which are face-centered cubic (FCC), body-centered cubic (BCC), or hexagonal close packed (HCP), are different. Face-centered cubic metals such as copper, aluminum, nickel, and platinum are inherently ductile because of the availability of a relatively large number of slip systems. A large number of slip systems results in low yield strengths, which causes plastic flow at stresses less than those required to cause fracture. Moreover, the availability of slip systems allows for relaxation of plastic strain incompatibilities that are often the reason for initiation of micro cracks. This behavior of FCC metals holds true both for low and high temperatures.

BCC metals such as iron, molybdenum, tungsten, and chromium have a tendency for brittle fracture as dislocation intersection and jog formation are more pronounced than in FCC metals. Also their yield strength has a strong dependence on temperature. At low temperatures BCC metals have a high yield strength, which results in plastic flow stress that is higher than the fracture stress, and hence failure by brittle fracture is more likely over failure by yielding. At higher temperatures, greater than $0.3T_{\text{melting}}$, failure occurs by ductile fracture.

HCP metals such as cadmium, zinc, titanium, and magnesium also exhibit brittle fracture at low temperatures due to the low number of slip systems. At higher temperatures, more slip systems are activated, and HCP metals then exhibit fracture behavior similar to that of FCC metals.

In addition to the inherent dependence of fracture behavior on the nature of chemical bonding and crystal structure, other parameters such as chemical composition, material processing, microstructure, and loading rate can significantly affect fracture toughness and hence fracture behavior. In the case of BCC metals, fracture toughness abruptly changes over a small temperature range.

A rapid transition from ductile to brittle behavior is especially dominant in steels with ferritic-pearlitic and martensitic structures. This *ductile to brittle transition* with temperature change in BCC metals is shown schematically in Figure 7-11. At lower temperature fracture occurs by cleavage, which is the direct separation along crystallographic planes due to simple breaking of atomic bonds. *Cleavage fracture* is usually associated with a particular crystallographic plane, for example (100) planes in iron.

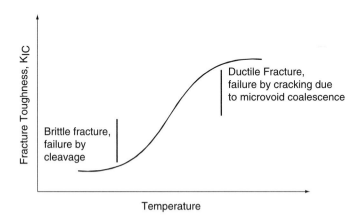

FIGURE 7-11 *Ductile to brittle transition in ferritic steels as the temperature decreases.*

Thus fracture occurs along a flat plane within one grain, and then on another flat plane of a slightly different orientation in the neighboring grain as shown in Figure 7-12. Cleavage fracture is also known as *transgranular fracture*. The flat cleavage facets through the grains have a high reflectivity, giving a cleavage fracture surface a bright, shiny appearance.

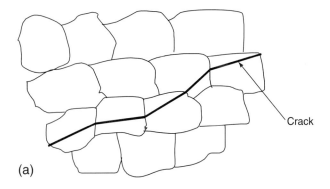

(a)

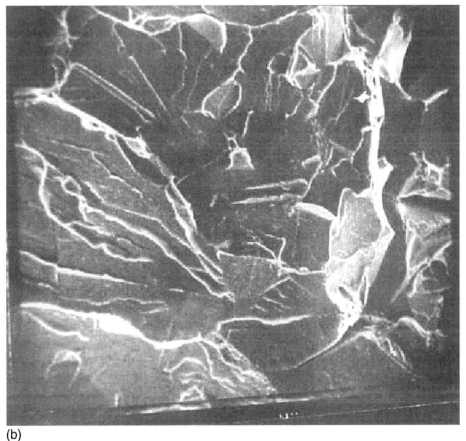

(b)

FIGURE 7-12 *(a) Schematic of a cleavage crack spreading through a polycrystalline metal. (b) Scanning electron microscope view of cleavage fracture in a steel sample at −196° C (original magnification 750×). (Courtesy of Dr. Robert A. Winholtz, University of Missouri–Columbia, MO.)*

Ductile fracture occurs by a sequence of microvoid initiation, growth, and coalescence at the tip of a preexisting crack. Typical engineering materials used for structural application are alloys that may contain significant amounts of second- and possibly third-phase particles. Pure metals are not often used and will fail under a monotonically increasing tensile load or due to a tensile overload by a slip deformation process resulting in necking and ultimately separation. In engineering metal alloys, second-phase particles are generally of the size 0.5 to 0.05 μm (carbides in steels) or of 0.05 to 0.005 μm [GP (Guinier-Preston) zones in Al–Cu precipitation hardened alloy]. These second-phase particles are generally harder than the matrix. Because of their greater hardness these particles cannot deform coherently with the matrix and thus separation can occur at the interface during plastic flow, causing a microscopic void to be formed. A large number of such microvoids may be formed at the location of the second-phase particles. These voids continue to grow by a slip and necking process until adjacent microvoids begin to coalesce to form a macroscopic crack. The process of necking and coalescence of microvoids results in the formation of dimples. Since the second-phase particles are randomly located, microvoids are also formed at random, which gives the fracture surface a dimpled and irregular shape (Figure 7-13). Another model for microvoid formation and growth is based on dislocation movement around the microscopic second-phase particles (Broek, 1973).

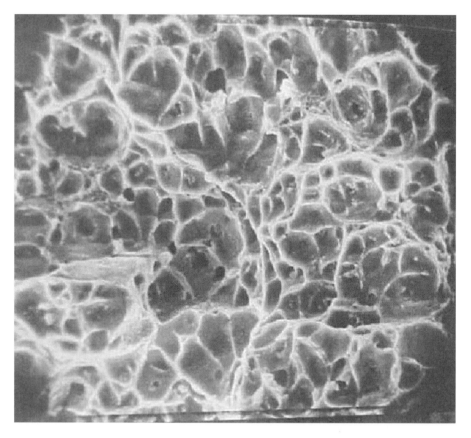

FIGURE 7-13 *Scanning electron microscope view of ductile fracture (original magnification 590×). (Courtesy of Dr. Robert A. Winholtz, University of Missouri–Columbia, MO.)*

Generally brittle metals fail by transgranular cleavage while ductile metals fail by coalescence of voids formed at second-phase particles. However, in certain cases cracks form and grow along grain boundaries. This phenomenon is called *intergranular fracture*, as shown in Figure 7-14. This type of fracture could occur for a variety of reasons, some of which are:

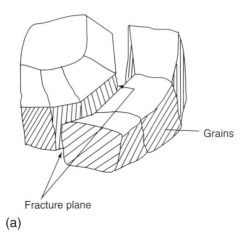

(a)

(b)

FIGURE 7-14 *(a) Schematic of intergranular fracture occurring along grain boundaries. (b) Scanning electron microscope view of intergranular fracture in an ASTM A 36 (0.25% C) steel sample at −196° C (original magnification 750×). (Courtesy of Dr. Robert A. Winholtz, University of Missouri–Columbia, MO.)*

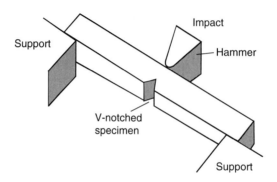

FIGURE 7-15 *Schematic of a Charpy V-notch impact test.*

- Precipitation of a brittle phase on the grain boundary
- Environmentally assisted cracking or stress corrosion cracking
- Hydrogen embrittlement
- Liquid metal embrittlement (for example, embrittlement of steel when placed in contact with liquid metals of lower melting point such as lithium or sodium)

All these mechanisms that cause intergranular fracture lower the fracture toughness of the affected material.

A high rate of loading (or dynamic loading) also usually lowers the fracture toughness. One of the most commonly encountered dynamic loading situations is that during an impact. Hence impact tests, such as the *Charpy V-notch impact test*, as shown in Figure 7-15, are often conducted to assess a material's tendency to brittle fracture. The impact is associated with a high rate of loading and introduction of a triaxial state of stress in front of the crack tip. These conditions promote fracture rather than plastic flow of the material around the crack tip. Temperature changes also dramatically affect fracture toughness during impact loading of BCC-type metals (see Figure 7-16) and the type of fracture surface

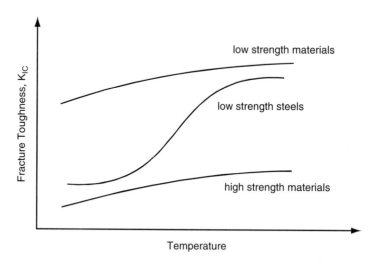

FIGURE 7-16 *Schematic of the variation of fracture toughness with temperature for different types of metals when subjected to impact.*

morphology (Figure 7-17). However, FCC and HCP materials are much less sensitive to temperature changes.

FIGURE 7-17 *Charpy impact samples showing ductile (rightmost sample) to brittle transition (leftmost sample) in ASTM A36 (0.25% C) steel. Testing temperatures from right to left are 200°C, 100°C, 25°C, and −196°C. (Courtesy of Dr. Robert A. Winholtz, University of Missouri–Columbia, MO.)*

7.7 Materials Selection for Fracture Control ⬟

One of the most important material parameters for resisting fracture due to crack propagation is the fracture toughness K_{IC} of the material. Fracture toughness varies considerably across the large classes of materials as shown by the Ashby charts in Figures 7-18 to 7-20. Note that the fracture toughness varies by nearly three orders of magnitude, with polymers being the weakest and metals being the toughest. Generally, materials having low strength and high toughness yield before they fracture, whereas materials with high strength and low toughness fracture before yielding. From the shape of the balloons representing metals in Figure 7-19, it can be inferred that as the yield strength increases the fracture toughness decreases. Different classes of materials exhibit different relationships between the yield strength and fracture toughness. For example, the toughness and strength of porous ceramics and foams is shown for compressive loading and these parameters also scale with the relative density. Note that in the Ashby chart the strengths displayed for metals and polymers are yield strengths, for ceramics and glasses these are the compressive strengths, and for composites these are the tensile strengths.

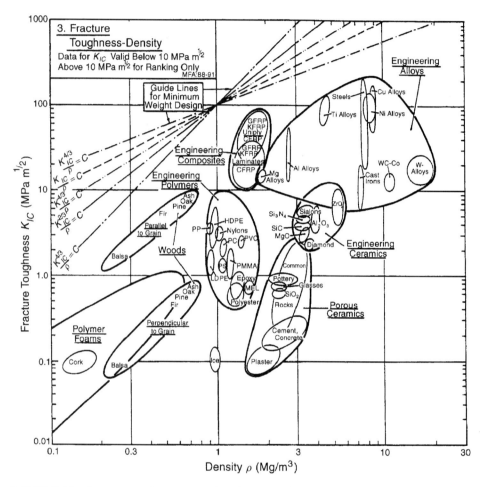

FIGURE 7-18 *Fracture toughness, K_{IC}, plotted against density, ρ. The guidelines are used in selecting materials to obtain minimum weight and fracture-limited design. (With permission from Michael F. Ashby.)*

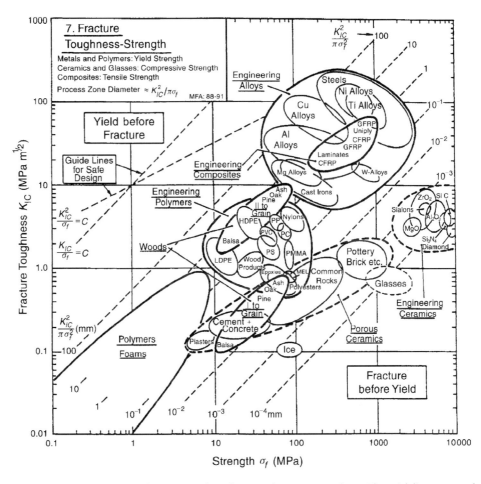

FIGURE 7-19 *Fracture toughness, K_{IC}, plotted against fracture strength, σ_f. The guidelines are used in selecting materials for damage tolerant design. (With permission from Michael F. Ashby.)*

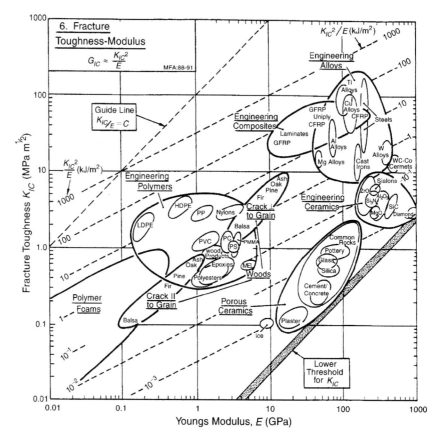

FIGURE 7-20 *Fracture toughness, K_{IC}, plotted against elastic modulus, E. The guidelines are used in selecting materials for deflection limited design. (With permission from Michael F. Ashby)*

EXAMPLE 7-1:

Consider a square beam of cross-sectional area $A = d^2$ (say), as shown in Figure E7-1. It contains a small crack of length $a < d$. Using the material index method of Ashby, determine the index that would optimize the mass and optimize the fracture toughness of the material of the beam.

Answer: The mass of the beam is given by

$$m = \rho L A. \tag{a}$$

The average stress at the crack plane is given by

$$\sigma = \frac{K_{IC}}{\sqrt{\pi a}} = \frac{Mc}{I} \tag{b}$$

where M is the bending moment at the crack plane and $c = d/2$.

Continued

EXAMPLE 7-1: *Cont'd*

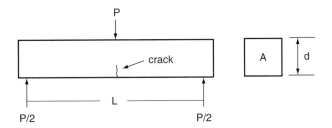

FIGURE E7-1

For a square beam the moment of inertia I is given by

$$I = \frac{A^2}{12}.$$ (c)

Eliminating A between Eqs. (a) and (c), and using Eq. (b), one can obtain the following expression for the mass m:

$$m = \sqrt{12 M c L^2 \sqrt{\pi a}} \frac{\rho}{\sqrt{K_{IC}}}.$$ (d)

From Eq. (d), if the factor $\frac{\sqrt{K_{IC}}}{\rho}$ (also known as the material index) is maximized, the mass m of the beam is minimized. Thus using the Ashby chart shown in Figure 7-18, suitable fracture resistant materials can be selected.

Consider the following numerical values for a problem: $m = 0.2\,\text{kg}$, $P = 20,000\,\text{N}$, $a = 0.02\,\text{m}$, $L = 2.0\,\text{m}$, $d = 0.1\,\text{m}$. It can be calculated that $c = 0.05\,\text{m}$ and $M = 10,000\,\text{Nm}$. Then the material index is from Eq. (d)

$$\frac{\sqrt{K_{IC}}}{\rho} = 387.8 \frac{Pa\sqrt{m}}{Kg/m^3} = 0.3878 \times 10^3 \frac{Pa\sqrt{m}}{Kg/m^3}$$ (e)

(Note: The right most term in Eq. (e) has units containing a 10^3 term to obtain units consistent with the Ashby chart in Figure 7-18.)

A materials selection line corresponding to the material index and parallel to the $\frac{\sqrt{K_{IC}}}{\rho}$ construction line is now drawn as shown in Figure 7-18, through the point 0.0388 on the y-axis (since the origin of the x-axis is at $0.1\,\text{kg/m}^3$). All materials falling on the line are possible candidate materials for this problem. Materials above the line will be somewhat superior to the list of materials falling on the selection line. The possible materials are:

- Polymer foams
- Polypropylene
- High-density polyethylene
- Nylon
- Magnesium alloys
- Aluminum alloys
- Titanium alloys
- Carbon fiber reinforced plastics (CFRP)

EXAMPLE 7-2:

In Chapter 3, Section 3.9, the solved design example is further extended here to include fracture toughness K_{IC} as an additional criteria for materials selection. In order to ensure that failure does not occur due to the presence of crack-like defects, we will use the K_{IC} versus strength S_y (Figure 7-19 where σ_f is shown as yield strength S_y) and the K_{IC} versus E (Figure 7-20) material selection charts. The results are shown in Table E7-2.

The K_{IC} versus S_y rankings are obtained by finding the ratio (K_{IC}/S_y) for each material; the material with the highest ratio is ranked as 1, and so on. Similarly, the K_{IC} versus E rankings are obtained by finding the ratio (K_{IC}/E) for each material; the material with the highest ratio is ranked as 1, and so on.

TABLE E7-2   Materials selection chart for the design example

Possible materials from E versus S_y chart	Weighted subtotal (from Table 3.2)	K_{IC} versus S_Y rank	Weighted Rank (Weight = 0.2)	K_{IC} versus E rank	Weighted Rank (Weight = 0.8)	Weighted total
		Additional considerations of material rankings from Ashby charts (Raw rankings and weighted rankings are given)				
Steel	1.25	1	0.2	2	1.6	3.05
Cast Iron	1.75	4	0.8	4	3.2	5.75
Titanium alloy	4.5	3	0.6	1	0.8	5.9
Zinc alloy	4.0	N/A	—	N/A	—	N/A
Aluminum	3.5	2	0.4	3	2.4	6.3

Note that fracture toughness relative to stiffness is ranked more important than relative to strength.

Notice that by including fracture toughness as an additional requirement, titanium is now a better alternative than aluminum. The final choice for the optimum material for the link remains as steel.

7.8   Design of Structures for Fracture ⵙ🏯💻

Development of fracture mechanics in Section 7.4 was based on externally loaded plates of infinite size containing finite-size center cracks. In actual practice, most applications will require the use of a plate of finite size that may contain finite cracks. Closed-form solutions for finite-size fracture problems are generally not available because of the nature of the boundary conditions, which have significant influence on crack tip stresses in this case. Thus finite element method or other approximate numerical solutions are used to obtain the stress intensity factor. Typically, the result is presented in the form

$$K_I = F\sigma_0\sqrt{\pi a} \tag{7.33}$$

where F is a correction factor that depends on the geometry of the specimen and mode of loading of the crack. It should again be emphasized that the condition of elastic stresses around the crack tip (except in a very small plastic region or zone around the crack tip) must be met and thus the material overall behaves in a linear elastic fashion.

Expressions for the correction factor F are typically written as a dimensionless polynomial of the form $F(a/w)$ or $F(\alpha)$, where a is the crack length and w is the width of the plate, and $\alpha = a/w$. Several commonly used fracture specimen geometries and the corresponding dimensionless correction factors are shown in Figures 7-21a–g (Tada, Paris, and Irwin, 1985).

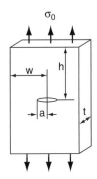

K Value for small a/w and 10% accuracy

$$K = \sigma_0\sqrt{\pi a} \quad \text{for (a/w} \leq 0.4)$$

Correction factor F for any $\alpha = a/w$

$$F = \frac{1 - 0.5\alpha + 0.326\alpha^2}{\sqrt{1 - \alpha}} \quad \text{(for h/w} \geq 2)$$

(a)

FIGURE 7-21A *Opening mode stress intensity factor for center cracked plate under tension.*

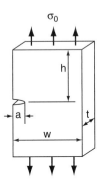

K value for small a/w and 10% accuracy

$$K = 1.12\sigma_2\sqrt{\pi a} \quad \text{for (a/w} \leq 0.13)$$

Correction factor F for any $\alpha = a/w$

$$F = 0.265(1-\alpha)^4 + \frac{0.857 + 0.265\alpha}{(1-\alpha)^{3/2}} \quad \text{(for h/w} \gg 1)$$

(b)

FIGURE 7-21B *Opening mode stress intensity factor for single edge cracked plate under tension.*

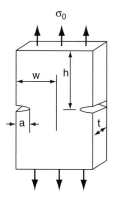

K value for small a/w and 10% accuracy

$$K = 1.12\sigma_0 \sqrt{\pi a} \quad \text{for } (a/w \leq 0.6)$$

Correction factor F for any $\alpha = a/w$

(c) $\quad F = (1 + 0.122\cos^4 \frac{\pi\alpha}{2})\sqrt{\frac{2}{\pi\alpha} \tan\frac{\pi\alpha}{2}} \quad \text{(for } h/w \geq 2)$

FIGURE 7-21C *Opening mode stress intensity factor for double edge cracked plate under tension.*

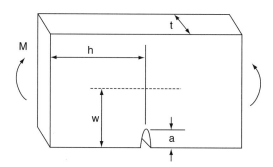

K value for small a/w and 10% accuracy

$$K = 1.12\sigma_0 \sqrt{\pi a} \quad \text{for } (a/w \leq 0.4)$$

Correction factor F for any $\alpha = a/w$

(d) $\quad F = \sqrt{\frac{2}{\pi\alpha} \tan\frac{\pi\alpha}{2}} \left[\frac{0.923 + 0.199(1-\sin\frac{\pi\alpha}{2})^4}{\cos\frac{\pi\alpha}{2}} \right] \text{(for } h/w \gg 1)$

FIGURE 7-21D *Opening mode stress intensity factor for single edge cracked plate in pure bending.*

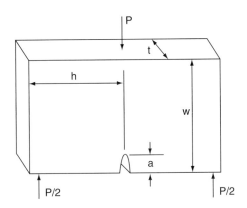

K value for small a/w and 10% accuracy

$$K = 1.12\sigma_2 \sqrt{\pi a} \quad \text{for } (a/w \leq 0.4)$$

Correction factor F for any $\alpha = a/w$

(e) $$F = \frac{1.99 - \alpha(1-\alpha)(2.15 - 3.93\alpha + 2.7\alpha^2)}{\sqrt{\pi}\ (1 + 2\alpha)(1-\alpha)^{3/2}} \quad \text{(for } h/w = 2)$$

FIGURE 7-21E *Opening mode stress intensity factor for edge cracked plate in three-point bend loading.*

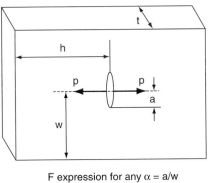

F expression for any $\alpha = a/w$

(f) $$F_P = \frac{1.297 - 0.297 \cos \frac{\pi\alpha}{2}}{\sqrt{\sin \pi\alpha}} \quad \text{(for } 0 \leq a/w \leq 1)$$

FIGURE 7-21F *Opening mode stress intensity factor for a center crack under internal loading.*

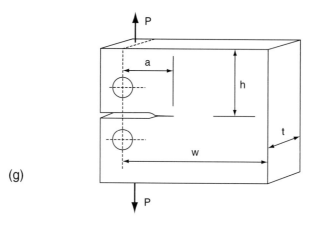

(g)

F expression for any α=a/w

$$F_P = \frac{(2 + \alpha)}{(1 - \alpha)^{3/2}} \, (0.886 + 4.64\alpha - 13.32\alpha^2 + 14.72\alpha^3 - 5.6\alpha^4) \quad \text{(for a/w} \geq 0.2)$$

FIGURE 7-21G *Opening mode stress intensity factor for a compact tension specimen.*

For elliptic cracks in a large plate, shown in Figure 7-22, the stress intensity factor varies along the crack front and the variation is given below for each of the two cases. It may be noted that at the end of the minor axis (at $\phi = 90°$) the stress intensity factor is the largest. At the end of the major axis (at $\phi = 0°$) it is the lowest.

Case (a): Embedded elliptic crack

$$K_I = \sigma_0 \sqrt{\frac{\pi a}{Q}} f(\phi) \tag{7.34}$$

where the flaw shape factor Q is given approximately by the expression

$$Q = 1 + 1.464 \left(\frac{a}{c}\right)^{1.65} \text{ and } a < c, \tag{7.35}$$

and the function $f(\phi)$ is

$$f(\varphi) = \left\{ \sin^2 \phi + \left(\frac{a}{c}\right)^2 \cos^2 \phi \right\}^{\frac{1}{4}}. \tag{7.36}$$

Note again that the value of K_I varies along the crack front as the angle φ changes and the maximum occurs for $\phi = 90°$.

Case (b): Semi-elliptic surface crack

$$K_I = F\sigma_0 \sqrt{\frac{\pi a}{Q}} f(\phi) \tag{7.37}$$

where F is given by

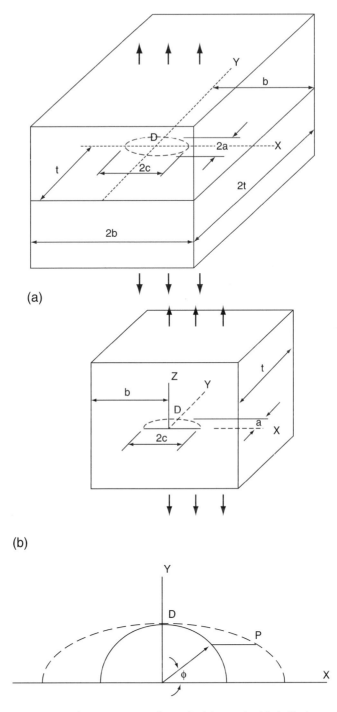

FIGURE 7-22 *Opening mode stress intensity factor for (a) an embedded elliptic crack and (b) semi-elliptic surface crack, in a large plate with dimensions b and t ≫ a and c.*

$$F = \left\{ 1.13 - 0.09\frac{a}{c} \right\}\{1 + 0.1(1 - \sin\phi)^2\} \quad \text{and} \quad a \leq c. \tag{7.38}$$

Most materials are more susceptible to failure under Mode I loading, that is, under normal tensile stresses, than due to the other two modes of fracture, that is, under the action of shear stresses. Thus in this treatment of fracture the emphasis is on design of structural components subjected to Mode I crack loading conditions only. Mixed-mode loading conditions have not been considered as they are beyond the scope of this book.

EXAMPLE 7-3:

A large center cracked panel, shown in Figure 7-21a, contains a crack of length $2a = 3.0$ in. The fracture toughness of the material is $K_{IC} = 30$ ksi(in.)$^{1/2}$, and the yield strength is 65 ksi. What is the maximum tensile stress σ_0 the panel can sustain before it fractures?

Answer: The large size of the panel implies that $w \gg 2a$, and also the length and thickness are much larger than 2a. When the remotely applied stress reaches a level to cause the stress intensity factor under Mode I (or opening mode) condition to become equal to or greater than K_{IC}, the panel will undergo brittle fracture.

Therefore,

$$K_I = K_{IC} = \sigma_0(\pi a)^{1/2}$$

$$30\,\text{ksi(in.)}^{1/2} = \sigma_0(\pi 1.5)^{1/2}$$

$$\Rightarrow \sigma_0 = 30/(\pi \times 1.5)^{1/2} \text{ ksi} = \underline{13.82\,\text{ksi}}.$$

Note that the stress required for causing fracture in the panel is much less than the yield strength of the material of the panel.

EXAMPLE 7-4:

An edge cracked plate as shown in Figure 7-21b has the dimensions $w = 100$ mm, $t = 5$ mm, $h = 300$ mm, and a remote tensile load of 100 kN is applied. Determine (a) the stress intensity factor K_I for a crack length of 15 mm, and (b) the critical crack length, a_c, for fracture if the material of the plate is AISI 4130 steel, which has a fracture toughness of $K_{IC} = 100$ MPa$\sqrt{m}$.

Answer:

(a) For a crack of length $a = 15$ mm, the crack length to width ratio is $\alpha = a/w = 15/100 = 0.15$, which is greater than $\alpha = 0.13$ (see Figure 7-21b). Hence we use the following expression for the correction factor F:

$$F = 0.265(1-\alpha)^4 + \frac{0.857 + 0.265\alpha}{(1-\alpha)^{3/2}}$$

$$\Rightarrow F = 0.265(1-0.15)^4 + \frac{0.857 + 0.265(0.15)}{(1-0.15)^{3/2}}$$

$$= 1.283$$

Continued

EXAMPLE 7-4: *Cont'd*

The stress caused by remote load is given by

$$\sigma = \frac{100 \times 10^3 (\text{N})}{100 \times 10^{-3}(\text{m}) \times 0.005(\text{m})} = 200 \text{ MPa}.$$

Thus, the stress intensity factor K_I is obtained as

$$K_I = F\sigma\sqrt{\pi a} = 1.283 \times 200\sqrt{\pi(0.015)} = 55.7 \text{ MPa}\sqrt{\text{m}}$$

(b) In this case the crack length is to be determined and hence the correction factor F cannot be obtained a priori. Hence we have to resort to a trial method as shown below.

First obtain an approximate value of crack length by assuming an infinite plate (though the plate is of finite size) by using the expression

$$K_{IC} = 1.12\sigma\sqrt{\pi a_c}$$

where a_c is the critical crack length being sought.

$$\Rightarrow 100(\text{MPa}\sqrt{\text{m}}) = 1.12 \times 200(\text{MPa})\sqrt{\pi a_c}$$
$$\Rightarrow a_c = 0.0634 \text{ m} = \underline{63.4 \text{ mm}}$$

However, this is an approximate value, and the correct value of the critical crack length is found as follows:

Calculate values of K_I in the neighborhood of a_c obtained earlier until a value of $K_I = K_{IC} = 100 \text{ MPa}\sqrt{\text{m}}$ is reached, as shown in the accompanying table.

Trial a_c (mm)	$\alpha = \frac{a}{w}$	F	K_I (MPa$\sqrt{m}$)
50	0.5	2.815	223.1
40	0.4	2.106	149.3
30	0.3	1.66	101.9
29.5	0.295	1.646	100.2

Hence, the desired value of the critical crack length is 29.5 mm.

EXAMPLE 7-5:

A billboard is supported on a tall rectangular post as shown in Figure E7-5. A semielliptic crack develops at the bottom of the post as shown and has the dimensions of $2c = 25 \text{ mm}$ and $a = 6 \text{ mm}$. The cross-sectional dimensions of the post are $125 \text{ mm} \times 50 \text{ mm}$. Because of a wind gust the wind load on the billboard produces a moment of 10,000 Nm at the bottom of the post. If the material of the post has a fracture toughness $K_{IC} = 50 \text{ MPa}\sqrt{\text{m}}$, determine the factor of safety against brittle fracture.

Answer: The tensile stress at the cross-section containing the crack is

EXAMPLE 7-4: *Cont'd*

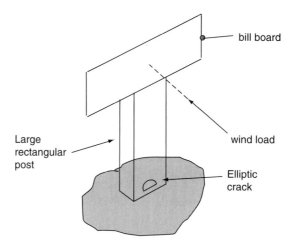

FIGURE E7-5

$$\sigma = \frac{Mc}{I} = \frac{10,000(\text{N.m}) \times 0.025(\text{m})}{\dfrac{0.125(\text{m}) \times (0.05\ \text{m})^3}{12}} = 192\ \text{MPa}.$$

Considering that the dimensions of the post are much larger than that of the crack, we use Eq. (7.37) to determine the value of the opening mode stress intensity factor K_I, in which $f(\phi)$ is obtained from Eq. (7.36) for $\phi = 90°$ at which angle the value of K_I is maximum. For this case $f(\phi) = 1$.

The value of Q is obtained from Eq. (7.35) as

$$Q = 1 + 1.464\left(\frac{6}{12.5}\right)^{1.65} = 1.436$$

and the value of the correction factor F is obtained from Eq. (7.38) as

$$F = \left\{1.13 - 0.09\frac{6}{12.5}\right\}\{1 + 0.1(1 - \sin 90°)^2\} = 1.087.$$

Thus from Eq. (7.37) the stress intensity factor is

$$K_I = 1.087 \times 192(\text{MPa})\sqrt{\frac{\pi(0.006\ \text{m})}{1.436}} \times 1.0 = 23.9\ \text{MPa}\sqrt{\text{m}}.$$

Thus the factor of safety, *FS*, against brittle fracture is given by

$$FS = \frac{K_{IC}}{K_I} = \frac{50}{23.9} = \underline{2.09}.$$

The factor of safety is much greater than 1 and hence the post is safe against brittle fracture.

Key Points to Remember

- Failure by yielding is the preferred mode of failure and fracture is not.
- Fracture mechanics provides an essential quantitative tool for prediction of failure in a structure by growth of crack-like defects.
- Fracture in a material is dependent on how failure manifests itself at the microscopic level (e.g., grains) and at the atomic level (e.g., dislocations).
- Fracture mechanisms depend on the type of material and the type of loading.
- Stress concentration effects of flaws cannot be used to describe failure due to very sharp cracks.
- Linear elastic fracture mechanics (LEFM) is applicable if the plastic zone created around the crack tip is small compared to all other characteristic dimensions of the specimen and the crack length.
- In LEFM, for a given mode of crack loading, a single parameter called the 'stress intensity factor' is sufficient to quantify the stress field in the vicinity of the crack tip.
- Fracture toughness is the material's resistance to failure due to the presence of and growth of a crack.
- Under certain conditions, such as plane strain, when the plastic zone size is small, if a specimen is loaded till the crack extends, the stress intensity factor value (e.g., K_I in Mode I loading) obtained at this load level is a property of the material and is called the fracture toughness, represented by K_{IC}.
- Three primary factors that control the susceptibility of a structure to brittle fracture are: material fracture toughness K_{IC}, crack size a, and stress level σ_o.

References

Anderson, T. L. (1995). *Fracture Mechanics: Fundamentals and Applications*, 2nd ed. CRC Press, Boca Raton, FL.

Ashby, M. F. (1999). *Materials Selection in Mechanical Design*, 2nd ed. Butterworth–Heinemann, Oxford.

Broek, D. (1973). The role of inclusions in ductile fracture and fracture toughness. *Engineering Fracture Mechanics*, vol. 5, pp. 55-66.

Broek, D. (1984). *Elementary Engineering Fracture Mechanics*. Martinus Nijoff, Boston.

Dowling, N. E. (1999). *Mechanical Behavior of Materials: Engineering Methods for Deformation, Fracture, and Fatigue*. Prentice Hall, Upper Saddle River, NJ.

Griffith, A. A. (1920). The phenomena of rupture and flow in solids. *Philosophical Transactions of the Royal Society*, Vol. A221, pp. 163-198.

Hertzberg, R. W. (1989). *Deformation and Fracture Mechanics of Engineering Materials*. Wiley, New York.

Inglis, C. E. (1913). Stresses in a plate due to the presence of cracks and sharp corners. *Transactions of Institution of Naval Architects*, Vol. 55, pp. 219-241.

Knott, J. F. (1981). *Fundamentals of Fracture Mechanics*, revised ed. Butterworths, London.

Tada, H., Paris, P. C., and Irwin, G. R. (1985), *The Stress Analysis of Cracks Handbook*, 2nd ed., Paris Productions, St. Louis, MO.

Problems

7-1. A large flat panel has a through thickness crack as shown in Figure 7-21a. It is subjected to a 150 MPa tensile stress and the plate material has a fracture toughness K_{IC} of 75 MPa(m)$^{1/2}$. Determine the critical crack length for this plate.

7-2. Calculate the opening mode stress intensity factor K_I for a rectangular specimen shown in Figure P7-2 containing an edge crack. $P = 35$ kN (7870 lb), $W = 50.9$ mm (2.0 in.), $t = 25.45$ mm (1.0 in.), $L = 203.6$ mm (7.0 in.), $a = 20.46$ mm (0.8 in.).

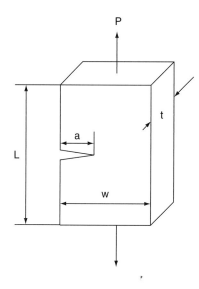

FIGURE P7-2

7-3. A lifting hook is made by cutting a flat plate of 50 mm thickness; its shape is shown in Figure P7-3. It supports a load of 50 kN. An edge crack develops because of a collision with a sharp object, and the crack length is 20 mm. If the hook is made of AISI 1045 steel, is it safe to continue using the hook?

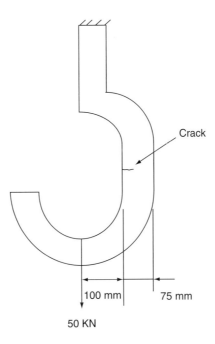

Crack

100 mm

75 mm

50 KN

FIGURE P7-3

7-4. If the fracture specimen shown in Figure P7-2 is made from AISI 4340 steel, Al 2024-T3, Al 6061-T6, or alumina, which specimen would have the highest failure load? Which specimen has the lowest fracture load?

7-5. A flat plate with a through circular hole is subjected to a tensile stress (σ), as shown in Figure P7-5. If a small through-the-thickness crack develops from the edge of the hole, estimate the stress intensity factor for this crack.

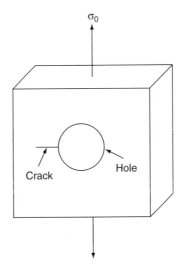

σ_0

Crack

Hole

FIGURE P7-5

7-6. An edge cracked plate shown in Figure 7-21b has dimensions $W = 50\,\text{mm}$, $t = 5\,\text{mm}$, and a large length. A load $P = 50\,\text{kN}$ is applied.
   (a) What is the stress intensity factor K_I for a crack of length $a = 10\,\text{mm}$?
   (b) For $a = 25\,\text{mm}$?
   (c) What is the critical crack length ac for brittle fracture if the material of the plate is 6061-T6 aluminum?

7-7. A cylindrical pressure vessel is made of AISI 4130 steel. The internal diameter is 100 mm and thickness is 40 mm. A semielliptic crack is found on the outer surface with a length of $2c = 40\,\text{mm}$ and depth of $a = 10\,\text{mm}$. Would it be safe to pressurize the vessel to an internal pressure of 300 psi? Would it be safe at 600 psi pressure?

7-8. High-pressure fluid at 50 MPa leaks into a through thickness center crack in a large panel, as shown in Figure P7-8. What is the stress intensity factor due to the internal pressure in the crack? Crack length $2a = 50\,\text{mm}$.

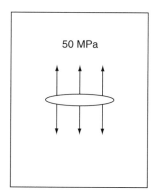

50 MPa

FIGURE P7-8

7-9. Consider a compact tension specimen, shown in Figure 7-21g. The specimen dimensions are $w = 100\,\text{mm}$, $a = 50\,\text{mm}$, $t = 50\,\text{mm}$. Crack propagation occurs when a load of 100 kN is applied. Calculate K_I for the material of the specimen. Is K_I the fracture toughness K_{IC} of the material, given that the material yield strength is 500 MPa?

7-10. A large and thin-walled pressure vessel is periodically inspected using a nondestructive technique. The smallest crack that can be detected is 1 mm in dimension. If the hoop stress in the vessel is 1200 MPa, what should be the fracture toughness of the material for the vessel to protect against fracture due to elliptic cracks on the outer surface along the axial length of the cylinder? The length of the elliptic crack is two times the depth of the crack.

7-11. A center crack of 60 mm length in a large plate is subjected to a remote tensile stress of 70 MPa. If the crack grows at a rate of 0.3 mm/day, when will the crack become unstable if the fracture toughness of the plate material is $K_{IC} = 40\,\text{MPa}\sqrt{\text{m}}$?

7-12. A beam is subjected to a three-point bending load as shown in Figure P7-12. The thickness of the beam is 5 cm. Find the maximum (or critical) load P that can be applied to the beam before the crack extends in an unstable manner, given that $K_{IC} = 100\,\text{MPa(m)}^{1/2}$.

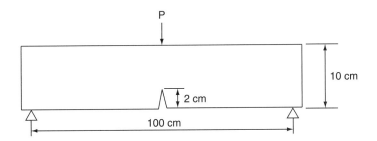

FIGURE P7-12

7-13. Which type of metal (BCC, FCC, HCP) should be chosen for a ship's hull that has to travel through very warm and very cold water to protect against brittle fracture?

7-14. Which type of metal (BCC, FCC, HCP) should be chosen for the container of very cold liquid nitrogen to protect against brittle fracture during normal use and during accidental impact of the container?

7-15. A double-edge cracked plate of 7075-T6 aluminum has the following dimensions: crack length $a = 6$ mm, thickness $t = 6.5$ mm, width $w = 16$ mm, and a large length. What is the maximum tensile load that can be applied without catastrophic failure of the plate?

7-16. Consider the rectangular specimen of Problem 7-2 with the given dimensions. Determine the stress intensity factor due to the applied tensile load of 35 kN for (a) center crack of length $2a = 13.0$ mm, (b) edge crack of length $a = 6.5$ mm, (c) double-edge crack each of length $a = 6.5$ mm, and (d) embedded elliptic crack of dimensions $2c = 13.0$ mm and $a = 3.25$ mm. Which cracked member is the most susceptible to fracture?

7-17. A pressure vessel weighing 200,000 lb is slowly lifted with a crane using two metal straps. Each strap is 2.5 in. wide and 0.35 in. thick. The straps are made of steel with a fracture toughness of 100 ksi$\sqrt{\text{in}}$. One of the straps has a center crack 0.5 in. wide, while the other strap has an elliptic edge crack of dimensions $2c = 0.5$ in. and $a = 0.1$ in. Both cracks are oriented perpendicular to the length dimension of the straps. Which strap, if any, would you recommend for replacement?

7-18. Give as many examples as you can of:
(a) Mode I or opening mode crack loading
(b) Mode II or in-plane shear mode crack loading
(c) Mode III or out-of-plane shear mode crack loading

7-19. If your subordinate presents you with a stress analysis of a machine component containing a sharp crack whose dimensions are known from ultrasound inspection, based on the Von Mises theory of yielding, will you accept the recommendation? Why or why not?

7-20. In Problem 7-3, if the crack grows at a rate of 0.05 mm for every cycle of loading and unloading, how many times can the hook be used before it must be repaired or replaced?

7-21. Solve Example 7-5 if the post is a solid cylindrical rod of the same area of cross-section as the rectangular post.

7-22. A mass of 50 kg is attached to the end of a solid circular shaft of a light material (that is, the weight of the shaft is negligible compared to the weight of the end mass). The shaft material has a fracture toughness of $10 \, \mathrm{MPa}\sqrt{\mathrm{m}}$. The shaft develops a surface semielliptic crack of size $2c = 20 \, \mathrm{mm}$ and $a = 4 \, \mathrm{mm}$. What is the maximum speed at which the shaft can be rotated in the vertical plane to prevent brittle fracture? Consider that the shaft dimensions are much larger than the crack size.

7-23. Solve Problem 7-22 if an elliptic crack of the same size is embedded in the center of the shaft.

8   Slender Compressive Axial Structures

*Nothing whatsoever takes place in the universe in which some
relation of maximum and minimum does not appear.*

—Leonhard Euler (1707–1783)

Objective: This chapter will introduce two design degrees of freedom—P_{cr} and ηL—in the context of the design of slender compressive axial structures.

What the student will learn:

- The concept of stability of equilibrium
- Definition and configuration of slender compressive axial structures
- Equilibrium and deformation of slender compressive axial structures
- 1-D elastic constitutive relation
- Linkage of mechanics and materials concepts through buckling tests
- Materials selection for slender compressive axial structures
- How to design slender compressive axial structures

8.1   Introduction ⍦🏛💻

In Chapter 2, we discussed two general categories of structural failure: strength and stiffness, focusing up to now on strength failures. In Chapter 3, we considered axial compression structures, principally of the compact kind, so as to avoid issues of stability or buckling. In this present chapter, we now concentrate on buckling failure of slender columns. Columns have been used as important structural elements since the earliest days of human engineering (see for example the columns in the Parthenon shown in Chapter 5). Carrying roof loads or upper story loads down to ground are a standard function of columns. Figure 8-1 shows a typical column in a building.

Before we get into the details of column buckling, we want to review the concept of stability (recall that we discussed this briefly in Chapter 3). For lack of better words, the concept of stability is related to the "quality" or "robustness" of an equilibrium state. The concept of stability is much more easily presented by a simple example.

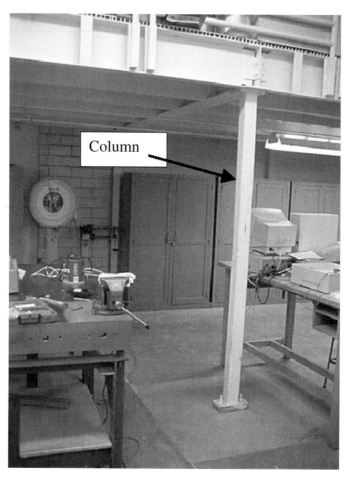

FIGURE 8-1 *A typical column in an industrial building. The column clearly is connected to a beam at its top end and then carries that load to the floor.*

Imagine a small, smooth ball (like a ball bearing) resting on a smooth surface (as shown in Figure 8-2). If the surface is concave (Figure 8-2a), the ball is in equilibrium at the bottom of the trough if no net force is acting on it. If we now displace the ball slightly from this equilibrium position, it inherently returns to the original equilibrium position (after perhaps a few oscillations of decaying amplitude). This condition is simply called *stable equilibrium*. The equilibrium position at the bottom of the trough is a *strong attractor*, such that small departures away from the bottom wind up decaying back to the equilibrium position.

Now if the surface is oppositely curved or concave (Figure 8-2b), the ball sitting at the apex of the curved surface is also in equilibrium if no net force is acting. But this time as we displace the ball slightly from this equilibrium position, the ball rolls away! This is called *unstable equilibrium*. The equilibrium position in this case is a *weak attractor* such that small departures away from the attractor do not return.

Finally, if the ball rests on a flat surface (Figure 8-2c), the displaced ball neither returns nor rolls away. This is a transitional configuration between the stable and unstable equilibria,

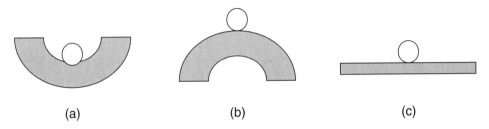

(a)   (b)   (c)

FIGURE 8-2 *Simple examples of the concept of stability.*

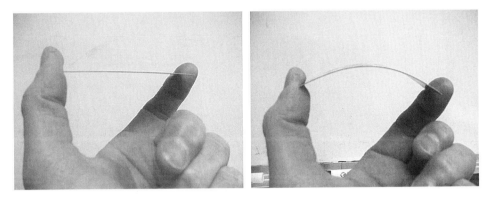

FIGURE 8-3 *A plastic strip acting as a simply supported column with the thumb and finger providing the compressive load. The first buckling mode, which will later be seen to be sinusoidal, is easily recognized in the right-hand picture.*

called *neutral stability*, and is also an example of the important concept of nearby or *adjacent equilibrium states*.

It is this issue of the quality of an equilibrium state under a small disturbance (formally called a *perturbation*) that we want to understand with respect to structures in general, and in slender axial compression structures (or columns) in particular. In what follows, we want to understand the load and other issues for a column to display the kind of behavior as the ball in Figure 8-2b. The load in this case is called the *critical buckling load*, that is, the load just required to cause the departure from the equilibrium position. The plastic strip shown in Figure 8-3 is a simple example of a buckled column.

We note here that although the *onset of buckling* may represent a significant failure state for the structure, it may not be catastrophic. That is, the column may snap to an adjacent equilibrium position and still carry load (albeit of reduced capacity)—this is called the *post-buckled state*. Although post-buckling is of great practical interest in structural design, the topic is beyond the scope of this text.

8.2   Equilibrium and Deformation: Compressive Loading ⓨ

In this chapter, we consider relatively slender, straight, prismatic, and uniform axial elements that are subject to axial compressive loads. In terms of configuration, these are not unlike the elements studied in the first part of Chapter 3 except that the loading is reversed, or the flexural

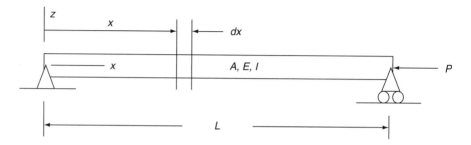

FIGURE 8-4 *Configuration of a long, straight, prismatic and slender column. Simply supported boundary conditions are shown. Note that a column does not have to be a vertical member as in Figure 8-1, but can be any straight slender element principally carrying axial compression.*

elements studied in Chapter 5 (except the loading is axial not transverse). Figure 8-4 shows the configuration geometry of these columns.

In Figure 8-4, simply supported boundary conditions are arbitrarily shown. Other boundary conditions we are already familiar with—clamped and free—could also be applied, and we will discuss these in more detail later. The term *slender* will be discussed later in this chapter. Also in what follows, the load **P** is assumed not to change direction during deformation; in Figure 8-4 then, the load continues to remain in the *x*-direction during deformation and not necessarily normal to the column end.

As mentioned earlier, our goal is to analyze the column just at the onset of buckling. However, to get a useful relation to work with, we must imagine the column has slightly buckled or displaced from its initially straight configuration. (Remember that free-body diagrams and their associated equilibrium equations must be written in the deformed configuration.) This departure from the initially straight configuration represents a curved or flexural deformed configuration; hence, internal shear forces and moments will exist.

We consider a free-body diagram of a section of the column of length ds, and prepare to sum forces and moments acting on the section in order to obtain the equilibrium expressions. Figure 8-5 shows the free-body diagram with internal forces (axial **N** and shear **V**), moments (**M**), and rotation angles (α). We allow each quantity to vary generally across the section from left to right in the positive axial coordinate direction, that is, d/dx; hence, quantities at the right end of the section are given by $(d/dx)dx$. Since we only consider small departures from equilibrium $ds \sim dx$, the rotation is small, and we use small-angle approximations $\sin \alpha \approx \alpha$ and $\cos \alpha \approx 1$.

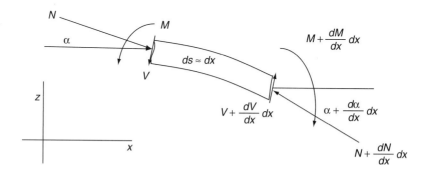

FIGURE 8-5 *Free-body diagram of a deformed infinitesimal element of the column in Figure 8-4.*

Summing forces in the x-direction:

$$N - \left(N + \frac{dN}{dx}dx\right) + V\alpha - \left(V + \frac{dV}{dx}dx\right)\left(\alpha + \frac{d\alpha}{dx}dx\right) = 0. \qquad (8.1a)$$

Upon simplification we get

$$\frac{dN}{dx}dx + V\frac{d\alpha}{dx}dx + \alpha\frac{dV}{dx}dx + \frac{dV}{dx}\frac{d\alpha}{dx}dxdx = 0. \qquad (8.1b)$$

Summing forces in the z-direction:

$$V - \left(V + \frac{dV}{dx}dx\right) - N\alpha + \left(N + \frac{dN}{dx}dx\right)\left(\alpha + \frac{d\alpha}{dx}dx\right) = 0. \qquad (8.2a)$$

or

$$\frac{dV}{dx}dx - N\frac{d\alpha}{dx}dx - \alpha\frac{dN}{dx}dx - \frac{dN}{dx}\frac{d\alpha}{dx}dxdx = 0. \qquad (8.2b)$$

Summing moments about the right end of the section:

$$-M + \left(M + \frac{dM}{dx}dx\right) + Vdx = 0. \qquad (8.3a)$$

That is,

$$\frac{dM}{dx} + V = 0. \qquad (8.3b)$$

A Note on Modeling

We are now at a critical step—put the model [Eqs. (8.1b), (8.2b), and (8.3b)] into a form that can provide useful results at a reasonable cost, that is, effort. Equations (8.1b) and (8.2b) are *nonlinear ordinary differential* equations, even given the small-angle assumptions already made. There are two reasons for this. First, V is a function of the rotation angle α, both of which are *solution dependent variables*. Second, the terms dxdx are quadratic. It will be seen that N is not a function of α.

As you can probably guess, nonlinear equations are usually harder to solve than linear equations, so there is strong motivation to try to linearize these nonlinear equations. An important consideration, however, is: What will be lost in the linearization process? That is a tough question, which may require nonlinear analysis for an answer! However, if the nonlinear terms can be argued to be small compared to the linear terms, it may be reasonable to ignore them in a *first-order* analysis, and the results may be fine for a large regime of the material/structural system response. This is the approach we will follow.

To simplify the analysis in our case, we make two additional assumptions. First, terms containing products of differentials, namely $dx \cdot dx$, are ignorably small compared to the other terms (prove to yourself that the square of a very small number is even smaller). Second, since no transverse loads are applied, and since deformations are limited to be small, terms made up of products of the small shear force and small rotation, $V\alpha$, are ignorably small as well. The following three linear second-order ordinary differential equations (with constant coefficients) in four unknowns (N, M, V, and α) then result from a linearization of Eqs. (8.1b), (8.2b), and (8.3b):

$$\frac{dN}{dx} = 0 \tag{8.4a}$$

$$\frac{dV}{dx} - N\frac{d\alpha}{dx} - \alpha\frac{dN}{dx} = 0 \tag{8.4b}$$

$$\frac{dM}{dx} = -V. \tag{8.4c}$$

Using (8.4a) in (8.4b), and using (8.4c) to eliminate V in (8.4b), the three equations in (8.4) reduce to

$$\frac{dN}{dx} = 0 \tag{8.5a}$$

$$\frac{d^2M}{dx^2} + N\frac{d\alpha}{dx} = 0. \tag{8.5b}$$

We now have two equations in three unknowns: N, M, and α. However, two additional simplifications can be made. First, (8.5a) simply says that N is constant along the length of the column, or $N = P$. Second, we know from our study of flexural elements (Chapter 5) that the moment in a beam is related to the beam curvature. The slope of the column in Figure 8-5 is $\alpha = dw/dx = w'$, and for small deflection the curvature is approximately equal to $d^2w/dx^2 = w''$. Then the moment in the column is equal to (see Chapter 5) $M = EIw''$. Using both of these simplifications in (8.5b) gives

$$\frac{d^2}{dx^2}\left(EI\frac{d^2w}{dx^2}\right) + P\frac{d^2w}{dx^2} = 0. \tag{8.6a}$$

For homogeneous prismatic columns, $EI = $ constant, and finally we get

$$\boxed{\frac{d^4w}{dx^4} + \left(\frac{P}{EI}\right)\frac{d^2w}{dx^2} = 0.} \tag{8.6b}$$

This equation governs the column deflection (buckling) given the earlier assumptions. It is a fourth-order, ordinary linear differential equation with constant coefficients; as such, the solution will require four boundary conditions. A list of typical boundary conditions is given below.

- Simply supported

$$w = 0, \ d^2w/dx^2 = 0 \tag{8.7a}$$

- Clamped

$$w = 0, \ dw/dx = 0 \qquad (8.7b)$$

- Free

$$d^2w/dx^2 = 0, \ d^3w/dx^3 = -(P/EI)dw/dx \qquad (8.7c)$$

[This last result is derived as follows: Equation (8.4b) gives $V = Pw'$, while $M = EIw''$, then from Eq. (8.4c) $V = -M'$, or $Pw' = -EIw'''$.]

Let's summarize where we are at this point. We have a linear theory of column buckling consisting of:

- The two equilibrium equations (8.5a) and (8.5b) (in fact we started with three equilibrium equations—summation of forces in two coordinate directions and one summation of moments—then combined them)
- The moment–curvature relation $M = EIw''$ from Chapter 5
- the kinematic equation $\alpha = dw/dx$

These four equations were combined into one fourth-order linear ordinary differential equation (8.6b). The solution $w(x)$ requires four constants of integration to be determined from four column boundary conditions, two at each end. We will pursue such solutions shortly.

Last but not least, it is important to pay tribute to the man who first developed this theory, the great French mathematician Leonhard Euler (1707–1783). The theory, developed in the 18th century, is still widely used today and referred to as the *Euler buckling theory* in honor of this great achievement.

8.3 Constitution

In this chapter we consider only columns whose material behavior would be adequately described by a linear elastic constitutive relation. Other material models may be invoked; one of the more common is an elastic-perfectly plastic constitutive model, but only the linear elastic model will be discussed here.

As our columns are considered "slender," a one-dimensional relationship will suffice. Just as in Chapter 3, this is simply stated as (with respect to the coordinate system shown in Figure 8-3):

$$\sigma_x = E\varepsilon_x. \qquad (8.8)$$

Bear in mind that we deal primarily in this chapter from a stiffness analysis perspective, and consequently the stress–strain relation does not play the same role it does in a strength analysis. Recall that we arrived at Eq. (8.6a) by invoking the *moment–curvature relation* of Chapter 5. That result directly used the constitutive model (8.8), so the constitutive nature of the material is still necessary in stiffness design. (We will come back to the question of column "strength" later in this chapter, however, principally to enforce the idea of column slenderness.)

It will be instructive to look at a load-deflection diagram for a perfect column, say the simply supported column of Figure 8-4. In Figure 8-6, we see the lateral deflection at the mid-length of the column to be zero up to a critical value of the load P_{cr} (which we will define and

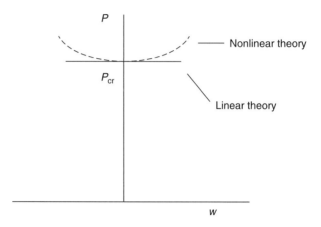

FIGURE 8-6 *Load–deflection curve for a column.*

study shortly); at this point the column buckles. In the post-buckled region, the linear theory (solid line) predicts no additional load being carried and increasing large deflections. The nonlinear theory (dashed line), which we do not discuss in this text, does predict some additional post-buckled load-carrying capacity. Note also the indeterminacy about to which side the column will buckle.

8.4 Buckling Tests

Up to this point, the theoretical treatment has considered only ideal, perfect columns. However, the real-world testing of columns for critical buckling loads, or for post-buckled load-carrying capacity, is notoriously difficult for one particular reason: imperfections. Nowhere is the sensitivity to imperfections greater than in buckling tests. For example, critical buckling load measurements for real columns (as well as for plates and shells) usually are less, often considerably less, than the predicted values from the corresponding buckling theory. Imperfections run the gamut from crookedness and loading eccentricity to nonideal boundary conditions and residual stresses from manufacturing.

The engineer goes to great lengths in buckling tests to ensure that the columns are initially straight and the loads are applied axially, with known boundary conditions. Even so, the scatter in buckling test data is usually large (large standard deviation). This uncertainty due to initial imperfections flows down into design through significant factors of safety or resistance factors. We will discuss the sensitivity to initial imperfections further a little later in the chapter.

8.5 Energetics

Just as we discussed with respect to constitutive relations in Section 8.3, the strain energy relations for a column are the same as they were for the compact axial compressive members discussed in Section 3.6. We merely repeat a useful form of the strain energy relation here for completeness:

$$U = \frac{1}{2} \int_{\text{volume}} \sigma_x \varepsilon_x dV. \tag{3.37}$$

We remind the student that we began the present chapter from the energy point of view, as energy considerations remain central to the analysis and design of slender compressive axial structures.

8.6   Analysis of Slender Compressive Structures Ϋ

In previous chapters we have relied in part on the direct stiffness method to aid our analysis. We could do that here as well (and this is in fact the method finite element codes use for column buckling analysis). However, the additional concepts needed—geometric stiffness and eigenvalue extraction—are just a little beyond the scope of this text. The interested reader is referred to other sources for details (see, for example, Przemieniecki, 1985). In this section we will rely on the classical treatment of column buckling, using the solution to the governing ordinary differential equation.

We return to our governing relationship Equation 8.6b, which has a general solution (Hildebrand, 1962)

$$w(x) = C_1 \sin \lambda x + C_2 \cos \lambda x + C_3 x + C_4 \qquad (8.9)$$

where C_1, C_2, C_3, and C_4 are constants of integration, and $\lambda^2 \equiv P/EI$ (see Problem 8-3).

Continuing with the simply supported column defined in Figure 8-3, we apply boundary conditions (8.7a):

$$
\begin{aligned}
w(0) &= (0)C_1 + C_2 + (0)C_3 + C_4 = 0 \\
w(L) &= \sin(\lambda L)C_1 + \cos(\lambda L)C_2 + LC_3 + C_4 = 0 \\
w''(0) &= -(0)C_1 - \lambda^2 C_2 = 0 \\
w''(L) &= -\lambda^2 \sin(\lambda L)C_1 - \lambda^2 \cos(\lambda L)C_2 = 0.
\end{aligned}
\qquad (8.10)
$$

The above represents a homogeneous system of linear algebraic equations, the form of which looks like $[\text{coeff.}]\{C_i\} = \{0\}$; in matrix form Eq. (8.10) is called the *characteristic equation*. The solution (*roots*, *characteristic values*, or *eigenvalues*) of this system of equations is nontrivial (the trivial solution is the vector $\{C_i\} = \{0\}$) if and only if the determinant of the coefficient matrix equals zero, i.e., $\det[\text{coeff.}] = 0$. Performing the determinant operation (see Problem 8-4) gives

$$-\lambda^4 L \sin(\lambda L) = 0. \qquad (8.11)$$

Clearly λ and L cannot be zero, so (8.11) is only nontrivially satisfied if $\sin(\lambda L) = 0$, that is,

$$\lambda L = n\pi, \quad n = 1, 2, \ldots. \qquad (8.12)$$

Hence the eigenvalues of (8.10) are given by (8.12), with the *eigenmode* being indicated by the number n.

Finally, note from the third of Eqs. (8.10) that $C_2 = 0$, from the first of (8.10) that $C_4 = -C_2 = 0$, and from the second of (8.10), using (8.12), that $C_3 = 0$. Then the solution for the column with both ends simply supported is

$$w_n(x) = C \sin(\lambda_n x) = C \sin\left(\frac{n\pi x}{L}\right), \quad n = 1, 2, \ldots \qquad (8.13)$$

FIGURE 8-7 *Critical buckling mode for the simply supported column is a half sine wave of undetermined amplitude.*

where we replace C_1 simply by C since there is only one nonzero constant. [Equation (8.13) is formally said to give the *eigenvectors* $w_n(x)$ of (8.10).]

The *critical* solution is the smallest of these, that is, $n = 1$, and then the *critical buckling load* and corresponding *critical buckling mode* are, respectively,

$$P_{\text{critical}} = P_1 = \lambda_1^2 EI = \frac{\pi^2 EI}{L^2}$$

$$w(x)_{\text{critical}} = w_1(x) = C \sin\left(\frac{\pi x}{L}\right).$$

(8.14a,b)

The buckled shape as given by Eq. (8-14b) is a half sine wave as seen in Figure 8-7.

Let's summarize where we are at this point:

- We have developed a solution for the linear theory for column buckling, for the specific case of simply supported boundary conditions.
- The solution consists of the *Euler buckling load* P_{cr} and the associated critical buckling mode.
- Note that the absolute value of the transverse column displacement is known only to within an undetermined constant C. This is an artifact of the linear theory, since the deformed configuration Figure 8-5 is considered to be infinitesimally close to the original configuration Figure 8-4 in the linear theory.

8.7 Material Selection for Slender Compression Structures

The material index for a slender compression structure can be determined in a manner similar to other structural components. For example, let us formulate the index for a lightweight, stiff column. The weight is to be minimized, that is,

$$W = mg = \rho A L g \leq W_{\text{max}}$$

(8.15)

where W_{max} is the maximum weight of the finished column. But reducing the cross-sectional area A of the column can lead to buckling if we're not careful. So the function (8.15) is subject to the *constraint*

$$P \leq P_{\text{cr}} = \frac{B\pi^2 EI}{L^2}$$

(8.16)

where B is a constant that depends on the column boundary conditions and that will be described in more detail shortly ($B = 1$ for the simply supported column considered above).

8.7.1 Material Index for a Lightweight, Stiff Column with Circular Cross-Section

We expect that the design load P and column length L are specified, as well as the maximum column weight. By way of example, we can imagine the design of a column with solid circular cross-section of radius r_0, such that $I_0 = \pi R_0^4/4 = A_0^2/4\pi$. Now eliminating the free variable A_0 from (8.15) and (8.16) at the equality condition results in

$$W_{\max} = \left(\frac{4}{\pi B}\right)^{1/2} (P)^{1/2} g L^2 \left(\frac{\rho}{E^{1/2}}\right). \tag{8.17}$$

As is usual in the Ashby method, the parameter groups represent functional requirements, geometry, and material properties. The mass of the column is minimized by choosing materials with the largest value of the material index:

$$\frac{E^{1/2}}{\rho} = \left(\frac{4}{\pi B}\right)^{1/2} (P)^{1/2} g \frac{L^2}{W_{\max}}. \tag{8.18}$$

8.7.2 Material Index for a Low-Cost, Stiff Column with Other Cross-Sections

Just as we did in earlier chapters, we can include the effect of shape into the material selection by defining a *shape factor* β. The shape factor is then incorporated in the material index, resulting in "effective material properties." As before, we compare various shapes on the basis of equivalent cross-sectional areas, so the shape factor for columns is

$$\beta = \frac{I}{I_0} \tag{8.19}$$

where I is the minimum area moment of inertia of the shape to be compared and I_0 is the moment of inertia of a solid circular section of the same area. Since the lowest critical buckling load is associated with the minimum area moment of inertia (assuming the same boundary conditions for all axes), it is conservative to compare shapes relative to their minimum area moment of inertia.

Returning to Eq. (8.16), we replace I with $\beta I_0 = \beta A_0^2/4\pi$. Then, as before, eliminating $A = A_0$ from Eq. (8.15) and the modified (8.16) results in a modified Eq. (8.17):

$$W_{\max} = \left(\frac{4}{\pi B}\right)^{1/2} (P)^{1/2} g L^2 \left(\frac{\rho}{\sqrt{\beta E}}\right). \tag{8.20}$$

Solving for the material index as before results in

$$\frac{E^{1/2}}{\rho} = \frac{1}{\sqrt{\beta}} \left(\frac{4}{\pi B}\right)^{1/2} (P)^{1/2} g \frac{L^2}{W_{\max}}. \tag{8.21}$$

Assuming $I > I_0$ (which would be desirable), $\beta > 1$ and $1/\sqrt{\beta} < 1$, which has the effect of moving the selection line down on the Ashby chart, thus allowing more materials in as candidates. An example of using the column shape factor will be given in the next section.

8.8 Design of Slender Compression Structures ⅄🏠🖥

8.8.1 Effect of Boundary Conditions

In Table 8-1, we present the results for combinations of the three common column boundary conditions in Eqs. (8.7).

When looking at Table 8-1, it is apparent that the critical buckling load looks like

$$P_{cr} = B\frac{\pi^2 EI}{L^2} \tag{8.22}$$

where the constant B is a function of the boundary conditions. Sometimes B has been called the *end fixity factor*. Another way to write Eq. (8.22) is to subsume B into the denominator with the length L as

$$P_{cr} = \frac{\pi^2 EI}{(\eta L)^2} \tag{8.23}$$

where $\eta = 1/\sqrt{B}$ and ηL is called the *effective column length*. The effective column length gives a measure of the relative stiffness of the various column configurations; by examination, that stiffness depends on the boundary conditions.

By definition, $\eta = 1$ for the simple supported–simple supported column, and this case becomes the reference case with a length L. If we compare the clamped–clamped case to this reference case, we note that the critical buckling load is four times as great ($B = 4$), so the clamped–clamped column must be effectively stiffer (with the same E, I, and L, of course). $\eta = 1/2$ for the clamped–clamped case, and the effective length is therefore one-half that of the reference case; that is, the clamped–clamped boundary conditions stiffen the column compared to the reference case, meaning that it would take a clamped–clamped column twice as long as the reference case to exhibit the same critical buckling load. Table 8-1 shows that the clamped–free column is the least stiff (has the greatest effective length). Does that make intuitive sense?

8.8.2 Real versus Ideal Boundary Conditions

The boundary conditions explored thus far are ideal mathematical constructs. Whether they are exactly represented in a real column is an open and important question for the engineer to consider. Given the effect of boundary conditions on the column response as shown in

TABLE 8-1 Critical Buckling Load, Mode Shape, and Effective Column Length for Several Combinations of Common Boundary Conditions

Boundary conditions	Critical load P_{cr}	Mode shape $w(x)$	Effective length ηL
Simple supported–Simple supported	$\dfrac{\pi^2 EI}{L^2}$	$C\sin\dfrac{\pi x}{L}$	L
Clamped–Clamped	$4\dfrac{\pi^2 EI}{L^2}$	$C\left(1 - \cos\dfrac{2\pi x}{L}\right)$	$L/2$
Clamped–Simple supported	$2.04\dfrac{\pi^2 EI}{L^2}$	$C[\sin \lambda x - \lambda L\cos \lambda x + \lambda(L-x)]$ $\lambda = 4.49/L$	$0.70L$
Clamped–Free	$\dfrac{1}{4}\dfrac{\pi^2 EI}{L^2}$	$C\left(1 - \cos\dfrac{\pi x}{2L}\right)$	$2L$

Table 8-1, the critical buckling could easily be misjudged with serious consequences, particularly since buckling is evidence of instability—catastrophic collapse may follow! Imagine that a vertical column is connected at its upper end to a beam, which would be a very common structural configuration (see, e.g., the picture of the Parthenon in Chapter 5). The beam is itself an elastic member, and thus is not likely to provide ideal restraint against translation or rotation.

Even if a boundary condition is ideal in one plane of the column, there may be a different condition in another plane. For example, the column may be simply supported in one plane by a pin connection (the plane cutting the pin transversely), but rigidly clamped in the orthogonal plane containing the pin.

8.8.3 Slenderness Ratio

As mentioned earlier, we have thus far ignored stress in the column, because the primary mode of column failure is due to stiffness. However, examining the column stress will give us insight into what we mean by a "slender" column, as well as when an axial compression member becomes a column or a compact compression structure (as studied in Chapter 3).

Given the 1-D geometry and constitutive relationship of a column, the stress can readily be given by

$$\sigma_{cr} = \frac{P_{cr}}{A}. \tag{8.24}$$

At this point, it will be convenient to introduce the concept of *radius of gyration* R_g:

$$R_g = \frac{I}{A}. \tag{8.25}$$

Substituting Eqs. (8.23) and (8.25) into (8.24) gives

$$\sigma_{cr} = \frac{\pi^2 E}{\left(\dfrac{\eta L}{R_g}\right)^2}. \tag{8.26}$$

The quantity $\eta L/R_g$ is called the *slenderness ratio*. The slenderness ratio gives a measure of the (effective) length of the column relative to the cross-sectional properties. The higher the slenderness ratio, the more pronounced the column length is relative to its cross-sectional dimensions.

Of course, good design would require σ_{cr} to be less than (or equal to) some design strength. For illustrations purposes, let's take the yield strength as the design strength (this would be consistent with the linear elastic assumption we have invoked), i.e., $\sigma_{cr} \leq S_{Yield}/FS$ or (with the factor of safety equal to unity for illustration purposes)

$$\left(\frac{\eta L}{R_g}\right)^2 \geq \frac{\pi^2 E}{S_{Yield}}. \tag{8.27}$$

Then, realizing we have a uniaxial stress state,

$$\left(\frac{\eta L}{R_g}\right) \geq \pi\sqrt{\frac{E}{S_{Yield}}} = \pi\sqrt{\varepsilon_{Yield}}. \tag{8.28}$$

Now, for many metals a reasonable estimate of strain at yield might be 0.2% or 0.002 m/m. Substitution of this value into Eq. (8.28) gives a value for the slenderness ratio equal to (or greater than) about 0.04. This means that keeping the slenderness ratio greater than, say, about 0.1 ensures that the column failure mode is buckling not strength (for the assumed value of yield strain). This was the basis for our claim in Chapter 3.

8.8.4 Initial Imperfections

As a simple example of sensitivity to imperfections, let us consider the slightly crooked column. We imagine our simply supported column of Figure 8-4 to have an initial shape of its first buckling mode:

$$w^*(x) = C^* \sin(\pi x / L) \tag{8.29}$$

where C^* is the magnitude of the initial imperfection. The total buckling mode deflected shape after loading must be $w^*(x) + w(x)$, where $w(x)$ is the additional deflection from the initial shape. The total deflected slope of the column is now $\alpha = d(w^* + w)/dx$. Returning now to Equation (8.5b), and realizing that the moment M is only related to the additional deflection after loading $w(x)$, we get a revised Eq. (8.6b):

$$\frac{d^4 w}{dx^4} + \left(\frac{P}{EI}\right) \frac{d^2(w^* + w)}{dx^2} = 0 \tag{8.30}$$

or

$$\frac{d^4 w}{dx^4} + \left(\frac{P}{EI}\right) \frac{d^2 w}{dx^2} = -\left(\frac{P}{EI}\right) \frac{d^2 w^*}{dx^2}. \tag{8.31}$$

Equation (8.31) is now nonhomogeneous, that is, there is a nonzero "forcing" term on the right-hand side. Given the assumed shape (8.29), we can simplify (8.31) to

$$\frac{d^4 w}{dx^4} + \left(\frac{P}{EI}\right) \frac{d^2 w}{dx^2} = \left(\frac{PC^* \pi^2}{EIL^2}\right) \sin\left(\frac{\pi x}{L}\right). \tag{8.32}$$

The solution can be shown to be (Brush and Almorath, 1975)

$$w(x) = \left(\frac{\zeta C^*}{1 - \zeta}\right) \sin\left(\frac{\pi x}{L}\right) \tag{8.33}$$

where $\zeta = P/P_{cr} = PL^2/\pi^2 EI$.

Equation (8.33) shows that as $P \to P_{cr}$, $\zeta \to 1$ and the solution becomes unbounded. Thus the initially crooked column buckles well before the ideal critical buckling load is reached, and that is a large source of the discrepancy between test and prediction. Our previous Figure 8-6 is now modified in Figure 8-8 for the case of initial crookedness.

8.8.5 Other Design Considerations

- Nonsymmetric cross-sections will have different buckling loads for different section planes. For example, a rectangular cross-section will have different critical buckling loads for the two planes of the cross-section. The design critical buckling load is the *smallest* load for the cross-section.

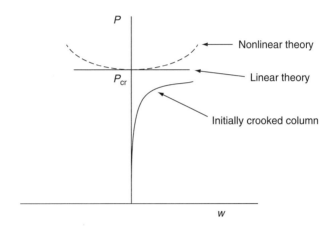

P

P_{cr}

Nonlinear theory

Linear theory

Initially crooked column

w

FIGURE 8-8 *Load-deflection curve for a column, now including initial crookedness.*

- A simple way to decrease the effective column length, thereby increasing its stiffness and critical buckling load, is to provide one or more intermediate supports. Mode shapes can give insight into optimal placement of the intermediate supports.
- Many of the uncertainties of column design are accounted for in practice by significant factors of safety, resistance factors, or other so-called *knock down factors* (for example, see AISC, 1999).

EXAMPLE 8-1: Design Example

An elevated trolley in a factory is used to carry loads from one location to another as shown in Figure E8-1A. The slow moving trolley wheels ride on rails that are in turn supported by beams. Each 10 m long beam is supported by 2 columns, with one beam placed within a few centimeters of the factory wall. The bottoms of the beams are 6 m above the factory floor with the intervening space open to factory personnel. The design load for the trolley is 20 kN, but little safeguards are in place to ensure the load carried. A small crane is available to lift the columns into place that has a lifting capacity of 350 N. Design a set of four columns to support the trolley that can be placed with the existing crane. Check at least two different column cross-section configurations.

I. *Problem Definition*

Performance requirements:
- Factor of safety $= 3$. This specification is chosen purposely high due to the potential for loss of human life should the columns fail catastrophically. There is significant uncertainty in the amount of load the trolley might accidentally carry, so we apply the FS to the load.
- No other performance requirements are given.

Continued

EXAMPLE 8-1: Design Example–*Cont'd*

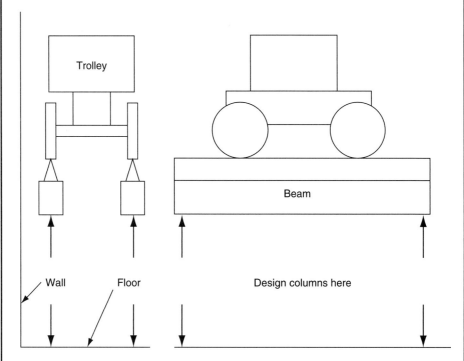

FIGURE E8-1A *Trolley configuration sketch.*

Service environment:
- Loads—are assumed to act statically. The load is 20 kN assumed to be distributed evenly between the two beams, thus the nominal load per column is:

$$P = FS \times \text{load per beam/number of columns per beam}$$
$$= 3(10\,\text{kN})/2 = 15\,\text{kN}.$$

(The assumption of static loads is made to keep the problem simple enough for educational purposes. In the real situation, the loads could easily be dynamic and thus require additional consideration.)
- The nominal column length is required to be 6 m.
- No other service environment specifications are given.

Project constraints:
- The column weight must not exceed 350 N so as to use the existing crane to lift the columns into place.
- We are to trade off two different cross-section configurations.
- No other constraints are given.

EXAMPLE 8-1: Design Example–*Cont'd*

II. *Preliminary Design*

- Identify DDOFs:
  1. Geometry ⇒ minimize volume to keep weight low ⇒ area (A) and cross-section shape (I)
  2. Density ⇒ keep weight low (ρ)
  3. Stiffness ⇒ modulus (E)
  4. Boundary conditions ⇒ this is an important variable that requires more discussion below
  5. Strength ⇒ not a primary design variable but good design suggests we check for yielding (S_y) just to be safe
- Trade study on boundary conditions: A review of Table 8-1 suggests that making both column end conditions simply supported gives the maximum critical buckling load. Making the upper boundary condition simply supported in this case also makes sense because the beam itself may rotate under the trolley load (maximum when the trolley is at mid-span). That rotation does not want to be translated to the column as an additional moment load; hence we should allow the beam to freely rotate relative to the column. If we make both ends simply supported, however, we will have made a four-bar linkage (floor, two columns, and beam) that will be highly flexible and require additional constraint against rigid body motion! So for this design we choose clamped (at the floor)—simply supported at the beam, which has the second highest critical bucking load in Table 8-1.
- Trade study on cross-section configuration: The desire to keep costs low (and, as always, to keep things as simple as possible!) suggests that we start with a simple solid circular cross-section of diameter d that will provide the same buckling resistance in all planes. We will also consider a hollow circular cross-section as well, to see what performance improvement can be had by modifying the cross-sectional shape.

  Material considerations suggest simple, readily available materials to keep costs low, but analysis needs to confirm the specific choice. Keeping material density low will also lead to less total weight and potentially less cost.
- FMEA: Failure expected to be from simple Euler buckling of the columns, but we will check for yielding just to be sure.

III. *Detailed Design*

From Eq. (8.21), the material index for a lightweight, solid circular ($\beta = 1$) stiff column is

$$\frac{E^{1/2}}{\rho} = \frac{1}{\sqrt{\beta}} \left(\frac{4}{\pi B}\right)^{1/2} (P)^{1/2} g \frac{L^2}{W_{\max}}$$

$$= \left(\frac{4}{2.04\pi}\right)^{1/2} (15 \times 10^3 \, \text{N})^{1/2} (9.8 \, \text{m/s}^2) \frac{36 \, \text{m}^2}{350 \, \text{N}}.$$

Now in order to plot this on the Ashby Young's modulus versus density chart, we need units of $\sqrt{\text{GPa}}/\text{Mg/m}^3$. A slight manipulation gets us there:

Continued

EXAMPLE 8-1: Design Example–*Cont'd*

$$\frac{E^{1/2}}{\rho} = \left(\frac{4}{2.04\pi}\right)^{1/2} (15 \times 10^3 \, \text{N})^{1/2} \frac{\sqrt{10^6 \, \text{m}^2}}{\sqrt{10^6 \, \text{m}^2}} (9.8 \, \text{m/s}^2) \frac{36 \, \text{m}^2}{350 \, \text{kg} \cdot \text{m/s}^2}$$

or

$$\frac{E^{1/2}}{\rho} = 3.09 \sqrt{\frac{\text{GPa}}{\text{Mg/m}^3}}.$$

We can plot this material index on the Ashby modulus versus density chart as follows:

$$3.09 \frac{(\text{GPa})^{1/2}}{\text{Mg/m}^3} = \frac{1000^{1/2}}{\rho} \Rightarrow \rho = 10.2 \approx 10$$

$$3.09 \frac{(GPa)^{1/2}}{\text{Mg/m}^3} = \frac{E^{1/2}}{0.1} \Rightarrow E = 0.0955 \approx 0.1.$$

Thus we plot a straight line between the coordinates $(0.1, 0.1)$ and $(10, 1,000)$. All materials above this line satisfy the light-stiff criteria as specified. Note, for example, that steel at this point is not an option.

To size our column, we can go back to either Eq. (8.15) or (8.16), the former being a little quicker. At this point, a reasonable choice might be the wood group. Choosing a midpoint density of $0.4 \, \text{Mg/m}^3$ and back solving for area, and then diameter, from Eq. (8.15) gives a solid wood column 13.8 cm (5.43 in.) in diameter.

Let's now investigate the potential for alternative cross-sectional shape by considering a hollow circular tube for our columns. Keeping in mind that we want to compare in this problem not only on equivalent section areas, but on equivalent section weight (remember that total weight is one of our design constraints), the section weight of the tube ρA is related to the solid column by

$$\rho A = 2\pi r_m t \rho = \rho_0 A_0 = \pi R_0^2 \rho_0.$$

Let's choose a factor of 3 improvement in area moment of inertia, that is, $\beta = I/I_0 = 3$ or

$$I = \pi r_m^3 t = 3I_0 = \frac{3\pi}{4} R_0^4.$$

Solving these last two equations simultaneously (this is left as an exercise for the student) shows that (for this case) $r_m = 1.22(\rho/\rho_0)^{1/2} R_0$ and $t = 0.41$ $(\rho_0/\rho)^{3/2} R_0$. Next we calculate a new material index with $\beta = 3$ in Eq. (8.21). We can do this by simply using our previous result above but now with the shape factor included, or

$$\frac{E^{1/2}}{\rho} = \frac{3.09}{\sqrt{3}} \frac{\sqrt{\text{GPa}}}{\text{Mg/m}^3} = \underline{1.78}$$

EXAMPLE 8-1: Design Example–*Cont'd*

Plotting as we did before, we draw a straight line between (18, 1000) and (0.1, 0.03) (the dashed line in Figure E8-1B). We immediately recognize that steel now appears as a candidate material. Using a midpoint density of $8 \, \text{Mg/m}^3$, we can calculate the mean radius and thickness of the hollow section as

$$r_{\text{m}} = 1.22(8/0.4)^{1/2}(6.9 \, \text{cm}) = \underline{37.6 \, \text{cm} \ (14.8 \, \text{in.})}$$

and

$$t = 0.41(0.4/8)^2(8/0.4)^{1/2}(6.9 \, \text{cm}) = \underline{0.31 \, \text{cm} \ (0.124 \, \text{in.})}$$

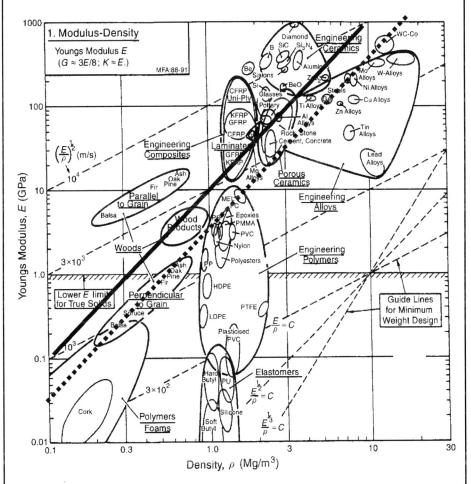

FIGURE E8-1B *Ashby material selection chart for this example. Solid line is for the default solid circular cylinder, while the dashed line is for the alternative hollow cross-section shape (with permission from Professor Michael Ashby and Granta Design).*

Continued

EXAMPLE 8-1: Design Example–*Cont'd*

Finally, let's check for simple axial stress:

$$\sigma = P/A = (15\,\text{kN})/(2\pi r_\text{m} t) = \underline{2.05\,\text{MPa}\ (0.293\,\text{ksi})}.$$

Clearly this is well below the yield stress of even standard structural steel (252 MPa or 36 ksi).

One other design feature we could incorporate is intermediate supports to the columns, especially along the wall. Providing intermediate connections between the columns and the wall has the result of reducing the "effective length" of the column and thus providing more load-carrying capability, as well as providing lateral bracing of the entire beam–column system (see Wempner, 1995).

We can now provide the initial column design specifications:

- 4 each COTS steel tube, 6 m long
- Nominal 76 cm (30 in.) outside diameter by 3 mm (0.125 in.) wall thickness

Key Points to Remember

- All equilibrium states are not equal—they may be stable, unstable, or neutrally stable.
- Slender axial compressive structures fail by an instability called buckling (stiffness failure) rather than strength failure.
- The linear theory of column buckling only considers the "onset" of buckling and not the post-buckled state (which requires a nonlinear theory).
- For a given configuration, a column will buckle when the axial compressive load reaches a certain critical value called the critical buckling load.
- The critical buckling load depends on the column "stiffness" EI, the column length L, and the boundary conditions of the column (the latter two often being combined into a description of the "effective column length" ηL).
- Real columns almost always differ from ideal columns, particularly with respect to boundary conditions and initial imperfections.
- A column will tend to buckle in the plane of least stiffness, so each symmetry plane should be checked in the design.

References

AISC (1999). *Load and Resistance Factor Design Specification*. American Institute of Steel Construction.

Brush, D. O., and Almorath, B. O. (1975). *Buckling of Bars, Plates, and Shells*. McGraw-Hill, New York.

Euler, L. (1759). On the strength of columns. *Academy Royal Society Belles Lettres Berlin Mém* 13, p. 252. (English translation by Van den Broek, J. A. (1947), *American Journal of Physics*, Vol. 15, pp. 315-318.)

Hildebrand, F. B. (1962). *Advanced Calculus for Applications*. Prentice-Hall, Upper Saddle River, NJ.

Przemieniecki, J. S. (1985). *Theory of Matrix Structural Analysis*. Dover.

Wempner, G. (1995). *Mechanics of Solids*. PWS Publishing, Boston.

Problems

8-1. Look around your local environment and identify columns with the following "apparent" boundary conditions:

- Clamped–clamped
- Simply supported
- Clamped–simply supported
- Fixed–free

Draw a sketch of each example, including a detail of the column cross-section.

8-2. For the wood table on the right side of Figure 8-1, discuss the likely boundary conditions, intermediate supports, and effective column length.

8-3. Show that λ^2 must equal P/EI to satisfy Eq. (8.6b). [Hint: take two and four derivatives of Eq. (8.9) and substitute into (8.6b).]

8-4. Place Eqs. (8.10) in the form [coeff.] $\{C_i\} = \{0\}$, and then show that the determinant of [coeff.] gives Eq. (8.11). Show also that for both ends simply supported, $C_1 = C_2 = C_3 = 0$.

8-5. Determine the critical buckling load and mode shape for a column with both ends clamped. Predict the location of the maximum transverse column displacement $w(x)|_{max}$, plot the mode shape, and verify.

8-6. Determine the critical buckling load and mode shape for a column with one end clamped and one end simply supported. Predict the location of the maximum transverse column displacement $w(x)|_{max}$, plot the mode shape, and verify.

8-7. Determine the critical buckling load and mode shape for a column with one end clamped and one end free. Predict the location of the maximum transverse column displacement $w(x)|_{max}$, plot the mode shape, and verify.

8-8. Derive the material index for a stiff, light column with hollow tubular cross-section of wall thickness t.

8-9. Derive the material index for a stiff, light column with solid rectangular cross cross-section.

8-10. Derive the material index for a stiff, light column with a wide-flange cross cross-section.

8-11. In the Design Example, verify the results that $r_m = 1.22(\rho/\rho_0)^{1/2} R_0$ and $t = 0.41(\rho_0/\rho)^{3/2} R_0$.

Design Problem

8-12. In order to protect a utility trench from caving in, a column needs to be placed between the two sidewall plates as shown in Figure P8-12.
   Assume that the soil load on the wall plates goes as $p(z) = Cz$, where $C = 10\,lb/in^2/in$. If the trench is 6 ft wide by 4 ft deep by 6 ft long, design a stiff, light column for the trench. What is a good depth z for the column to be placed?

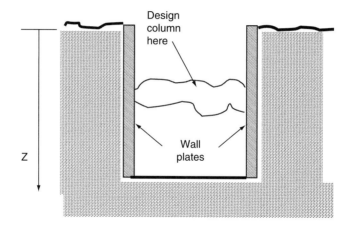

FIGURE P8-12

9   Materials for the Twenty-First Century

The future ain't what it used to be.

—Pogo

Objective: To acquaint the reader with currently used novel materials and potential materials of the future.

9.1   Introduction ⊛

Major advances in science and engineering have often occurred when disparate areas, such as materials science, manufacturing, engineering design, chemistry, biology, and computational power, converge to address certain needs or produce new knowledge that spurs innovation and new application. For example, high-refractive-index polymers with refractive indices of 2.0 and higher are expected to revolutionize optical switching and transmission. The recently developed carbon nanotubes of 2- to 3-nm diameter and tens of microns in length are expected to be the ultimate fiber needed as reinforcement in composite materials. In general, the fabrication of nanomaterials in the form of spheres or fibers or flakes with dimensions of the order of nanometers are revolutionizing materials development in a vast majority of applications ranging from structural materials to human implants. Though research is occurring across a wide range of length scales ranging from angstroms (10^{-10} m, atomic separation distance) to tens of meters (as in structural members), the focus on nanoscience and nanotechnology has been very prominent lately.

After studying the previous chapters in this book one should have a basic idea of mechanics of materials under the action of a variety of external loading conditions (Chapters 3–6, 8). Basic materials science concepts and engineering design concepts have been treated in Chapters 1 and 2 and Review Module R1. Throughout the chapters the mechanical response of materials (that is, deformation and stress and strain) has been explained on the basis of materials science concepts wherever possible. Various material responses such as tensile, compressive, shearing, yielding, work hardening, and fracture have been explained using fundamental mechanisms at the atomistic or microscopic level, which in turn can be explained by materials science concepts. In this effort we have limited ourselves to homogeneous isotropic materials, and all our discussion on mechanics has been mostly limited to metals. However, considering that 69 elements (with atomic numbers 3–5, 11–13, 19–31, 37–51, 55–83, 87–106) out of 106 in the periodic table of elements are inherently metallic in nature, that is not a very restrictive assumption. We have not yet discussed the properties and applications of other important material classes such as polymers, ceramics, engineered materials, multifunctional materials, nanomaterials, energy and power materials, electronic materials, and biological and bio-inspired materials. Even this list is by no means exhaustive and many other materials exist. In this chapter we will only briefly present the properties, the mechanics–materials link, and structural or other usage of some of these materials. It is our hope that with this exposure engineering and science students will better understand material behavior, design more efficient and safer structures using new/existing materials, will be able to design materials concurrently with the needs of the system design (and not design the system around the available materials), and will be able to shorten the design cycle—and, most importantly, will take up careers that will help to further the frontiers of materials development, characterization, and application.

9.2 Novel Materials 🕸

9.2.1 Polymers

In common spoken English, the word *plastic* is generally used for polymers. A *polymer* is a compound of high molecular weight that is formed from repeating subunits called *monomers*. The monomer generally consists of a parent compound with a double chemical bond. An example of a vinyl chloride monomer is shown in Figure 9-1.

During a chemical process called *polymerization* the double bond opens up to form a single bond and these single bonds join with adjacent mers to form long-chain molecules that constitute the polymer (Figure 9-2).

Details on the polymerization process may be obtained from materials science books or polymer chemistry books (such as Askeland and Phule, 2003; Allock and Lampe, 1990).

The long-chain molecules mainly exist in four configurations: linear, branched, crosslinked, and ladder polymers (Figure 9-3).

Two major classes of polymers are thermoplastic and thermosets. *Thermoplastic* polymers are generally linear or branched polymers with weak van der Waals bonds between atoms of different chains. They soften or melt on heating and can flow. Examples include polyethylene, polystyrene, polyvinyl chloride, and polyimide. Thermoplastics can have an *amorphous* structure (that is, there is no apparent order among molecules and the polymer chains are arranged randomly) or a *semicrystalline* structure (small, platelike crystalline regions are obtained by precipitation of the polymer from a dilute solution).

FIGURE 9-1 *Vinyl chloride* (C_2H_3Cl) *monomer.*

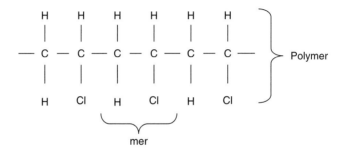

FIGURE 9-2 *Polymerization of vinyl chloride monomers to form a polymer.*

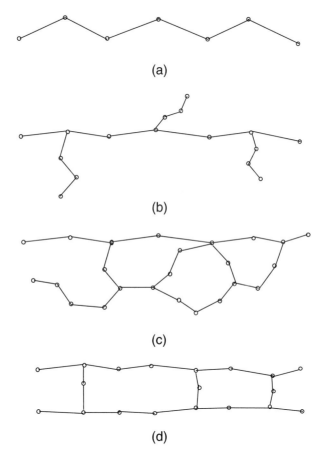

FIGURE 9-3 *Configurations of molecules in polymers: (a) linear, (b) branched, (c) crosslinked, (d) ladder.*

Thermosetting polymers are composed of a three-dimensional crosslinked network of chains and do not soften on heating. Thermosets are strong and rigid as crosslinking makes it difficult for molecules to slide past one another. Examples include polyester, polyurethane, phenolic, and epoxy.

9.2.1.1 Conducting Polymers

The general conception is that plastics, unlike metals, do not conduct electricity. Plastics are used as insulation around ordinary electrical cables. However, that is not necessarily true. Plastics can, after certain modifications, be made electrically conducting, and this discovery was considered so revolutionary that the inventors Alan Heeger, Alan G. MacDiarmid, and Hideki Shirakawa were rewarded with the Nobel Prize for chemistry in 2000.

Conducting polymers are conjugated polymers, namely organic compounds that have conjugated double bonds in a molecule. Conjugated double bonds mean that the single and double bonds alternate. This enables the electrons to be delocalized over the whole system and so be shared by many atoms. This means that the delocalized electrons may move around the whole system, causing an electric current to flow. The most common conducting polymers are polyaniline (PAni) and polypyrrole (PPY).

Polypyrrole has been used for the development of micro-muscles. Polyaniline films and ions sandwiched around ion-conducting film are considered as material for artificial muscles for robots. A current flow reduces one side and oxidizes the other. Ions are transferred. One side expands and the other contracts, resulting in a bending of the sandwich. Electrical and chemical energies are thus transformed into mechanical energy in the artificial muscle.

9.2.1.2 Polymer Foams

Cellular polymers or *foams* are multiphase material systems that consist of a polymer matrix, a fluid phase, generally air or gas, and filler material (only if used). Most polymers can be converted into a cellular material, but only a few polymers such as polyurethane, polystyrene, and polyolefins have been widely used. Cellular or foamlike materials are also very common in nature. Some examples are wood, cork, leaves, sponge, proteins, and bones. Many foods such as bread, meringue, and chocolate have a foamlike structure. Humans have tried to mimic their structure using both natural and simulated materials and have used these new materials in many engineering and biological applications. The most common example is the white coffee cup used in fast-food restaurants. However, lately more of these materials are being used for other structural or load-bearing applications and energy-absorbing applications. Some examples are metal foams, ceramic foams (Reddy and Schmitz, 2002), and high-porosity hydroxyapatite foam scaffolds for bone substitute (Ebaretonbofa and Evans, 2002). Foams are ideal energy absorbers as they can undergo large deformations at nearly constant stress. Also, if fillers, such as glass fibers or hollow microspheres of glass, are dispersed throughout the polymer matrix, the specific energy of absorption in a composite structure can be increased.

Cellular polymers or *polymer foams* have a cellular structure produced by gas bubbles formed during the polymerization process. The properties of the foam depend on the size, shape, and topology of the cells that constitute the foam. A variety of cell shapes, sizes, and connectivities can be observed in both synthetic and natural foams, though the latter show more variety. Natural foams are observed in materials such as cork, wood, cancellous bone, iris leaf, stalks of various plants, and coral.

Foams are classified according to the nature of the cell structure (open, closed, or mixed) or according to their stiffness (rigid or flexible). Historically, rigid-type foam materials have been used to increase stiffness and provide extra energy absorption when applied, but their contribution diminishes with aging of the structure. Rigid polymer foams have found limited use in automobile bodies and aerospace structures for increased crash resistance. Semiflexible and flexible foams have been used for cushioning and vibration damping, but not for improving crash resistance. Thus if flexible or semiflexible foams having a relatively higher load-bearing capacity with high recovery properties (similar to a coiled metal spring) can be fabricated, they could be used for energy absorption in lightweight structures.

Polymeric foams have many uses depending upon their physical characteristics. The most popular application of polymeric foams is for thermal insulation (Gibson and Ashby, 1997). Polymeric foams are used as insulators because of their low thermal conductivity. Applications include things as simple as Styrofoam coffee cups and refrigerated trucks, and complicated structures such as the insulation for the booster rockets on the Space Shuttle (Gibson and Ashby, 1997).

Polymeric foams have been found to be of good use in marine applications, filters, water-repellent membranes that allow air to permeate whatever is underneath the membrane for burn victims, and coatings for the roofs of houses to keep moisture away from wood. Foam-coated surfaces are often rough and have a high coefficient of friction (Gibson and Ashby, 1997). This high value makes the foam able to serve well as a nonslip surface. These types of foams are used on the soles of shoes, floors, mats, trays, and many other types of surfaces (Gibson and Ashby, 1997). Polymeric foams have been found to have a great ability to damp sound and vibration (Gibson and Ashby, 1997; Titow and Lanhoam, 1975; Ratcliffe and Crane, 1996). This property of the foam, combined with its low weight and easy moldability to a specified shape, allows the foam to be used in walls and ceilings of buildings. These types of foams are also used in automobiles and houses to absorb sound. Often a layer of foam padding is applied to the floors before carpet is laid so that there is less noise heard on one side or the other of the floor (Ratcliffe and Crane, 1996).

Polymeric foams, especially polyurethane foams, are used in the medical industry as well. In the area of medicine, polyurethane foams are used as cushioning in wheelchairs, splints, braces, and some types of bandaging. One of the features that make polyurethane foams appealing to the medical field is that the open cells in the foam allow the user a comfortable way to get better. The foam actually absorbs any impact that may occur, preventing further damage to an injured area. The foam also is a good material for use in this manner because the foam allows the skin to "breathe," because fresh air can pass through the open cells that are in the foam. Also, depending on the foam, it may also be used as a waterproof barrier, preventing the injured area from getting wet.

The other most popular use for polymeric foams is in packaging (Gibson and Ashby, 1997). In packaging, especially of delicate materials, it is necessary to use a material that will absorb the energy if the package is dropped or falls. Polymeric foams have been found to be great at absorbing energy.

Polymeric foams also play an important role in structures. The most common uses for polymeric foams in a structural application are in sandwich panels. A sandwich panel is usually made of two plates, usually a metal (such as aluminum), separated by a specific width (defined by the application) of polymeric foam or some other cellular solid (such as wood). Some examples of sandwich structures include the skin of airplanes, on and inside space vehicles, skis, boats (especially racing yachts), and buildings (Gibson and Ashby, 1997).

Polyurethane foam is made by mixing a polyol with a polyisocyanate. The polyol is usually either a polyether or polyester and sometimes contains a catalyst to improve the reaction rate. The polyol hydrates when mixed with the isocyanate group, which is found in the polyiso-cyanate, creating a foaming action. A specific polyisocyanate is methylene diphenylisocyanate (MDI). The polyisocyanate is responsible for creating the urethane structure (Salamone, 1999). However, the polyol is also responsible for making the foam flexible, or rigid. If the polyol has a low molecular weight, it will create rigid foam, whereas a polyol with a high molecular weight will create a flexible foam (Szycher, 1999). The reason for this drastic difference in foam after fabrication is due to the crosslinking that occurs during foam formation. In flexible foams, there is less crosslinking than in rigid foams (Szycher, 1999). So, the lower the

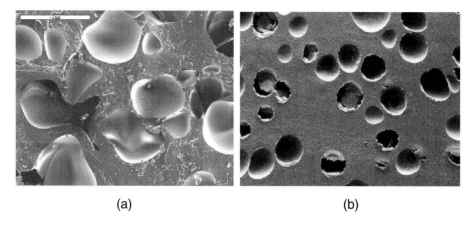

(a)   (b)

FIGURE 9-4 *Scanning electron microscope views of the microstructure of (a) flexible polyurethane foam (original magnification 20×), and (b) rigid polyurethane foam (original magnification 50×).*

molecular weight of the polyol, the more crosslinking occurs. The microstructures of flexible and rigid polyurethane foams are shown in Figure 9-4.

9.2.1.3 Polyvinylidene Fluoride (PVDF) Polymer

Piezoelectric materials are a class of materials that have electromechanical coupling. This means that if the material is mechanically strained, the material produces corresponding electrical voltage (piezoelectric effect), or if an electrical voltage is applied across the material, the material undergoes mechanical strain (inverse piezoelectric effect). The piezoelectric effects can be considered as a transfer of electrical and mechanical energy. Such transfer can only occur if the material is composed of charged particles and can be polarized. Two main classes of piezoelectric materials are used: ceramics and polymers. The best known piezoceramic is lead zirconate (PZT). It has a recoverable strain of 0.1% and is widely used as an actuator and sensor for a wide range of frequencies, including ultrasonic applications; it is well suited for high-precision applications as well (Piefort, 2001). Piezopolymers are mainly used as sensors; the best known is polyvinylidene fluoride (PVDF). PVDF was first studied by Kawai at the end of the 1960s (Srinivasan and Mcfarland, 2001) and has been commercially available (known by the trade name Kynar) since the early 1980s. Piezoelectricity is obtained by orientating the molecular dipoles of polar polymers such as PVDF in the same direction.

Because of their light weight and low stiffness, PVDF actuators can be used in complex structural engineering with minimal effect on the structure being measured. Moreover, since PVDF is a polymer, it is inherently compliant and hence PVDF actuators can be used in situations requiring less stiffness and less weight. Piezoelectric polymers such as PVDF and PZT can be used both as sensors (the direct piezoelectric effect) and as actuators (the inverse piezoelectric effect) (Tzou and Ye, 1994). Piezoelectric actuators have been used for active shape, vibration and acoustic control of structures (Lalande, Chaudhry, and Rogers, 1994). PVDF is resistant to strong acids, solvents, and reducing agents and is used in several industries, including electrical and electronic manufacturing, chemical processing, pulp and paper, and transportation.

9.2.2 Ceramics

Ceramics (including glasses) are inorganic, nonmetallic materials consisting of metallic and nonmetallic elements bonded primarily with ionic and covalent bonds. These high-strength bonds give rise to the special characteristics of these materials. They offer many desirable alternatives to the metals and polymers in common usage, as listed in Table 9-1.

There are a large number of ceramic materials, and they have found a wide spectrum of usage. They are used in floor tiles, refractories, spark plugs, sensors, abrasives, as thin coatings, etc. The Space Shuttle *Columbia* crashed in 2003 because of damage to external ceramic tiles that protected the aluminum frame from the heat generated during reentry of the shuttle into the earth's atmosphere. More details and information on ceramic materials processing, structure, and uses can be obtained from Budinski and Budinski (2004), Askeland and Phule (2003), and McColm (1983). Some more examples are given below.

Aluminum oxide (Al_2O_3) is used in rocket nozzles, pump liners, refractory tiles, check valves, reinforcement in composites, as an insulator in spark plugs, and in electronic and electrical devices. It has a Young's modulus of 360 GPa, tensile strength of 210 MPa, compressive strength of 2600 MPa, and fracture toughness of $1.75 \, MPa\sqrt{m}$.

Silicon carbide (SiC) is used as an abrasive in grinding wheels and bonded abrasive papers. It has excellent oxidation resistance up to very high temperatures and is used for coating metals, carbon–carbon composites, and other ceramics to provide protection at high temperatures. It is also used for making heating elements for furnaces. It has a Young's modulus of 410 GPa, tensile strength of 300 MPa, compressive strength of 2000 MPa, and fracture toughness of $3.0 \, MPa\sqrt{m}$. Silicon carbide has a higher hardness and lower density than aluminum oxide. Silicon carbide made by a chemical vapor deposition process has the highest thermal conductivity among ceramics.

Silicon nitride (Si_3N_4) is widely used as cutting tool materials, for gas turbine parts that must resist thermal cycling, and as balls in bearings. It is also finding application in diesel engines and hot extrusion dies because of its low wear rate at high temperatures. It has a Young's modulus of 200 GPa, tensile strength of 145 MPa, compressive strength of 1000 MPa, and a fracture toughness of $1.9 \, MPa\sqrt{m}$.

Sialon ($Si_3Al_3O_3N_5$) is an acronym that stands for silicon aluminum oxynitride. It is formed by blending silicon nitride, silica, aluminum oxide, and aluminum nitride and then subjecting

TABLE 9-1 Comparison of the Properties of Ceramic, Metal, and Polymer Materials

Property	Ceramic	Metal	Polymer
Hardness	High	Medium	Low
Melting point	High	Low–high	Low
Tensile strength	Low–medium	High	Low
Yield strength	No yielding occurs	Medium–high	Low
Compressive strength	High	High	Low–medium
Elastic modulus	Medium–high	Low–high	Low
Fatigue strength	Low in tension	Medium–high	Low
	Medium in compression		
Thermal expansion	Low–medium	Medium–high	High
Thermal conductivity	Medium	Medium–high	Low
Chemical resistance	High	Low–medium	Medium
Electrical resistance	High	Low	High
Fabricability	Low	High	Medium–high

the blend to pressureless sintering. A glassy phase forms during sintering that makes it possible to achieve 100% dense structures. It is being researched for applications in cutting tools, engine components, and other applications where high temperatures and wear are a concern. It has a Young's modulus of 300 GPa, tensile strength of 450 MPa, compressive strength of 3500 MPa, and a fracture toughness of $7.5 \, \text{MPa}\sqrt{\text{m}}$. Sialon is the lightest, strongest, and most fracture resistant among all the ceramics listed here.

9.2.3 Nanomaterials

The prefix *nano-* means one billionth. A nanometer (nm) is one billionth of a meter, or $10 \, \text{Å}$ (10×10^{-10} m), or about 10 times the size of a hydrogen atom. Nanoscale studies on materials implies the study of molecules and structures with at least one dimension between about 1 and 100 nm. Thus intuitively at the nanoscale we may be dealing with a few atoms or a very small molecule, and hence the measured properties would be very different from the bulk properties and closer to the properties exhibited by atoms and molecules, which can be explained by theories such as quantum mechanics. Familiar bulk properties such as electrical conductivity and hardness do not simply scale with size at the nanometer dimension. For example, Ohm's law relates current, voltage, and resistance in bulk metals, but it is not applicable when electrons have to move through a group of just a few atoms. Another example where Ohm's law fails is for carbon nanotubes (CNTs) and superconducting materials. It has also been observed that as the size of the *nanoparticles* becomes smaller the crystal structure may change from that of the bulk structure. For example, an 80-nm aluminum particle has a face-centered cubic (FCC) structure (the same as that of bulk aluminum), while a 5-nm particle has an icosahedral structure (one with 20 faces). This change in structure for very small nanoparticles (<10 nm) is believed to occur as that structure would have the lowest energy compared to other forms (Poole and Owens, 2003). At the nanoscale many other properties—magnetic, electrical, optical, reactivity, and so forth—also undergo dramatic changes from the bulk values. For example, if aluminum nanoparticles are exposed to air they violently react with oxygen and form a 3- to 5-nm-thick Al_2O_3 layer on the surface. Thus these particles should be made in a solution which protects from oxidation. Iron nanoparticles with 10 atoms and greater than 18 atoms are more reactive with hydrogen than others.

Besides nanoparticles there are other bulk materials with nanosized microstructure that are made with nanoparticles and are called *nanomaterials*. One of the most common methods of making bulk disordered nanostructured materials is by the method of compaction and consolidation. For example, nanostructured copper–iron alloys are made by compacting a mixture of Fe and Cu powder in the ratio of 85% Fe, 15% Cu in a mold at a pressure of 1 GPa for 24 hours, followed by heating to 400°C under a pressure of 870 MPa for one-half hour. The final compacted material has an average grain size of 40 nm and a density that is 99% of the maximum possible density. The fracture stress is 2.8 GPa, which is about five times the fracture stress of iron having larger grain sizes of the order of 100 μm. However, there is no significant change in the elastic modulus (Poole and Owens, 2003). It has been found that the elastic modulus begins to decrease if the grain size decreases to below 20 nm. The yield strength shows an increasing trend with reducing grain size. Thus the smaller the grain size the higher the yield strength, as materials with smaller grains have more grain boundaries, which blocks dislocation motion more effectively. It is postulated that materials with grain size smaller than 30 nm cannot exhibit dislocation-based deformation. Hence, such small-grain-size materials behave in a more brittle fashion compared to larger-grain-size polycrystalline metals. Nano-grain-size copper shows elongations of about 5% compared to 60% for coarse-grained copper. It should be noted that these properties can change depending on the manufacturing method and on the experimental technique used to measure them. Nanomaterials are still evolving and it may be years before the physical, mechanical, optical, electrical, and magnetic properties are fully understood.

9.2.4 Composite Materials

Composite materials are engineered materials that have triggered many new design applications over the past 30 years. Composite materials are obtained by binding together two or more distinct but compatible materials. One of the major advantages of composites is that they can be tailored or engineered as per specifications of an optimum design. Among the different composites, fiber-reinforced polymers are the most popular. The use of graphite fibers embedded in a polymer matrix is well known in sports equipment, such as tennis rackets and golf clubs. Among engineering structures, composite materials have had a major impact in the aerospace industry. Examples of composites also exist in nature—for example, wood (cellulose fibers in a lignin matrix) and bones (short and soft collagen fibers embedded in a mineral called apatite).

In all the examples mentioned above the common theme is that high-strength and generally high-stiffness fibers are embedded in a matrix of lower properties. Besides fibers, other reinforcements are whiskers, platelets, and particulates of varying sizes, which could be as small as a few nanometers. Composites are thus multiphase materials, and interfaces between reinforcement and matrix are inherent in these materials. In general the reinforcements are the principal load-carrying members, whereas the surrounding matrix keeps them in the desired location and orientation, acts as the load transfer medium, and protects them from environmental damage. Consider a fiber of length l and diameter d, embedded in matrix and an external load is applied to the matrix surface. The load on the fiber then builds from both ends. If the fiber is long enough, the load could build to the breaking strength of the fiber. The length of the fiber at which the load builds to the breaking stress at the middle of the fiber is called the critical length, l_c. The ratio (l_c/d) is called the critical aspect ratio of the fiber. A high aspect ratio, for a given bond strength between fiber and matrix, enables the reinforcing fiber to carry a greater portion of the applied load and thus lower the fraction of load carried by the matrix. This is also true for other types of reinforcements. Hence spheres, which have an aspect ratio of 1, are not well suited to carrying the external load in composites reinforced with them. More details on the mechanics of load transfer and mechanics of composites can be found in composite materials books (e.g., Broutman and Agrawal, 1980; Chawla, 1987). Mechanical properties of continuous fiber-reinforced matrix composites, with load applied parallel to the fiber direction, can be determined from the rule of mixtures as

$$E_c = V_m E_m + V_f E_f \qquad (9.1)$$

where
  E_c = elastic modulus of the composite,
  V_m, V_f = volume fraction of matrix and fibers, respectively,
  E_m, E_f = elastic modulus of matrix and fibers, respectively.
Similarly the strength σ_c and density ρ_c of the composite can be determined from the rule of mixtures as

$$\sigma_c = V_m \sigma_m + V_f \sigma_f \qquad (9.2)$$

$$\rho_c = V_m \rho_m + V_f \rho_f. \qquad (9.3)$$

When the load is applied perpendicular to the fibers, each component acts independently of each other, and the elastic modulus of the composite is given by

$$\frac{1}{E_c} = \frac{V_m}{E_m} + \frac{V_f}{E_f}. \tag{9.4}$$

Many different combinations of reinforcements and matrix materials exist. A few of the most common types of composites are briefly mentioned next. A new class of composite materials called nanocomposites is described in the next section.

Polymer matrix composites: Typically glass, Kevlar, graphite, and boron fibers are embedded in a polyester or epoxy matrix. The tensile strength of E-glass fiber/epoxy composite can be the same as that of 7075-T6 grade aluminum, while boron/epoxy composite can greatly exceed the strength of aluminum. The tensile elastic modulus of boron fiber/epoxy composite also exceeds that of 7075-T6 aluminum, while the E-glass fiber/epoxy composite has a modulus of about 50% that of aluminum.

Metal matrix composites: These materials consist of a metal matrix reinforced with metal or ceramic fibers and are generally used for high-temperature applications. Metal matrix composites when compared to polymer matrix composites have higher temperature capability, greater tensile, shear and compressive strength, and greater thermal and electrical conductivity. Aluminum is the most common matrix material. When strengthened with aluminum oxide fibers it is used in the pistons of diesel engines, and with silicon carbide fibers and whiskers in aerospace applications; boron fiber/Al composite was used in the tubular truss members of the Space Shuttle orbiter. Copper-reinforced with SiC fibers is used to produce high strength propellers for ships. Tunsten fiber/copper matrix composites are used as electrical contacts

Ceramic matrix composites: Ceramic materials are attractive in composite materials for their high strength and stiffness at high temperatures. However, they seriously lack in fracture toughness. Thus the use of fibers to toughen a ceramic matrix has been a vigorous area of research. Nevertheless, ceramic matrix composites differ from other composite materials in important ways. In general the fibers in a composite bear a greater portion of the applied load, because the load portioning depends on the ratio of the fiber to matrix elastic moduli, and this ratio is high. In ceramic matrix composites this ratio is low and can also approach 1. Hence the matrix will have to bear a relatively larger portion of the applied load. Other reasons for the distinctive behavior of these composites is the low ductility of matrix, high fabrication temperatures, thermal mismatch between fibers and matrix, and chemical compatibility. Examples of ceramic matrix composites are SiC fibers/LAS (lithium aluminosilicate) matrix and SiC whisker/alumina composite. Ceramic matrix composites have found some application in cutting tools, engines, energy conversion, electrical applications, and military equipment, especially where resistance to aggressive environments is required.

9.2.4.1 Nanocomposites

Nanocomposites, which are a new class of composites, are particle-filled matrices in which at least one dimension of the dispersed particles is of the order of a nanometer. The matrix material can be a polymer, metal, or ceramic. The filler nanoparticles can be of three types depending on how many dimensions of the particles are in the nanometer range. When all three dimensions are in the nanometer range, they are called *isodimensional nanoparticles;* when two dimensions are in the nanometer range and the third is larger, they are called *nanotubes* or whiskers, such as carbon nanotubes and cellulose whiskers. When only one dimension of the particle is in the nanometer range and the other two dimensions are two to three orders of

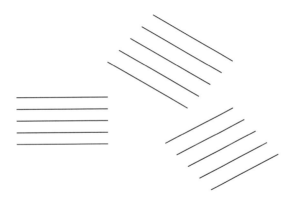

FIGURE 9-5   *Agglomerate of layered silicate nanoparticles.*

magnitude larger, we obtain *flakelike nano particles*, such as clay and layered silicates. We briefly describe nanoclay or *layered silicate nanocomposites* and carbon nanotube reinforced (filled) polymer nanocomposites next.

Layered silicates are stacked face-to-face in an agglomerate, as shown in Figure 9-5.

The high aspect ratio of the silicate nanolayers is ideal for reinforcement. However, special techniques have to be employed to separate the nanolayers to enable them to be adequately dispersed in the polymer (Alexander and Dubois, 2000). The preferred dispersion modes are intercalation or exfoliation, as shown in Figure 9-6.

In an intercalated structure polymer chains are intercalated between silicate layers resulting in a well-ordered multilayer morphology. In an exfoliated structure the silicate layers are uniformly dispersed in a continuous polymer matrix. The greater and more uniform the dispersion, the greater the influence in improving mechanical properties of the composite. Also, the exfoliated particles have a more favorable aspect ratio and larger surface area compared to intercalated particles, and hence the former have a more favorable effect on properties of the nanocomposite. This is possible due to good coupling between the large surface area of clay and polymer, which facilitates stress transfer from the matrix to the reinforcement. Addition of 5 to 10 wt. % of filler to polymers has been found to result in at least a twofold increase in Young's modulus and tensile strength, and significant improvements in dimensional stability, thermal stability, resistance to ignition, and so forth. However, in some cases brittleness in the nanocomposite may result.

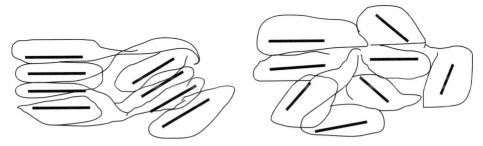

FIGURE 9-6   *Interaction of layered silicate reinforcement with a polymer matrix resulting in (a) intercalated nanocomposite and (b) exfoliated nanocomposite.*

Carbon nanotubes have recently received a lot of attention due to their exceptional mechanical and functional properties and are being billed as one of the most promising reinforcements for the next generation high-performance nanocomposites. *Single-wall carbon nanotubes* (SWNTs) have a high aspect ratio (2 nm diameter, $20-100\,\mu m$ length), very high Young's modulus (about 1200 GPa), and high strength (about 100–200 GPa). Compared to glass fibers SWNT has a Young's modulus that is about 18 times higher and a tensile strength that is about 50 to 100 times higher. Thus in theory, adding SWNTs to a polymer matrix should result in a composite with exceptional mechanical properties (Thostenson and Chou, 2002). However, SWNT-reinforced epoxy composites made by casting or injection processes have been found to be weaker or only slightly stronger than neat epoxy resin (Thostenson and Chou, 2003). This is because the nanotubes do not disperse uniformly in the resin and tend to conglomerate into ropes or other nanostructures, and have weak interfacial bonding with the matrix. Vigorous research is currently underway to overcome these drawbacks in the use of SWNTs in polymer composite fabrication.

9.2.5 Multifunctional Materials

When a material performs at least one or more functions besides its structural function of carrying loads and defining shape, it is called a *multifunctional material*. Composite materials are among the best candidates for inculcating multifunctionality. For example, in a fiber-reinforced polymer matrix composite, two types of fibers could be embedded, one that provides strength and another that produces galvanic current when embedded in a conducting polymer matrix, which may also contain microspheres filled with colored pigments that break at a critical stress level and cause a color change in the highly stressed or damaged area of a structure. In addition the composite material could be engineered to act as a smart material. For example, fiber optic sensors could be incorporated at discrete locations that can activate embedded piezoelectric actuators to produce macroscopic changes in the shape or position of an object. The optical fibers could also serve as damage sensors and could initiate a damage healing process.

Another possible multifunctional material could be carbon nanotube reinforced polymer composites. Carbon nanotubes can exhibit either metallic or semiconducting properties depending upon their diameter and structure (chirality). It is believed that in the metallic state CNTs can carry 10^9 amps/cm^2 compared to 10^6 amps/cm^2 for copper. This is due to the fact that CNTs have very few defects that can scatter electrons, and thus have a very low resistance. Also the thermal conductivity of CNTs is two times that of diamond, which is the best thermal conductor among bulk materials. Thus CNT-reinforced polymers will possess higher tensile strength and elastic modulus and could also make the polymer a conductor of heat or electric current.

9.2.6 Energy and Power Materials

We all use energy and power on a daily basis, and it is almost like second nature to us. However, most of us do not pay any attention to the materials technology behind the success of the energy and power industry and the future challenges. Some of the needs for new or improved materials in this sector are in the areas of energy storage (batteries), energy conversion (fuel cells), electric power generation (higher temperature gas turbines), and propellants. We briefly discuss the materials and issues in each of these areas, mainly with the intent of creating awareness among the readers.

Energy storage in batteries depends on the properties of the materials used for electrodes, which in turn determines the cell potential, capacity, and energy density. The life of the battery depends on the stability of the interfaces between reactive materials. In this regard micro-structured or layered nanomaterials, nanocomposites, nanoporous materials, and conducting polymers are some examples of materials that may provide high-power, long-lasting batteries functioning in a wide range of service conditions.

Nanocomposites could offer highly energetic materials that are resistant to accidental explosion for use as propellants in rockets or missiles. For example, nanoscale aluminum powder could possibly increase the burn rate of the propellant and increase energy density significantly over other existing materials.

Conversion of energy in one form to electric energy is a vital part of many industrial sectors, for example automobiles. Presently one of the most active areas of research is the development of fuel cells that directly convert chemical energy into electrical energy. Because there is no combustion, the fuel cell is not limited by the Carnot cycle and thus has the potential for high efficiency. The performance of a fuel cell is dependent on the resistance of the electrolyte and the nature of the electrodes (to enhance reaction kinetics by increasing electrode surface area). In addition the choice of the fuel is fundamental to operation of the fuel cell. High-temperature fuel cells (500–1000°C) operate on hydrocarbon fuels, whereas low-temperature cells (70–250°C) operate on hydrogen.

9.2.7 Biological and Bio-Inspired Materials

New, synthetic materials are being developed using the biological structures, systems, natural fabrication processes, and natural building blocks. The resulting materials have greatly improved physical and mechanical properties and are often multifunctional as well. For example, *biological materials* such as bones and teeth are hard, strong, and tough, with hierarchically organized structures from the nanometer scale to the macroscopic scale. Hard biological materials have hierarchically organized organic/inorganic composites in which soft materials (for example, proteins, membranes, fibers) organized at lengths of 1 nm to 100 nm are used as framework for the growth of specifically oriented and shaped inorganics. The high-modulus inorganic phase provides stiffness and the organic phase improves toughness. For example, the structure of an abalone shell consists of layered plates of calcium carbonate ($CaCO_3$) of about 200 nm thickness held together by about 10 nm thick organic mortar. This laminated biological material has been mimicked to create microlaminated ceramic–metal and ceramic–organic composites with significantly superior mechanical properties. Some armor panels have been developed using ceramic–metal and ceramic–organic microlaminates.

Possibly the oldest and best known naturally occurring biological material is silk fiber. It is one of the strongest and toughest fibers with a very high strength/weight ratio. Silk fibers perform the same whether they are subjected to tensile or compressive forces, a property that differentiates them from synthetic high-performance fibers. Efforts are under way to synthesize the protein structures found in silk in polymers and possibly improve their strength-to-weight ratio.

Structural members that only need to carry tensile loads can be made highly weight efficient—they can be formed as simple strings or cables. In essence, a network made of cables (a *cable-net*) is a special case of a *tensegrity* structure (a structure made only of separate tensile and compression members that has tensile integrity; Connelly and Back, 1998). Due as well to their superior chemistry, which allows for high strength and elasticity, *spiderwebs* are a perfect embodiment of a tensegrity structure. Early human fishing nets were likely copied from the spider web.

There is considerable literature on the role of silk and web construction by spiders in relation to prey capture, evolution, behavior, communication, site selection, and architecture (Denny, 1976). For some time we have been aware of the basic elements of orb-web construction.

The evolution of silk production in spiders was as important as the evolution of flight in insects. With its unique properties and enormous potential for material science, it has attracted a comprehensive research effort, particularly in the last several years. Spider silk, while its tenacity is slightly less than nylon, is twice as elastic but with a tensile strength superior to tendon, rubber, bone cellulose, and it is still 50% as strong as the finest steel. It has to be 40 miles long before it breaks under its own weight. Secreted as heterogeneous water-soluble liquid of proteinaceous fibroin, the molecules of the polypeptide chains are arranged as alpha helixes (soluble configuration with intramolecular hydrogen bonds). Beta-pleated sheets are formed by applying shearing forces (which occur during extrusion through spigots) and liquid silk becomes solid (insoluble configuration with intermolecular bonds). This composite of alpha-helixes (which supply elasticity) and the accordion-like crystalline beta-sheets (which supply strength) gives spider silk its unique properties. Strength and elasticity also depend on water content; when high, silk becomes highly viscoelastic and can be stretched more than 300% before breaking occurs.

9.3 Closure

Since prehistoric times and especially in the last century, material advances have been the key to significant technology breakthroughs. The electronic revolution, the information technology boom, the nuclear age, the aerospace era, and new energy storage and generation possibilities have all been possible because of breakthroughs in materials technology. For example, if a new material is developed for gas turbines enabling an increase of $100°C$ in the turbine inlet gas temperature, the operating efficiency of the turbine will increase by 10%. This increase in efficiency will result in enormous savings for the power generation industry and aircraft carriers. As another example, if nuclear power plants can be operated in the temperature range of $1000°$ to $1200°C$ (compared to the current $400–500°C$) the power output efficiency can be significantly increased. In addition large quantities of hydrogen gas can be obtained by direct thermal cracking of water ($2H_2O \rightarrow 2H_2 + O_2$), which will be a more cost-effective process compared to the current process of generating hydrogen from steam using fossil fuels (mainly methane gas). The increased power output can reduce dependence on natural oil resources, and the hydrogen can be used as an alternative fuel for power generation or as fuel in automobiles, resulting in very significant economic progress. However, new high-temperature materials will have to be developed to achieve this objective.

Materials have always been the basic ingredient of all physical systems. Materials in conjunction with design processes affect system cost throughout the system life cycle in terms of cost of raw materials, processing and fabrication, assembly, inspection, repair and maintenance, and recycling or disposal. Figure 9-7 shows the factors affecting material behavior and their effect on design parameters. It is well recognized now that nearly 80% of the cost associated with a system is locked in very early in system development by the materials and design decisions made. However, at this early decision stage an engineer or scientist must keep an open mind with regard to materials selection and not let simple familiarity with one material or a material class influence the choice.

Every material has strengths and weaknesses. In the majority of applications multiple requirements are placed upon the material, and more often than not these requirements are at odds with each other. Hence materials selection will often become a complex trade-off among different, often conflicting, factors.

A materials scientist generally focuses on one or a few aspects of materials structure or behavior. However, engineers and designers almost always design and produce systems that

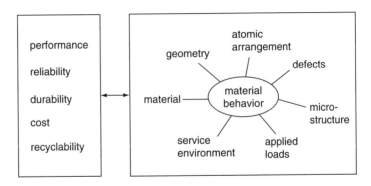

FIGURE 9-7 *Effect of material behavior and properties on design.*

are quite complex and require a variety of materials in the various subsystems that comprise the system. It is possible that introduction of or changing one material can have consequences for other materials used in the system. Thus engineers must be cognizant of different materials, their properties and behavior under different working conditions, their compatibility, and other factors. This textbook and this chapter have been an endeavor to introduce the reader to this systems approach, and it is sincerely hoped that this objective has been achieved.

References

Alexander, M., and Dubois, P. (2000). Polymer-layered silicate nanocomposites: Preparation, properties and uses of a new class of materials. *Materials Science and Engineering*, Vol. 28, pp. 1-63.

Allock, H. R., and Lampe, F. W. (1990). *Contemporary Polymer Chemistry*, 2nd ed. Prentice-Hall, Upper Saddle River, NJ.

Askeland, D. R., and Phulé, P. P. (2003). *The Science and Engineering of Materials*, 4th ed. Thomson Engineering, Boston.

Broutman, L. J., and Agrawal, B. D. (1980). *Analysis and Performance of Fiber Composites*. Wiley, New York.

Budinski, K. G., and Budinski, M. K. (2004). *Engineering Materials: Properties and Selection*, 8th ed. Prentice-Hall, Upper Saddle River, NJ.

Chawla, K. K. (1987). *Composite Materials: Science and Engineering*. Springer Verlag, Berlin.

Connelly, R., and Back, A. (1998). Mathematics and tensegrity. *American Scientist*, Vol. 86, pp. 142-151.

Denny, M. (1976). The physical properties of spider's silk and their role in the design of orb-webs. *Journal of Experimental Biology*, Vol. 65, pp. 483-506.

Ebaretonbofa, E., and Evans, J. R. G. (2002). High posrosity hydroxyapatite foam scaffolds for bone substitute. *Journal of Porous Materials*, Vol. 9, No. 4, pp. 257-263.

Gibson, L., and Ashby, M. (1997). *Cellular Solids: Structure and Properties*, 2nd ed. Cambridge University Press, New York.

Lalande, F., Chaudhry, Z., and Rogers, C. A. (1994). Modeling considerations for in-phase actuation of actuators bonded to shell structures. AIAA/ASME Adaptive Structures Forum, Hilton Head, SC.

McColm, J. (1983). *Ceramic Science for Materials Technologists*. Chapman and Hall.

Piefort, V. (2001). *Finite Element Modeling of Piezoelectric Active Structures*. Thesis, Universite Libre de Bruxelles.

Poole, Charles P. Jr., and Owens, Frank J. (2003). *Introduction to Nanotechnology*. Wiley, New York.

Ratcliffe, C., and Crane, R. (1996). Optimizing reinforced polyurethane as a combined structural and damping component. *Advanced Materials for Vibro-Acoustic Applications*, NCA-Vol. 23. ASME.

Reddy, E. S., and Schmitz, G. J. (2002). Ceramic foams. *American Ceramic Society Bulletin*, Vol. 81, No. 12, pp. 35-37.

Salamone, J. (1999). *Concise Polymeric Materials Encyclopedia*. CRC Press, New York, pp. 1425-1426.

Srinivasan, A. V., and Mcfarland, D. M. (2001). *Smart Structures: Analysis and Design*. Cambridge University Press, Cambridge, UK.

Szycher, M. (1999). *Schyzer's Handbook of Polyurethanes*. CRC Press, New York.

Thostenson, E. T., and Chou, T.-W. (2002). Aligned multi-walled carbon nanotube-reinforced composites: Processing and mechanical characterization. *Journal of Physics D—Applied Physics*, Vol. 35, No. 16, pp. L77-L80.

Thostenson, E. T., and Chou, T.-W. (2003). On the elastic properties of carbon nanotube-based composites: Modelling and characterization. *Journal of Physics D—Applied Physics*, Vol. 36, No. 5, pp. 573-582.

Titow, W., and Lanhoam, B. (1975). *Reinforced Thermoplastic*. Wiley, New York, p. 266.

Tzou, H. S., and Ye, R. (1994). Pyroelectric and thermal strain effects of piezoelectric (PVDF and PZT) devices. *Adaptive Structures and Composite Materials: Analysis and Application*, International Mechanical Engineering Congress and Exposition, Chicago.

Appendix 1:   Material Properties

MATERIAL	Type	Cost ($/kg)	Density (ρ, Mg/m^3)	Young's modulus (E, GPa)	Shear modulus (G, GPa)	Poisson's ratio (v)	Yield stress (σ_Y, MPa)	UTS (σ_f, MPa)	Breaking strain (ε_f, %)	Fracture toughness (K$_c$, MN m$^{-3/2}$)	Thermal expansion (α, $10^{-6}/C$)
Alumina (Al$_2$O$_3$)	ceramic	1.90	3.9	390	125	0.26	4800	35	0.0	4.4	8.1
Aluminum alloy (7075-T6)	metal	1.80	2.7	70	28	0.34	500	570	12	28	33
Beryllium alloy	metal	315.00	2.9	245	110	0.12	360	500	6.0	5.0	14
Bone (compact)	natural	1.90	2.0	14	3.5	0.43	100	100	9.0	5.0	20
Brass (70Cu30Zn, annealed)	metal	2.20	8.4	130	39	0.33	75	325	70.0	80	20
Cermets (Co/WC)	composite	78.60	11.5	470	200	0.30	650	1200	2.5	13	5.8
CFRP Laminate (graphite)	composite	110.00	1.5	1.5	53	0.28	200	550	2.0	38	12
Concrete	ceramic	0.05	2.5	48	20	0.20	25	3.0	0.0	0.75	11
Copper alloys	metal	2.25	8.3	135	50	0.35	510	720	0.3	94	18
Cork	natural	9.95	0.18	0.032	0.005	0.25	1.4	1.5	80	0.074	180
Epoxy thermoset	polymer	5.50	1.2	3.5	1.4	0.25	45	45	4.0	0.50	60
GFRP laminate (glass)	composite	3.90	1.8	26	10	0.28	125	530	2.0	40	19
Glass (soda)	ceramic	1.35	2.5	65	26	0.23	3500	35	0.0	0.71	8.8
Granite	ceramic	3.15	2.6	66	26	0.25	2500	60	0.1	1.5	6.5
Ice (H$_2$O)	ceramic	0.23	0.92	9.1	3.6	0.28	85	6.5	0.0	0.11	55
Lead alloys	metal	1.20	11.1	16	5.5	0.45	33	42	60	40	29
Nickel alloys	metal	6.10	8.5	180	70	0.31	900	1200	30	93	13
Polyamide (nylon)	polymer	4.30	1.1	3.0	0.76	0.42	40	55	5.0	3.0	103
Polybutadiene elastomer	polymer	1.20	0.91	0.0016	0.0005	0.50	2.1	2.1	500	0.087	140
Polycarbonate	polymer	4.90	1.2	2.7	0.97	0.42	70	77	60	2.6	70
Polyester thermoset	polymer	3.00	1.3	3.5	1.4	0.25	50	0.7	2.0	0.70	150
Polyethylene (HDPE)	polymer	1.00	0.95	0.7	0.31	0.42	25	33	90	3.5	225
Polypropylene	polymer	1.10	0.89	0.9	0.42	0.42	35	45	90	3.0	85
Polyurethane elastomer	polymer	4.00	1.2	0.025	0.0086	0.50	30	30	500	0.30	125
Polyvinyl chloride (rigid PVC)	polymer	1.50	1.4	1.5	0.6	0.42	53	60	50	0.54	75
Silicon	ceramic	2.35	2.3	110	44	0.24	3200	35	0.0	1.5	6
Silicon Carbide (SiC)	ceramic	36.00	2.8	450	190	0.15	9800	35	0.0	4.2	4.2
Spruce (parallel to grain)	natural	1.00	0.60	9	0.8	0.30	48	50	10	2.5	4
Steel, high strength 4340	metal	0.25	7.8	210	76	0.29	1240	1550	2.5	100	14
Steel, mild 1020	metal	0.50	7.8	210	76	0.29	200	380	25	140	14
Steel, stainless austenitic 304	metal	2.70	7.8	210	76	0.28	240	590	60	50	17
Titanium alloy (6A14V)	metal	16.25	4.5	100	39	0.36	910	950	15	85	9.4
Tungsten Carbide (WC)	ceramic	50.00	15.5	550	270	0.21	6800	35	0.0	3.7	5.8

Appendix 2:   Section Properties

Section	Area properties	Area moments of inertia and torsion constants
Solid circle	$A = \pi R^2 = \pi D^2/4$	$I_{xx} = \pi D^4/64$ $J_{zz} = 2I_{xx} = \pi D^4/32$
Hollow circle	$A = \dfrac{\pi}{4}\,(D_o^2 - D_i^2)$	$I_{xx} = \dfrac{\pi}{64}\,(D_o^4 - D_i^4)$ $J_{zz} = \dfrac{\pi}{32}\,(D_o^4 - D_i^4)$

(Continues)

(*Continued*)

Section	Area properties	Area moments of inertia and torsion constants
Solid square 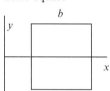	$A = b^2$	$I_{xx} = b^4/12$ $J_{zz} = 0.141b^4$
Solid rectangle 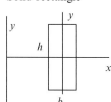	$A = bh$	$I_{xx} = \dfrac{bh^3}{12}, \; I_{y'y'} = \dfrac{b^3h}{12}$
Solid equilateral triangle 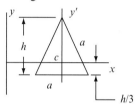	$A = \dfrac{\sqrt{3}}{4}a^2$ $h = \dfrac{\sqrt{3}}{2}a$	$I_{xx} = \dfrac{ah^3}{36}, \; I_{y'y'} = \dfrac{a^3h}{48}$ $J_{zz} = \dfrac{1}{874}a^4 \approx \dfrac{1}{26}h^4$

Appendix 3:   Material Property Charts

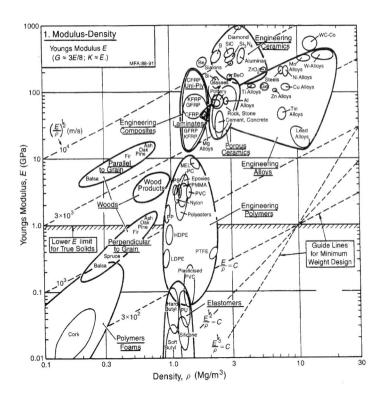

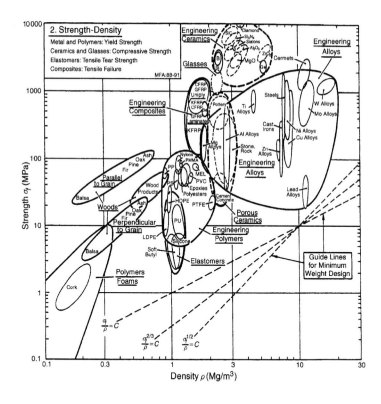

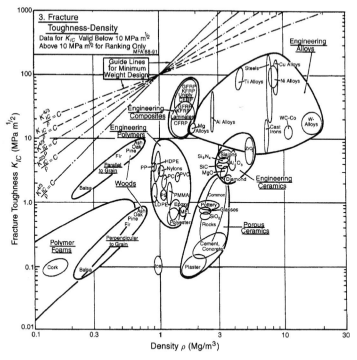

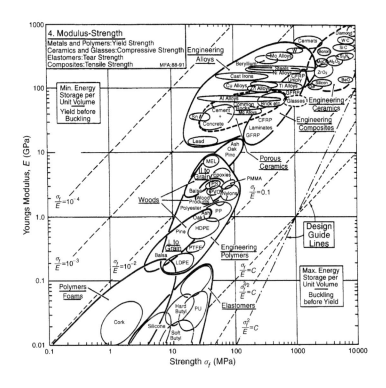

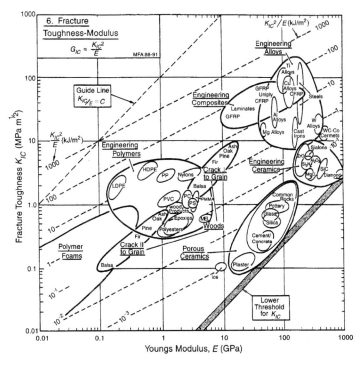

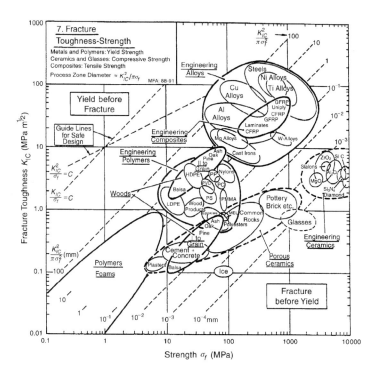

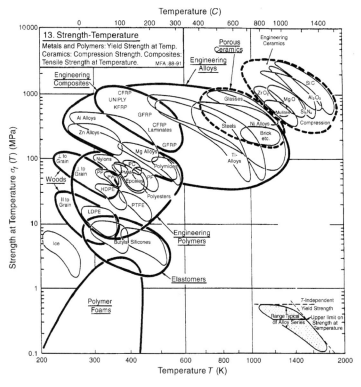

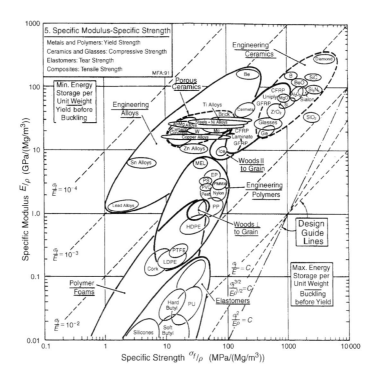

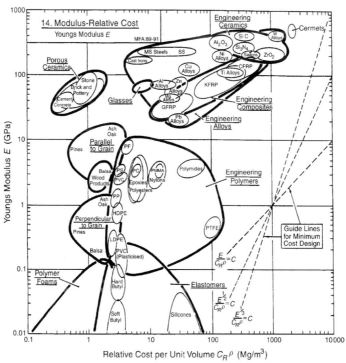

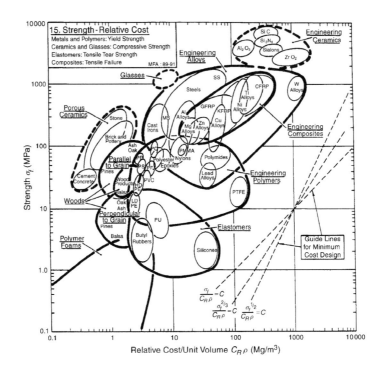

Review Module R1:   Fundamental Materials
   Science Concepts

1   Introduction

The existence of interatomic bonds can be explained on the basis of the forces of interaction between atoms. Atomic models relate these forces to the energy contained in the structure of the atoms. The consideration of the structure of the atom and of the nature of the interatomic bond is not only of theoretical but also of practical interest in the study of large-scale mechanical behavior. For example, all macroscopic fracture phenomena due to static or dynamic fatigue loads are initiated on the atomic or molecular scale and are essentially governed by atomic or molecular processes. Thus, to explain the fracture phenomena it is essential to understand the nature of the atomic bond and the conditions for its formation and disruption.

2   Atomic Structure

Quantum mechanics, wave mechanics, and complex mathematical tools are required to explain the basic nature of the various atomic phenomena. In this work we will not use these theories and concepts but rather use the results to explain and understand mechanical behavior of materials. Interested readers are referred to books on nuclear physics and quantum and statistical mechanics.

An atom consists of negatively charged, positively charged, and neutral particles, called the electrons, protons, and neutrons, respectively. The electrical charge carried by each electron and proton is 1.9×10^{-19} coulomb. The protons and the neutrons reside in the nucleus of the atom while the electrons revolve around the nucleus. The numbers of electrons and protons in an atom are equal and hence the atom as a whole is electrically neutral. The operating forces within an atom are the forces of electrostatic attraction between the electrons and the protons and the centrifugal forces of the electron, resulting from its acceleration.

Most of the mass of the atom is contained in the nucleus. The mass of each proton and neutron is about 1.67×10^{-24} g, but the mass of each electron is 9.11×10^{-28} g. The *atomic mass*, which is equal to the average number of protons and neutrons in the atom, is the mass of the Avogadro number (6.02×10^{23}) of atoms. Thus an Avogadro number of atoms of oxygen have an atomic mass of 16 g, while the same number of diatomic oxygen molecules will have a mass of 32 g. The *atomic number* of an element is equal to the number of electrons or protons in each atom. Thus, an oxygen atom, which contains eight electrons and eight protons, has an atomic number of 8.

2.1   The Periodic Table of Elements

The familiar periodic table of elements, which can be found in many textbooks on physics, chemistry, and materials science, is based on the distribution of electrons in nonradiating orbits around the nucleus. The distribution of electrons over the orbits is governed by *Pauli's exclusion principle*. There are four quantum numbers, namely the principal quantum number n, angular momentum quantum number l, magnetic quantum number m_l, and the spin quantum number m_s. The principal quantum number determines the energy level which is an integral multiple of the energy quantum ε. The angular momentum quantum number is related to the angular momentum of the motion of the electron relative to the nucleus. It is an integral multiple of the energy quantum and expresses the ellipticity of the orbit. The quantum number n varies from 1 to ∞, and l varies from 0 (circular orbit) to $n - 1$. Although the quantum numbers n and l are sufficient to describe the orbital motion of one electron, the remaining quantum numbers m_l and m_s are needed for atoms consisting of more than one electron. The magnetic quantum number defines the number of energy levels, or orbitals, for each angular momentum quantum number. The total number of magnetic quantum numbers for each l are $2l + 1$, that is, $+l$ to $-l$ including zero. The spin quantum number defines the spin motion superimposed on the orbital motion and can have the values $+1/2$ or $-1/2$ to reflect the two possible spins.

With the aid of these quantum numbers and Pauli's exclusion principle, the occupation of orbitals by any number of electrons can be explained. This principle states that no two electrons belonging to any individual nucleus may have the same set of four quantum numbers. After one orbit has been occupied by the maximum number of electrons that can be accommodated according to the exclusion principle, additional electrons occupy orbits of greater radii and smaller electrostatic attraction to the nucleus. Thus the impenetrability of matter, which is manifested by the small compressibility of solids, is essentially a consequence of the Pauli exclusion principle.

Thus the energy level to which an electron belongs is determined by the four quantum numbers. The maximum number of electrons in the innermost shell, the K shell (quantum number $n = 1$), is $2 (= 2n^2)$, the next shell, L (quantum number $n = 2$), has 8 electrons $(= 2 \times 2^2)$, and so on. Consider, for example, a sodium atom with an atomic number of 11. It has 11 electrons and protons and 12 neutrons. The K shell has 2 electrons with quantum numbers $n = 1$, $l = 0$, $m_l = 0$, $m_s = \pm 1/2$. The L shell has 8 electrons and two energy levels corresponding to the combinations $n = 2$, $l = 0$, and $n = 2$, $l = 1$. The remaining two quantum numbers have not been taken into consideration in determining the energy levels since they have a very small influence on the energy levels. The first of these energy levels has 2 electrons with the quantum numbers $n = 2$, $l = 0$, $m_l = 0$, $m_s = \pm 1/2$. The second energy level has 6 electrons with the quantum numbers $n = 2$, $l = 1$, $m_l = -1$, 0, $+1$, $m_s = \pm 1/2$. The M shell contains the remaining single electron. The M shell, with a principal quantum number $n = 3$ and angular momentum quantum number $l = 0$, 1, 2, has three possible energy levels. The last remaining electron can occupy any one of these three energy states depending on the conditions. A schematic representation of the electron orbits in a sodium atom, defined by the quantum numbers (n, l) are shown in Figure 1.

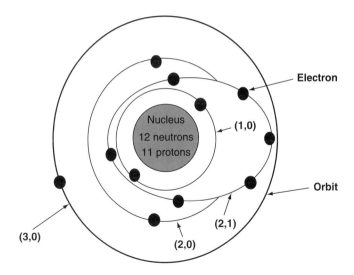

FIGURE 1 *Schematic representation of electron orbits in sodium atom.*

The order of succession in which the electrons occupy the various orbits is determined by the energy expended in the process of binding successive electrons. Thus, if, after the completion of a secondary shell of angular momentum number $l < (n - 1)$ within the total group of such shells associated with the principal quantum number n, the binding energy of an electron in the first l shell of the $(n + 1)$ level is higher than that of the successive l shell in the n level, the first l shell of the $(n + 1)$ level will be occupied by electrons before the still unoccupied i shells of the n level are occupied.

Elements with atomic numbers whose electrons completely occupy all orbits defined by a principal quantum number, such as krypton with atomic number 36, are called *inert*. Krypton has 8 electrons in the first two energy levels of the N shell, the maximum allowed. Such a shell is called *closed electron shell* because of the stability of the grouping of electrons. Elements with a closed outer electron shell are inert and do not participate in bonding or reacting with other elements under normal circumstances. Thus the *valence* of the element with a closed outer shell is zero. Electrons that occupy orbits outside of closed secondary shells and whose number is always less than 8, which is the minimum number of electrons forming a secondary shell, are called *valence electrons*. Hence, aluminum with 3 electrons in the outermost shell has a valence of three, and carbon has a valence of four. The valence can also depend on the nature of the chemical reaction. For example, phosphorus has 5 electrons in the outermost M shell. Of these, 2 electrons are in the first energy level and 3 electrons are in the second energy level within the M shell. When phosphorus reacts with oxygen it displays a valence of five, but when it reacts with hydrogen it displays a valence of three.

Only the valence electrons determine the interaction between atoms, and thus the chemical and mechanical properties of the material. The similarity of the chemical and mechanical behavior is related to the similarity of the structure of their outer electron shell. This is the reason for the occurrence of the "periods" in the periodic table of the elements. For example, column 1A in the periodic table consists of the alkali metals. Each of the alkali metals has one electron more than the preceding inert gas, that is, they have one electron outside of the last closed shell. Because of the similar nature of the outermost electron shell the alkali metals have similar chemical and mechanical behavior. For instance, all alkali metals are more compressible than other metals, all have high electrical conductivity, all are lighter than water, and they react rapidly with oxygen in a highly exothermic process.

The forces that two atoms exert on each other are the result of the electric charges of their valence electrons. The orbits in closed shells remain almost unperturbed by the influence of neighboring atoms. When two atoms approach each other closely enough, the valence electrons that are at the greatest distance from their own nuclei take up new common orbits around the nuclei of both atoms, thus establishing an atomic bond. The strength of this bond depends on the mutual influence between the valence electrons and the nuclei of the binding atoms.

3 Binding Energy

An atom consists of positively charged nucleus surrounded by electrons. The electrons revolve around the nucleus in a series of orbitals, which are regions in which the electrons are most likely to be found. The interactions of the electrons with each other and with the nucleus are responsible for the binding of the atoms into molecules and crystals. The interatomic or intermolecular forces can be conceptualized as being produced by the interaction of two potential fields of attractive and repulsive energy, or of two fields of repulsive and attractive electrostatic charges.

Four principal types of bonds are formed, namely ionic bond, covalent bond, metallic bond, and intermolecular bonds. The first three bonds are formed between atoms or ions, while the last bond type is formed between individual molecules. All four bond types are responsible for the cohesion of crystalline solids. In ionic bonds the attractive forces are the result of opposite electric charges; in covalent bonds they are the result of sharing of valence electrons; in metallic bonds they are due to the general attraction of the positive ions for the negatively charged free electron clouds; in intermolecular bonds they are the result of magnetic interaction between dipoles.

Assuming that the potential energy of interaction $U(r)$ has the form

$$U(r) = -\frac{a}{r^m} + \frac{b}{r^n}. \tag{1.1}$$

$n > m$ since the repulsive potential (b/r^n) must decrease more rapidly with increasing distance r than the attractive potential (a/r^m), as otherwise no equilibrium would be possible. The actual values of m and n vary with the nature of the bond. For ionic, covalent, and metallic bonds the value of $m = 1$. The value of n varies between 9 and 11 for ionic and covalent bonds, and between 6 and 9 for metallic bonds. For intermolecular bonds the values are $m = 6$ and $n = 9$ to 12.

The interaction force $F(r)$ is derived by differentiating the potential energy of interaction $\phi(r)$:

$$F(r) = \frac{d}{dr}U(r) = \frac{ma}{r^{m+1}} - \frac{nb}{r^{n+1}}. \tag{1.2}$$

The equilibrium position of the two interacting particles is attained at the distance of maximum stability, that is, of minimum energy at the bottom of the "potential trough" shown schematically in Figure 2. At the point where $U(r)$ reaches a minimum, its first differential $F(r) = 0$, indicating equilibrium between the interacting forces.

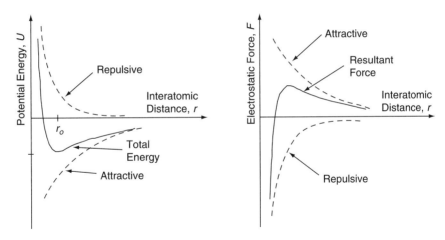

FIGURE 2 *Interatomic energy and force.*

The minimum (negative) potential energy or the maximum bond energy U_o defines the equilibrium position or atomic separation, r_o. The area of the function $F(r)$, which thus represents the bond energy, gives the energy required to remove the particle out of the field of mutual interaction. It is evident from this function that any displacement from the equilibrium position gives rise to restoring forces that, for small displacements, can be assumed to change linearly with displacement. The linearity fails, as soon as the displacements become large. The restoring forces differ for both directions of the displacement. Figure 2 shows that the restoration or the elastic modulus, which is determined of the tangent to the $F(r)$ curve, or the curvature of the $U(r)$ curve at the equilibrium position $r = r_o$, increases rapidly if the particles are crowded together but decreases more gradually with increasing separation. The maximum ordinate of the $F(r)$ curve gives the maximum force required to separate the particles. This ordinate represents the theoretical or atomic cohesive strength of an elastic substance of perfectly regular atomic arrangement.

4 Atomic Bonds

The three *primary bond* types are the ionic, covalent, and metallic bonds and are formed between atoms and ions. A relatively weaker secondary bond, called the *intermolecular bond*, is formed between molecules. The energy of the bond is generally an inverse function of the distance between the elementary particles. The primary bonds are stronger than the secondary bonds by about two orders of magnitude.

4.1 Primary Bonds
Ionic Bonds
Ionic bonds are formed due to the electrostatic attraction between positively charged ions (cations) and negatively charged ions (anions) of atoms of different chemical elements. This type of bond is generally found in the formation of inorganic salts such as sodium chloride and aluminum oxide. Metallic elements can easily lose their outermost or valence electrons, whereas the nonmetallic elements have a strong tendency to acquire electrons. This results in the formation of stable metallic cations and nonmetallic anions, which are attracted to each other and form strong ionic bonds. It may be noted that the smaller the number of valence

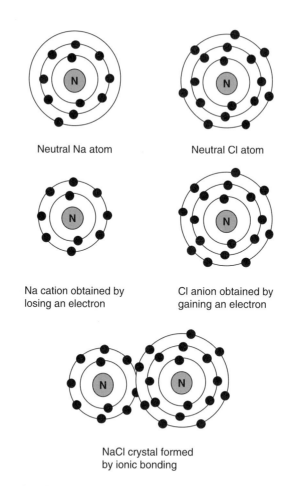

Neutral Na atom

Neutral Cl atom

Na cation obtained by
losing an electron

Cl anion obtained by
gaining an electron

NaCl crystal formed
by ionic bonding

FIGURE 3 *The ionic bond in a sodium chloride molecule.*

electrons in the interacting atoms, the easier it is to form an ionic bond. It is for this reason that monovalent alkali metals (which have one electron in the outermost shell) react vigorously with halogens (which need one electron to complete their outermost shell). For example, sodium chloride molecule formation by ionic bonding is shown schematically in Figure 3. However, Na^+ and Cl^- ions do not link up in pairs as shown in that figure, because then there would be strong attractive forces within the paired ions of a sodium chloride molecule but negligible attraction between the molecules. As a result, sodium chloride would not exist. Actually, each positive ion attracts all neighboring negative ions and vice versa. Thus the ions surround each other in such a way that the attraction between neighboring unlike charges exceeds the repulsion due to like charges. The ionic bond is sufficiently strong to result in hard ionic solids with high melting points (the melting point of sodium chloride is 801°C).

Covalent Bonds

These bonds are formed by the "sharing" of valence electrons by neighboring atoms in an attempt to complete their outer electron shells. A requirement for covalent bonding is that each atom have at least one half-filled electron shell. This arrangement allows the energy level to be

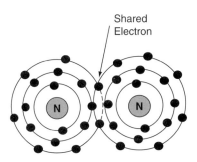

Shared Electron

FIGURE 4 *The chlorine molecule formed by covalent bonding.*

lowered substantially by having each bonding electron in the electron shells of the two atoms simultaneously. The more the bonding electron shells overlap, the more the energy is lowered, hence the stronger the bond. For example, chlorine requires 1 electron to complete its outer shell. This can be accomplished if two chlorine atoms join together and share their outer electrons, as shown in Figure 4. With 7 electrons in the outermost shell of each chlorine, one is shared or exchanged by both atoms. These atoms have alternately 6 and 8 electrons in the outer shell, becoming alternately positively and negatively charged. Through this sharing process the two atoms form a closely knit diatomic molecule. Because the mutual attraction within the molecule extends to a pair of atoms only, the cohesive strength is very low.

In elements with six electrons in the outer shell such as sulfur and selenium, one electron of each atom is shared with the two nearest neighboring atoms. Since each atom can form a bond with only two neighboring atoms, a linear chain of atoms can be formed. Such a one-dimensional long-chain molecular structure is typical of crystalline sulfur and selenium. Between the long-chain molecules only molecular interaction is possible and hence the cohesive strength of these materials is small.

If there are five electrons in the outer shell, as in arsenic and antimony, an atom shares one electron each with three immediate neighbors completing a stable eight electron outer shell. In these materials large plane sheet like molecules are formed. These layers or sheets of atoms are connected by molecular bonds. The molecular interaction produces a weak cohesion perpendicular to the plane of the molecules. Thus these materials are strong in the plane of the sheets but weak perpendicular to the sheet.

Atoms with 4 electrons in the outer shell, such as carbon and silicon, share 1 electron each with the four immediate neighbors, producing a three-dimensional structure in which the atoms occupy the center and the corners of a tetrahedron. Each of the four corner atoms shares 1 of its outer electrons with the central atom, thereby making a closed shell of 8 electrons. The central atom shares each of its 4 electrons with one of the corner atoms, thereby contributing to their closed shells. In this structure the covalent bonds are formed in three directions in space, resulting in materials with very high strength, rigidity, and melting point. The most popular example is diamond, which is the hardest material known and has the highest melting point of 3550°C. The diamond structure results in poor electrical conductivity, since the alternately closed outer shells of the constituent atoms produce configurations similar to that of the electrically inert gas neon.

Metallic Bonds

If the number of an atom's valence electrons is less than four it does not form covalent bonds, and if the electrostatic charges on the ions are not of opposite sign they cannot form ionic

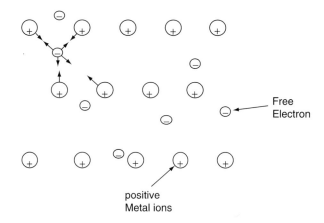

FIGURE 5 *Schematic representation of metallic bonding.*

bonds. In such a case the material atoms, for example, metals with one or two valence electrons, readily give up their valence electrons forming positive ions, while the freed electrons are shared by all the positive ions. The freed electrons are referred to as an *electron gas*, which permeates the whole crystal. The bonding is then an electrostatic attraction between the array of positive ions and the electron gas, as shown in Figure 5.

The free electrons move around freely between the positive ions. Each electron has an attraction between each of the neighboring metal ions. Thus the metal ions are nondiscriminating and very versatile in the formation of bonds. Hence there is no restriction on the number of bonds and the directions of the metallic bonds, which are characteristic of the ionic and covalent bonds. This versatility in bond formation produces the typical metal properties of ductility and malleability. The presence of the electron gas results in the properties of good electric conductivity and thermal conductivity.

4.2   Intermolecular Bonds

These are weak bonds, which produce loosely bound lattices. These bonds are produced by the magnetic interaction between dipoles, or temporarily induced in electrically neutral molecules by an external electric field surrounding an ion or another dipole. Intermolecular bonds are responsible for the formation of molecular crystals, such as paraffin, crystallized rubber, and gels. In substances formed by intermolecular bonds the shape of the molecules is the primary factor influencing their mechanical behavior and properties. The van der Waals bond is the most important bond in this category. Van der Waals forces are primarily electrostatic and arise as a result of the momentary shifts in the relative positions of the nucleus and electrons of an atom. Thus, the atom forms an electric dipole that in turn induces dipoles in neighboring atoms. The attraction between charges of opposite sign in the dipoles thus gives rise to the attractive forces. The van der Waals bond is responsible for binding atoms in solidified inert gases.

5   Crystal Structure

5.1   Crystal Structure

Generally solid materials are formed in either a crystalline or amorphous form, or as some combination of the two. The crystalline materials, such as metals and ceramics, consist of a

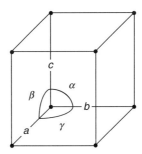

FIGURE 6

very ordered arrangement of atoms, while the amorphous materials, such as many plastics and glass, do not possess any symmetry or order in the arrangement of atoms. In this section we will concentrate on the common crystal structures formed by crystalline solids. (A more detailed study on crystal structure can be found in *Mechanical Properties of Matter*, by A. H. Cottrell, Wiley, 1964, and *Mechanical Properties of Solid Polymers*, by I. M. Ward, 2nd edition, Wiley, 1983.)

Studies using the experimental technique of X-ray diffraction (see W. H. Zachariasen, *Theory of X-ray Diffraction in Crystals*, Wiley, New York, 1945) have shown that there are 14 different ordered arrangements possible, called the Bravais lattices. A lattice in space is a three-dimensional array of points, infinite in extent, in which every point has identical surroundings. When atoms are placed at each of these points, a crystal structure results. Only a typical lattice with the sides as a, b, and c, and the angles between any two sides represented by the angles α, β, and γ, is shown in Figure 6. This figure shows the axes system and the lattice parameters. The lattice parameters describe the size and shape of the Bravais lattice. (The Bravais lattice is also known as the unit cell, which is a subdivision of the lattice that still retains the overall characteristics of the entire lattice.) The dimensions of the sides shown as a, b, and c define the size of the Bravais lattice, while the shape is defined by the angles between the sides. The lattice parameters or the axes and the angles are listed in Table 1 only for seven different crystal structures each associated with a different axes system. The crystal structures listed here are cubic, orthorhombic, tetragonal, monoclinic, rhombohedral, hexagonal, and triclinic. The cubic system has 3 orthogonal axes of equal length, orthorhombic has 3 orthogonal axes of unequal length, tetragonal has 3 orthogonal axes of which 2 are of equal length, monoclinic has 2 orthogonal and 1 inclined, rhombohedral has 3 nonorthogonal but

TABLE 1 Characteristics of the Seven Crystal Structures

Structure	*Axes*	*Angles between axes*
Cubic	$a = b = c$	All angles are 90°
Tetragonal	$a = b \neq c$	All angles are 90°
Orthorhombic	$a \neq b \neq c$	All angles are 90°
Hexagonal	$a = b \neq c$	$\alpha = 120°$, $\beta = \gamma = 90°$
Rhombohedral	$a = b = c$	All angles equal, but not 90°
Monoclinic	$a \neq b \neq c$	$\alpha = \gamma = 90°$, $\beta \neq 90°$
Triclinic	$a \neq b \neq c$	All angles unequal and not 90°

equally inclined axes of equal length, hexagonal has 2 nonorthogonal axes of equal length at right angles to the third axis, and triclinic has 3 unequal axes inclined to each other. Of these seven crystal structures, only cubic and hexagonal are particularly important in the study of behavior of metals.

5.2 Atomic Packing and Its Effect on Mechanical Behavior

To facilitate a discussion on the formation of crystal structures in terms of the packing of atoms into different arrangements, the atoms will be considered as hard spherical balls. This assumption is unrealistic, as it does not recognize electron clouds or charge distributions. However, it is convenient and useful if we are considering only the geometric arrangement of atoms in crystals. This concept will also be useful in interpreting the mechanical behavior of crystalline materials, especially metals.

There are a number of possible regular arrangements of "near-neighbor" atoms or ions, that is, of atoms connected by equivalent bonds and pictured as spheres of equal size, around any considered spherical atom or ion:

(a) Two symmetrical neighbors on a straight line through the atom
(b) Three neighbors in the corners of an equilateral triangle, the center of which is occupied by an atom
(c) Four neighbors in the corners of a tetrahedron, the center of which is occupied by an atom

The foregoing three arrangements result in the rather loose packing typical of crystal structures of nonmetals formed by covalent bonds. These arrangements are not limited to nonmetals; there are certain metals whose crystal structures are of the 2-, 3-, and 4-neighbor type, respectively, for instance: tellurium with 2 neighbors, antimony with 3 neighbors, and silicon with 4 neighbors. These metals most closely resemble the nonmetallic elements, which form covalent bonds. The 3-neighbor arrangement is also characteristic for the hexagonal structural formation of a large number of organic compounds (benzene ring).

(d) Six neighbors in the corners of an octahedron, the center of which is occupied by the atom. This type of packing produces the *simple cubic* lattice. No element is known to crystallize in this form. It is actually a transition structure between the loosely packed lattices produced by covalent bonds and the closely packed ionic and metal lattices.
(e) Eight neighbors in the corner of a cube, the center of which is occupied by the atom. This is the *body-centered cubic* structure, in which a number of metals crystallize.
(f) Twelve neighbors in the *closest packed* structure of spheres of identical size. This arrangement leaves the minimum of space between the spheres and may have either cubical symmetry, resulting in the *face-centered cubic* structure, or hexagonal symmetry, resulting in the *hexagonal* closed packed structure, which is slightly less regular than the cubic form of closest packing. An individual atom in the hexagonal closest packing is in contact with a hexagon of spheres arranged in the same plane and with two triangles of spheres above and below.

Closest packing is characteristic of the structure of the majority of metals and a number of ionic crystals. It expresses the tendency of the atoms to form the most stable structure, that is, a structure with the maximum bond energy. In ionic crystals the positive ions surround themselves with the maximum number of negative ions, whereas in metallic crystal structures free electrons surround the positive ions.

Actually the 8-neighbor (body-centered cubic) structure is also rather closely packed since, in addition to the 8 nearest neighbors connected by strong bonds, there are 6 neighbors at only slightly larger distances, connected by weaker bonds. For certain metals the total bond energy, and thus the stability of the structure resulting from the 14 bonds of unequal strength, appears to be higher than that produced by the 12 bonds of equal strength of the face-centered group.

Only as long as the structure is made of spheres of equal size are the cubic and hexagonal closest packings the two alternatives of maximum stability. An infinite variety of arrangements is possible for the closest packing and therefore for maximum stability of spheres of two or more number of different diameters. In such structures, mostly metal alloys, intermetallic compounds, and ionic crystals, the number of closest neighbors can be increased considerably beyond 12. Several packings are known with as many as 20 neighbors. This refers particularly to compounds of metals with the nonmetallic light atoms of smaller size, such as carbon, nitrogen, and hydrogen. In this case the metal ions are themselves arranged in closest packing locations while the carbon or nitrogen atoms are located in the interstitial solid solutions or interstitial compounds, and may require a slight expansion of the surrounding metal lattice to fit the light atoms into the interstices between the ions. Because of the high density of this arrangement and the large number of bonds formed, these compounds are unusually strong and hard and have extremely high melting points, for example as in the nitrides and carbides.

In the modern process of surface hardening of steel parts by nitriding, cyaniding, and carburizing, hard interstitial compounds are formed within a thin surface layer. In addition, since the slight volume expansion associated with the introduction of the carbon or nitrogen atoms into the metal lattice near the surface is prevented by the adjoining metal unaffected by the treatment, relatively high compressive stresses are introduced into the surface layer. Thus, both the high strength and high endurance of surface-treated steel parts are interpretable in terms of simple changes in the atomic structure conceived as densely packed spheres.

5.3 Slip Systems in Metals

Deformation in metals occurs by a process of slip on certain slip planes and along specific slip directions. The slip planes are those with the highest planar atomic fraction (the ratio of the area of atoms on the plane to area of the plane) while the slip directions are those with the highest linear atomic fractions (the ratio of the length of atoms along a direction to length of the direction). The various combinations of a slip plane and slip directions is called a slip system. Table 2 lists the slip systems in metallic crystal structures.

TABLE 2 Slip Systems in Metals

Crystal structure (typical metals)	Slip plane, (number of nonparallel planes, n)	Slip direction, (number of slip directions per plane, m)	Number of slip systems $(n \times m)$
BCC (Fe, MO, Ta, W)	$\{110\}$, (6)	$\langle \bar{1}11 \rangle$, (2)	$(6 \times 2 = 12)$
	$\{112\}$, (12)	$\langle 11\bar{1} \rangle$, (1)	$(12 \times 1 = 12)$
	$\{123\}$, (24)	$\langle 11\bar{1} \rangle$, (1)	$(24 \times 1 = 24)$
FCC (Al, Cu, Ni, Pt)	$\{111\}$, (4)	$\langle 1\bar{1}0 \rangle$, (3)	$(4 \times 3 = 12)$
HCP (Cd, Ti, Zn)	$\{0001\}$, (1)	$\langle 11\bar{2}0 \rangle$, (3)	$(1 \times 3 = 3)$

Review Module R2:   Systems of Units

TABLE 1 A Taxonomy of Systems of Units

	"The Good"	*"The Bad"*	*"The Ugly"*	*"The Dead"*
System	SI	U.S. Customary or British Gravitational	English Engineering	cgs (dyne, cm, g, s) fps (poundal, ft, lb_m, s)
Basis	Absolute	Gravitational	?	
Base (primitive, fundamental) mechanics dimensions and units	mass kg length m time s	force lb length ft time s	force lb_f mass lb_m length ft time s	
Derived dimension and unit (mechanics only)	force N	mass slug	—	
Newton's 2nd law $F \sim d/dt(mv)$	$F = (1)\,ma$ Proportionality constant $= 1$	$F = (1)\,ma$ *Actually:* $m = (1)\,F/a$	$F = (1/g_c)\,ma$ Since force and mass are defined, prop. const. $\neq 1$	
	$1\,\text{N} = (1\,\text{kg})(1\,\text{m/s}^2)$	$1\,\text{lb} = (1\,\text{slug})(1\,\text{ft/s}^2)$ *Actually:* $1\,\text{slug} = (1\,\text{lb})(1\,\text{ft/s}^2)$	$1\,\text{lb}_f = \dfrac{(1\,\text{lb}_m)(32.2\,\text{ft/s}^2)}{32.2\,\text{lb}_m\text{ft/lb}_{fs^2}}$	$1\,\text{dyne} = (1\,\text{g})(1\,\text{cm/s}^2)$ $1\,\text{poundal} = (1\,\text{lb}_m)(1\,\text{ft/s}^2)$
Magnitudes	Decades	—	—	

Notes:

- The SI system is *the* system of units in use by most of the world outside of the USA, and is the standard for nearly all current engineering and scientific publications. The SI system is an "absolute" system since mass is taken as a primitive or fundamental dimension (as compared to force or weight); force is a derived dimension. Force and mass have separate and distinct units. The constant of proportionality in Newton's law is nondimensional and of unit value. SI also uses order-of-magnitude descriptors (in decades).

- Other systems of units are less desirable in practice, but a working knowledge of them is required in the USA. The U.S. Customary system is a "gravitational" system ("relative" system) since it takes force (weight) as a primitive dimension (but weight obviously depends on the local gravitational acceleration). It does, however, have separate and distinct units for force and mass, and a constant of proportionality in Newton's law that is nondimensional and of unit value. Limited order-of-magnitude descriptors are used (e.g., ksi).

- The English Engineering system is the least desirable of commonly used systems of units. Both force and mass are taken as primitives. Hence, the constant of proportionality in Newton's law is neither nondimensional nor of unit value $(1/g_c)$. Force and mass do not have separate and distinct units. Limited order-of-magnitude descriptors are used (e.g., ksi).

- The cgs and fps are rarely used at the present time, which is a blessing!

Review Module R3:   Free-Body Diagrams

The *free-body diagram* (FBD) is a tool for formulating mathematical models via Newton–Euler mechanics. It provides for bookkeeping of all of the externally applied forces and moments on the system of interest (SOI), as well as internal forces and moments at the free-body boundaries.

There are three simple steps for developing FBDs:

1. Define the SOI (i.e., a particle, body, portion of a body, etc.).
   The definition must include a coordinate system. The SOI must be considered in its displaced configuration. For convenience in writing the equations of motion, displace the SOI in the positive coordinate direction(s).
2. Completely isolate the free body.
   Isolate the free body from the SOI, and show on the FBD all forces and moments:
   - Applied by external agents (e.g., gravity, concentrated forces)
   - Occurring internally as reactions at the boundaries when the SOI is isolated
3. Show on the FBD the relevant geometry and any other information required to do a complete accounting.

An example follows in Figure 1.

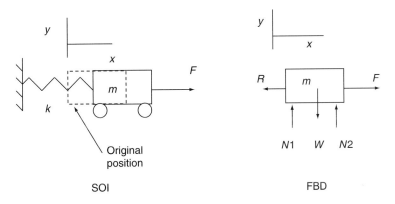

FIGURE 1   *Simple free-body diagram.*

Review Module R4:   Matrix Algebra

Consider a system of linear equations of the form

$$a_{11}x_1 + a_{12}x_2 + a_{13}x_3 + \ldots + a_{1n}x_n = f_1$$
$$a_{21}x_1 + a_{22}x_2 + a_{23}x_3 + \ldots + a_{2n}x_n = f_2$$

$$\vdots$$

$$a_{m1}x_1 + a_{m2}x_2 + a_{m3}x_3 + \ldots + a_{mn}x_n = f_m$$

where the a_{ij} are constant coefficients. It is remarkable that this system of equations can be arranged as an array of the form

$$[A]\{X\} = \{F\}$$

where

$$[A] = \begin{bmatrix} a_{11} & a_{12} & \cdots & a_{1n} \\ a_{21} & a_{22} & \cdots & a_{2n} \\ \vdots & \vdots & \vdots & \vdots \\ a_{m1} & a_{m2} & \cdots & a_{mn} \end{bmatrix}, \{X\} = \begin{Bmatrix} x_1 \\ x_2 \\ \vdots \\ x_m \end{Bmatrix}, \{F\} = \begin{Bmatrix} f_1 \\ f_2 \\ \vdots \\ f_m \end{Bmatrix}$$

and then solved for in this formulation by manipulating the arrays rather than the individual equations. The array $[A]$ is called a *matrix*, the quantities $\{X\}$ and $\{F\}$ are called *vectors*, and the manipulation is called *matrix algebra*.

Some general principles of matrices and matrix algebra are given here, sufficient for the scope of this text.

1 Order of a Matrix and Square Matrix

A matrix with m rows and n columns is said to be of *order* $m \times n$ ("m by n"). A *square matrix* is order $m \times m$, or simply "of order m."

Example

$$\begin{bmatrix} 1 & 2 & 3 \\ 4 & 5 & 6 \end{bmatrix} \text{ order } 2 \times 3$$

$$\begin{bmatrix} 1 & 2 \\ 3 & 4 \\ 5 & 6 \end{bmatrix} \text{ order } 3 \times 2$$

2 Coefficients of a Matrix and the Identity Matrix

The members or elements of a matrix $a_{ij}(i = 1, 2, \ldots, m, j = 1, 2, \ldots, n)$ are called the *matrix coefficients*. If the coefficients are such that they are defined by

$$\delta_{ij} = \begin{cases} 1, & i = j \\ 0, & i \neq j \end{cases}$$

then the matrix is called the *identity matrix* [1].

3 Multiplication of a Matrix by a Constant

Multiplication of a matrix by a constant is distributive.

Example

$$3 \begin{bmatrix} 1 & 3 \\ 0 & 2 \end{bmatrix} = \begin{bmatrix} 3 & 9 \\ 0 & 6 \end{bmatrix}$$

4 Submatrix

The *submatrix* $[M_{ij}]$ of a square matrix $[A]$ results from eliminating the ith row and jth column.

Example

$$[A] = \begin{bmatrix} 1 & 2 & 3 \\ 4 & 5 & 6 \\ 7 & 8 & 9 \end{bmatrix}$$

$$[M_{11}] = \begin{bmatrix} 5 & 6 \\ 8 & 9 \end{bmatrix}, \quad [M_{12}] = \begin{bmatrix} 4 & 6 \\ 7 & 9 \end{bmatrix}, \quad [M_{31}] = \begin{bmatrix} 2 & 3 \\ 5 & 6 \end{bmatrix}, \quad \ldots$$

5 Determinant of a Second-Order Square Matrix

The *determinant* of a second-order square matrix [A] is a single scalar value |A|, given by

$$|A| = a_{11}a_{12} - a_{21}a_{22}.$$

Example

$$\begin{vmatrix} 2 & 5 \\ 4 & 12 \end{vmatrix} = 2(12) - 4(5) = 4$$

6 Minors of a Square Matrix

The *minors* M_{ij} of a square matrix [A] are the determinants of any of the square submatrices formed by deleting the *i*th row and *j*th column of [A]. The minor M_{ij} is called the *minor of a_{ij}*.

Example

$$[A] = \begin{bmatrix} 1 & 2 & 3 \\ 4 & 5 & 6 \\ 7 & 8 & 9 \end{bmatrix}$$

$$[M_{11}] = \begin{bmatrix} 5 & 6 \\ 8 & 9 \end{bmatrix}, \; M_{11} = 45 - 48 = -3$$

$$[M_{12}] = \begin{bmatrix} 4 & 6 \\ 7 & 9 \end{bmatrix}, \; M_{12} = 36 - 42 = -6$$

7 Cofactor of a Square Matrix

A *cofactor* of the square matrix [A], α_{ij}, is a "signed minor" of [A] given by

$$\alpha_{ij} = (-1)^{(i+j)} M_{ij}.$$

Example

$$[A] = \begin{bmatrix} 1 & 2 & 3 \\ 4 & 5 & 6 \\ 7 & 8 & 9 \end{bmatrix}$$

$$[M_{11}] = \begin{bmatrix} 5 & 6 \\ 8 & 9 \end{bmatrix}, \; M_{11} = 45 - 48 = -3, \; \alpha_{11} = (-1)^{1+1}(-3) = -3$$

$$[M_{12}] = \begin{bmatrix} 4 & 6 \\ 7 & 9 \end{bmatrix}, \; M_{12} = 36 - 42 = -6, \; \alpha_{12} = (-1)^{1+2}(-6) = 6$$

8   Determinant of a Third-Order (or Higher) Square Matrix

The determinant of a third-order square matrix $[A]$ is found as the sum of the products obtained by multiplying each element of row (or column) of a $[A]$ by its cofactor. For example;

$$|A| = \alpha_{11}a_{11} + \alpha_{12}a_{12} + \alpha_{13}a_{13} = \alpha_{21}a_{21} + \alpha_{22}a_{22} + \alpha_{23}a_{23} = \ldots.$$

This is finding a determinant by the method called *expansion by minors*.

Example

$$\begin{vmatrix} 1 & 2 & 1 \\ 1 & 2 & 0 \\ 1 & 0 & 2 \end{vmatrix} = (-1)^{1+1}\begin{vmatrix} 2 & 0 \\ 0 & 2 \end{vmatrix} + (-1)^{1+2}\begin{vmatrix} 1 & 0 \\ 1 & 2 \end{vmatrix} + (-1)^{1+3}\begin{vmatrix} 1 & 2 \\ 1 & 0 \end{vmatrix} = 4 - 2 + (-2) = 0$$

9   Transpose of a Square Matrix

The *transpose* $[A]^T$ of a matrix $[A]$ is found by interchanging corresponding row and column elements, that is; $a_{ij} \rightarrow a_{ji}$, $i \neq j$. A *symmetric matrix* is defined by $a_{ij} = a_{ji}$.

Example

$$[A] = \begin{bmatrix} 1 & 2 & 3 \\ 4 & 5 & 6 \\ 7 & 8 & 9 \end{bmatrix}, \ [A]^T = \begin{bmatrix} 1 & 4 & 7 \\ 2 & 5 & 8 \\ 3 & 6 & 9 \end{bmatrix}$$

10   Product of Two (or More) Square Matrices

The product of two square matrices $[A][B]$, each of order m, is a square matrix $[C]$ also of order m, each element c_{ij} of $[C]$ being found by multiplying the ith row of $[A]$ by the jth column of $[B]$. For example, if $m = 3$, then

$$c_{11} = a_{11}b_{11} + a_{12}b_{21} + a_{13}b_{31}$$
$$c_{23} = a_{21}b_{13} + a_{22}b_{23} + a_{23}b_{33}$$

In general, $[A][B] \neq [B][A]$.

Example

$$\begin{bmatrix} 9 & 3 \\ 4 & 5 \end{bmatrix}\begin{bmatrix} 2 & 0 \\ 1 & 7 \end{bmatrix} = \begin{bmatrix} 9(2) + 3(1) & 9(0) + 3(7) \\ 4(2) + 5(1) & 4(0) + 5(7) \end{bmatrix} = \begin{bmatrix} 21 & 21 \\ 13 & 12 \end{bmatrix}$$

11 Inverse of a Square Matrix

The *inverse* $[A]^{-1}$ of a square matrix $[A]$ may be found from dividing the transpose of the matrix of cofactors of $[A]$, cofac$[A]$, by the determinant of $[A]$, that is,

$$[A]^{-1} = \frac{1}{|A|}(\text{cofac}[A])^{\text{T}}.$$

A matrix is *singular* when $[A]^{-1}$ does not exist. Also,

$$[A]^{-1}[A] = [A][A]^{-1} = [1].$$

Example

$$\begin{bmatrix} 1 & 0 & 1 \\ 2 & 1 & 1 \\ 1 & 1 & 2 \end{bmatrix}^{-1} = \frac{1}{\begin{vmatrix} 1 & 0 & 1 \\ 2 & 1 & 1 \\ 1 & 1 & 2 \end{vmatrix}} \begin{bmatrix} 1 & -3 & 1 \\ 1 & 1 & -1 \\ -1 & 1 & 1 \end{bmatrix}^{\text{T}} = \frac{1}{2}\begin{bmatrix} 1 & 1 & -1 \\ -3 & 1 & 1 \\ 1 & -1 & 1 \end{bmatrix}$$

12 Matrix Solution of a System of Linear Equations

Given the linear system of equations in matrix form $[A]\{X\} = \{F\}$, the solution may be found from

$$[A]^{-1}[A]\{X\} = [A]^{-1}\{F\}$$

$$[1]\{X\} = [A]^{-1}\{F\}$$

$$\{X\} = [A]^{-1}\{F\}$$

Answers to Selected Problems

Chapter 1

1-9   Point C: Negligible load
      Point B: Bending moment, torsion, shear
      Point A: Bending, shear

1-11 Point A: Bending and shear
      Point B: Bending, shear and torsion

1-25 (a) The negative term is attractive.

      (b) $\Rightarrow r_0 = \left(\dfrac{8\beta}{\alpha}\right)^{\frac{1}{7}}$

      (c) Attractive energy $= 8\times$ Repulsive energy

1-26 $< S_{UT} >= 1215\,\text{MPa}$

Chapter 2

2-3   There are 4 CDOFs (x and z translation, x and y rotation).

2-7   Beam is statically indeterminate; tripod is statically determinate.

2-10 Mode I and mode II

2-11 The $45°$ crack will be under mode I and mode II loading.

2-13 The ductile material will result in a longer fatigue life.

2-15 Plate with the elliptic hole exhibits a greater stress concentration on the x-axis.

Chapter 3

3-1   Diameter $= 0.4997$ inch

3-2   For a temperature change from $10\,°\text{C}$ to $40\,°\text{C}$, $\sigma = 37.8\,\text{MPa}$ (compressive).

3-4   12 inches

3-6   (a) $15{,}000\,\text{psi}$ (b) $-12{,}000\,\text{psi}$

3-7   (b) $0.06\,\text{mm}$

3-8   0.5

3-9   Diameter $= 11.88\,\text{mm}$

3-12 $532\,\text{GPa}$

3-18 Diameter $= 0.5003$ inch

3-22 2.89 inch

Chapter 4

4-1   (b) 3.063 kN m
4-2   7.64 kip in
4-3   106.9 mm
4-9   [0 1 1] and [$\bar{1}$ 0 1] are slip directions for (1 $\bar{1}$ 1). CRSS = 255 kPa
4-10 42.1 kN
4-14 8.99 lb·in
4-16 2.55 t/b
4-18 (a) 3.77 mrad; (b) $T_1 = -444\,\text{lb}\cdot\text{in}$, $T_3 = -3556\,\text{lb}\cdot\text{in}$

Chapter 5

5-4   (a) 999 milli-in; (b) 1.498 min
5-5   (a) 1.57 m; (b) 1.57 mm/mm
5-9 $y_{max} = -p_0 L^4 / 144EI$
5-11 (a) $M_{max} = 9 p_0 L^2 / 128$, $y_{max} = -0.01 p_0 L^4 / EI$

Chapter 6

6-4 $\sigma_n = 31.3\,\text{MPa}$
      $\tau_n = 51.1\,\text{MPa}$
6-6   (a) $\sigma_{x'x'} = 90.4\,\text{MPa}$
      $\sigma_{y'y'} = 9.0\,\text{MPa}$
      $\sigma_{z'z'} = 49.0\,\text{MPa}$
      $\tau_{x'y'} = -54.5\,\text{MPa}$
      $\tau_{y'z'} = -59.0\,\text{MPa}$
      $\tau_{z'x'} = 65.7\,\text{MPa}$
6-7   (a) $\sigma_n = 0\,\text{MPa}$
      $\tau_n = 45.0\,\text{MPa}$
      (b) $\sigma_n = 50.0\,\text{MPa}$
      $\tau_n = 44.1\,\text{MPa}$
6-10 $\sigma_1 = 300\,\text{MPa}$, $\sigma_2 = 50\,\text{MPa}$, $\sigma_3 = -200\,\text{MPa}$
      $\tau_{max} = 250\,\text{MPa}$
6-11 $\sigma_1 = 20\,\text{MPa}, \sigma_2 = -10\,\text{MPa}, \sigma_3 = -10\,\text{MPa}$
      For σ_1: $\cos(n, x) = 0.577$ $\theta_{1x} = 54.7°$
      $\cos(n, y) = 0.577$ $\theta_{1y} = 54.7°$
      $\cos(n, z) = 0.577$ $\theta_{1z} = 54.7°$
      For σ_2: $\theta_{2x} = 45.0°$ For σ_3: $\theta_{3x} = 65.9°$
      $\theta_{2y} = 90°$ $\theta_{3y} = 144.8°$
      $\theta_{2z} = 135°$ $\theta_{3z} = 114.1°$
6-14 $\sigma = \sigma_1 n_x^2 + \sigma_2 n_y^2 + \sigma_3 n_z^2$
      $\tau^2 = n_x^2 n_y^2 (\sigma_1 - \sigma_2)^2 + n_y^2 n_z^2 (\sigma_2 - \sigma_3)^2 + n_x^2 n_z^2 (\sigma_3 - \sigma_1)^2$
      $I_1 = \sigma_1 + \sigma_2 + \sigma_3$
      $I_2 = \sigma_1 \sigma_2 + \sigma_2 \sigma_3 + \sigma_3 \sigma_1$
      $I_3 = \sigma_1 \sigma_2 \sigma_3$
6-15 $\sigma_{oct} = 1/3(\sigma_1 + \sigma_2 + \sigma_3) = 1/3(I_1)$
      $\tau_{oct}^2 = 1/9[(\sigma_1 - \sigma_2)^2 + (\sigma_2 - \sigma_3)^2 + (\sigma_3 - \sigma_1)^2]$
6-17 Hydrostatic and pure shear states, respectively
      $$\begin{bmatrix} 2 & 0 & 0 \\ 0 & 2 & 0 \\ 0 & 0 & 2 \end{bmatrix}, \begin{bmatrix} -1 & 2 & 4 \\ 2 & 0 & 1 \\ 4 & 1 & 1 \end{bmatrix}$$

$\sigma_{oct} = 2\,\text{MPa}$

$\tau_{oct} = \frac{2\sqrt{33}}{3}\,\text{MPa}$

6-20 Maximum principal stress = $1758\,\text{lb/in}^2$

Minimum principal stress = $-332\,\text{lb/in}^2$

Maximum shear stress = $1045\,\text{lb/in}^2$

6-22

Stress state	σ_1 (MPa)	σ_2 (MPa)	θ (degree)	τ_{max} (MPa)
a	23.5	6.5	34.7	8.5
b	20	−14	14	17
c	23.5	6.5	−34.7	8.5
d	−6.5	−23.5	−34.7	8.5

6-23 (a) $\sigma_{x'} = 131.6\,\text{MPa}$, $\sigma_{y'} = 28.4\,\text{MPa}$, $\tau_{x'y'} = 6.7\,\text{MPa}$

(b) $\sigma_{x'} = 48.4\,\text{MPa}$, $\sigma_{y'} = 111.6\,\text{MPa}$, $\tau_{x'y'} = 41.3\,\text{MPa}$

6-24 5.1

6-25 $\varepsilon_{xx} = 0$, $\varepsilon_{yy} = -3(10^{-2})$, $\varepsilon_{zz} = 0$

$\gamma_{xy} = 6(10^{-2})$, $\gamma_{yz} = 9(10^{-2})$, $\gamma_{xz} = 24(10^{-2})$

6-26 $4\mathbf{i} + 21\mathbf{j} + 8\mathbf{k}$

6-27 $\varepsilon_{xx} = 0$, $\varepsilon_{yy} = 9(10^{-2})$, $\varepsilon_{zz} = 0$

$\gamma_{xy} = 4(10^{-2})$, $\gamma_{yz} = 6(10^{-2})$, $\gamma_{xz} = 12(10^{-2})$

6-30 (a) $\varepsilon_{x'x'} = 393.4\,\mu\varepsilon$, $\varepsilon_{y'y'} = 182.6\,\mu\varepsilon$, $\varepsilon_{z'z'} = 324\,\mu\varepsilon$

$\gamma_{x'y'} = 349\,\mu\varepsilon$, $\gamma_{y'z'} = 15.6\,\mu\varepsilon$, $\gamma_{z'x'} = 231.6\,\mu\varepsilon$

6-31 (a) $\varepsilon_1 = 546.7\,\mu\varepsilon$, $\varepsilon_2 = 274.6\,\mu\varepsilon$, $\varepsilon_3 = 78.7\,\mu\varepsilon$

$\frac{1}{2}\gamma_{max} = 234\,\mu\varepsilon$

Direction of ε_1: $\cos(n, x) = 0.844$, $\theta_{1x} = 32.4°$

$\cos(n, y) = 0.367$, $\theta_{1y} = 68.5°$

$\cos(n, z) = 0.388$, $\theta_{1z} = 67.2°$

6-35 $\sigma_{xx} = 179.1\,\text{MPa}$, $\sigma_{yy} = 155.3\,\text{MPa}$, $\sigma_{zz} = 131.4\,\text{MPa}$

$\tau_{xy} = 11.9\,\text{MPa}$, $\tau_{yz} = 11.9\,\text{MPa}$, $\tau_{xz} = 23.9\,\text{MPa}$

6-37 $-1.96\,\text{mm}^3$

6-38 (a) $400\,\text{MPa}$, (b) $346\,\text{MPa}$

6-39 (a) 0.75 in, (b) 0.75 in, (c) 0.63 in

6-42

	σ_1 (MPa)	σ_2 (MPa)	σ_3 (MPa)	τ_1 (MPa)	τ_2 (MPa)	τ_3 (MPa)	θ (°)
a	14	4	0	7	5	2	−26.6
b	17.8	7.2	0	8.9	5.3	3.6	24.4
c	14.7	0	−25.7	20.2	12.8	7.4	82.9

6-43 (1) b; (2) a; (3) c

6-46 (a) maximum normal stress theory $\tau = 9.2\,\text{mm}$

(b) maximum shear stress theory $\tau = 16.6\,\text{mm}$

(c) distortion energy theory $\tau = 11.5\,\text{mm}$

6-49 $\sigma_1 = \sigma_2 = \dfrac{pr}{2t}$, $\tau_{max} = \dfrac{pr}{4t}$

6-52 (a) $\sigma_1 = 28.8\,\text{MPa}$, $\sigma_2 = 14.4\,\text{MPa}$

$\tau_{max} = 14.4\,\text{MPa}$

(b) $\sigma_1 = \sigma_2 = 14.4\,\text{MPa}$

$\tau_{max} = 7.2\,\text{MPa}$

6-56 45° for both tensile and compressive loads

6-57 4710 Pa

6-61 $\sigma_{xx} = 6$ MPa, $\sigma_{yy} = \sigma_{zz} = 10$ MPa, safe

6-63 (a) $\varepsilon_1 = 480\,\mu\varepsilon$, $\varepsilon_2 = 20\,\mu\varepsilon$

     (b) $\gamma_{max} = 460\,\mu\varepsilon$, normal strain $= 250\,\mu\varepsilon$

Chapter 7

7-1 8.0 cm

7-2 14.44 MPa$\sqrt{m}$

7-3 Not safe

7-5 $0.56\sigma_0\sqrt{\pi a}$

7-7 Safe in both cases

7-10 Greater than 48 MPa$\sqrt{m}$

7-12 59.4 kN

7-15 18.7 kN

7-16 (a) 3.86 MPa$\sqrt{m}$, (b) 4.32 MPa$\sqrt{m}$, (c) 4.32 MPa$\sqrt{m}$, (d) 2.52 MPa$\sqrt{m}$

7-17 Center cracked strap

7-21 2.87

Chapter 8

8-2 Effective column length: $\sim 0.7L$

8-5 $-2 + 2 \cdot \cos(\lambda L) + \lambda L \cdot \sin(\lambda L) = 0$

8-6 $\tan(\lambda \cdot L) = \lambda \cdot L \Rightarrow \lambda \cdot L = 4.49$

Index

Crack tip (*contd.*)
  plastic zone, 289
  stress intensity factor, 285
Critical buckling load, 319, 326
Critical buckling mode, 326
Critical resolved shear stress, 125, 251
Critical shear stress, 121, 145
Critical stress intensity factor, 37
Cylindrical pressure vessel, 254

D

Decision matrices, 43
Deformation, 17, 198
  inelastic, 240
  plastic, 240
Density, 59
Design degrees of freedom, 2, 12
Detailed design, 43
Deviatoric stress state, 116, 242
Dilatational, 116
Direction cosines, 215, 218
Direct stiffness method, 82
Dislocations, 121
Dislocation glide, 121, 145
Dispersion hardening, 253
Displacement, 198
Distortion Energy Theory, 247, 248
Ductile fracture, 294
Ductile to brittle transition, 292

E

Edge vectors, 123
Effective column length, 328
Eigenmode, 325
Eigenvalues, 325
Eigenvectors, 325
Elastic curve, 165
Elastic energy release rate, 288
Elastic modulus, 63
Embodiment design, 42
End fixity factor, 328
Endurance limit, 39
Engineering strain, 60, 65
Engineering stress, 65, 66
Equations of constraint, 11
Equilibrium, 17
Euler-Bernoulli assumptions, 161
Euler buckling load, 326
Euler buckling theory, 323, 326

F

Factor of safety, 12
Fatigue, 38
Finite elements, 84
Finite element analysis, 9, 82
First-order, 321

Flake-like nanoparticles, 349
Flexural structures, 151
Flexural testing, 169
Foams, 342
  polymer, 342
  polyurethane, 343
Four-point bending, 169
Fracture, 275, 278
  cleavage fracture, 292
  ductile fracture, 294
  fracture mechanics, 275, 285
  fracture toughness, 37, 289
  fracture work, 280
  intergranular fracture, 295
  transgranular fracture, 293
Fracture stress, 66
Free body diagram, 56

G

Generalized design template, 2
General state of strain, 200
General state of stress, 198
Gradient, 58
Griffith's energy approach, 286

H

Hertzian stresses, 258
Hexagonal closed-packed, 20
Hooke's law, 60, 204
Hydrostatic, 75
Hydrostatic stress state, 242

I

Ionic crystals, 277
Inextensional, 160
Intergranular fracture, 295
Invariants of strain, 235
Isodimensional nanoparticles, 348
Isotropic, 19

K

Knockdown factor, 13, 331

L

Laminated, 164
Layered silicate nanocomposites, 349
Line forming elements, 9
Linear elastic behavior, 65
Link, 51
Load and resistance factor, 14
Load factor design, 12
Load path, 14
Local stiffness matrix, 80

M

Mapping, 84

Slenderness ratio, 55, 329
Slip lines, 121
Slip plane, 121
Small-scale yielding, 285
Solution dependent variables, 321
Solution hardening, 252
Spallation, 259
Specific strain energy, 209
Spider webs, 351
Spring stiffness, 80
Stable equilibrium, 318
Statically indeterminate, 11
Step function, 172
Stiffness, 60
Stiffness matrix, 137
Strain-curvature relation, 162
Strain-displacement equations, 59, 200
Strain energy, 209
Strain energy density, 209
Strain hardening, 241, 250
Strain invariant, 235
Strain tensor, 200
Stress concentration effect of flaws, 282
Stress couple, 165
Stress curvature relation, 165
Stress element, 57
Stress equilibrium equations, 204
Stress intensity factor, 37, 285, 302–305
Stress invariants, 217
Stress resultant, 165
Stress-strain relations, 17
Strong attractor, 318
Structural analysis, 11
Strut, 51
Surface-forming elements, 9
Surface forces, 195

T

Taxonomy of structures, 8
Tensegrity, 351
Tensile elastic modulus, 54, 60
Tensile strength, 54, 66
Tensor, 196

Tensorial strains, 200
Thermoplastic, 340
Thermosetting, 341
Three-point bending, 169
Torque, 112
Torsion constant, 132, 145
Torsion spring element, 137
Torsional stiffness, 117, 145
Torsional stiffness matrix, 137
Total potential energy, 77
Total structure design, 2
Toughness, 68
Transformation of stress, 214
Transgranular fracture, 293
Translational spring, 80
Triaxial, 208
True strain, 67
True stress, 67
True stress–true strain curve, 68
Twist angle, 114, 145
Twist gradient, 115
Twisting moment, 112
Types of stress-strain curves, 69

U

Ultimate tensile strength, 66
Unconditionally stable, 56
Unstable equilibrium, 56, 318

W

Weak attractor, 318
Weight, 59
Work hardening, 250

Y

Yield criteria, 241, 244
  distortion energy, 247
  maximum normal stress, 244
  maximum shear stress, 245
Yield strength, 63, 66
Yield stress, 66, 243
Young's modulus, 63, 65